THYMUS UPDATE 2

T Lymphocyte Differentiation in the Human Thymus

THYMUS UPDATE

This series aims to bring together an in-depth assessment of selected areas of current research on the thymus gland together with an annual update of general information on meetings, publications and products relevant to the field.

Edited by Marion D. Kendall and Mary A. Ritter

Volume 1
THE MICROENVIRONMENT OF THE HUMAN THYMUS

Volume 2
T LYMPHOCYTE DIFFERENTIATION IN THE HUMAN THYMUS

Additional Volumes in Preparation

Volume 3
ROLE OF THE THYMUS IN TOLERANCE INDUCTION

THYMUS UPDATE 2

T Lymphocyte Differentiation in the Human Thymus

Edited by

Marion D. Kendall

Division of Anatomy,
United Medical and Dental Schools of Guy's and St. Thomas's Hospitals, St. Thomas's Campus, London, UK

and

Mary A. Ritter

Department of Immunology,
Royal Postgraduate Medical School,
Hammersmith Hospital, London, UK

harwood academic publishers
chur • london • paris • new york • melbourne

Harwood Academic Publishers

Post Office Box 197
London WC2E 9PX
England

58, rue Lhomond
75005 Paris
France

Post Office Box 786
Cooper Station
New York, New York 10276
United States of America

Private Bag 8
Camberwell, Victoria 3124
Australia

ISBN 3-7186-4932-2

 Printed in the United Kingdom by Bell and Bain Ltd., Glasgow.

CONTENTS

Introduction

T lymphocyte differentiation requires an interaction between lymphoid progenitor cells and the thymic microenvironment. Volume 1 of *Thymus Update* concentrated on the microenvironmental aspects of the interaction ("The Microenvironment of the Human Thymus"). This second, companion volume of *Thymus Update* focuses on intrathymic T cell development from the lymphocytes' point of view. It addresses aspects such as how lymphoid progenitors are attracted and recruited into the thymus, the genetic events leading to functional T cell receptor genes, the development of the capacity to synthesise and express T cell receptor and associated accessory molecules at the cell surface, the role of lymphokines and their receptors, the much neglected area of innervation in thymocyte differentiation, and the deleterious effects of both intrinsic defects (MHC deficiency) and extrinsic (steroids) factors on thymocyte development.

This theme is continued in Part 2 which contains a valuable review of cytokines. Much current immunological research is devoted to the subtle roles that these substances play in immune responses, and it is now becoming clear that they are also vitally important in intrathymic T cell differentiation. In addition, as before, this second part of the volume contains reports from several meetings that have been held over the last twelve months. One of the more important of such events in the European field of thymology in 1988 was a meeting on *The Thymus: Histophysiology and Dynamics in the Immune System* in the Netherlands. This workshop provided a friendly and informal venue for the discussion of research data and exchange of ideas, and a detailed report from the chairmen of each of the 7 sessions is contained in this volume of *Thymus Update*. The editors would like to thank the many people who have become reporters on their behalf at these events.

As in the previous volume, the Appendix is a reference section that provides regular information on forthcoming meetings, PhD theses just completed, research reagents, new books and a comprehensive listing of recent (July 1987-

October 1988) research publications on the thymus. This and the previous section relies heavily on material from our readers and the editors are always pleased to receive such contributions. Our aim is that *Thymus Update* will be recognised as an intellectually lively and useful world-wide forum for thymologists that encourages "audience participation".

Finally, our next volume will focus on the highly topical and controversial area of the role of the thymus in the induction of tolerance to self antigens and the acquisition of MHC restriction. Review articles will discuss experimental data concerning the existence and nature of positive and negative selection, and the microenvironmental cells that are involved in these intrathymic processes. We look forward to many stimulating contributions.

Marion D. KENDALL
Mary A. RITTER

List of Contributors

Bauvois, Brigitte
Laboratoire de Physiopathologie du Développement, Centre National de la Recherche Scientifique et Ecole Normale Supérieure, 46 rue d'Ulm, 75230 Paris, Cédex 05, France

Campana, Dario
Department of Immunology, Royal Free Hospital School of Medicine, Rowland Hill Street, Hampstead, London, NW3 2QG, UK

Deugnier, Marie-Ange
Laboratoire de Physiopathologie du Développement, Centre National de la Recherche Scientifique et Ecole Normale Supérieure, 46 rue d'Ulm, 75230 Paris, Cédex 05, France

Dunon, Dominique
Laboratoire de Physiopathologie du Développement, Centre National de la Recherche Scientifique et Ecole Normale Supérieure, 46 rue d'Ulm, 75230 Paris, Cédex 05, France

Felten, David L.
Department of Neurobiology and Anatomy, University of Rochester, School of Medicine, Rochester, New York 14642,USA

Felten, Suzanne Y.
Department of Neurobiology and Anatomy, University of Rochester, School of Medicine, Rochester, New York 14642, USA

Gearing, Andrew J. H.
Immunobiology, National Institute for Biological Standards and Control, South Mimms, Hertfordshire, EN6 3QG, UK

Hougen, Hans P.
University Institute of Forensic Medicine, Frederik V's Vej 11, Copenhagen, Denmark

Imhof, Beat A.
Laboratoire de Physiopathologie du Développement, Centre National de la Recherche Scientifique et Ecole Normale Supérieure, 46 rue d'Ulm, 75230 Paris, Cédex 05, France

Janossy, George
Department of Immunology, Royal Free Hospital School of Medicine, Rowland Hill Street, Hampstead, London, NW3 2QG, UK

Kendall, Marion D.
Division of Anatomy, United Medical Schools of Guy's and St. Thomas's Hospitals, St. Thomas's Campus, London, UK

Ladyman, Heather M.
Royal Postgraduate Medical School, Hammersmith Hospital, Du Cane Road, London, W12 0NN, UK

Larché, Mark
Department of Immunology, Royal Postgraduate Medical School, Hammersmith Hospital, Du Cane Road, London, W12 0NN, UK

Lechler, Robert I.
Department of Immunology, Royal Postgraduate Medical School, Hammersmith Hospital, Du Cane Road, London, W12 0NN, UK

Ritter, Mary A.
Department of Immunology, Royal Postgraduate Medical School, Hammersmith Hospital, Du Cane Road, London, W12 0NN, UK

Schuurman, Henk-Jan
Division of Immunopathology, Departments of Internal Medicine and Pathology, University Hospital, Catharijnesingel 101, 3511 GV Utrecht, The Netherlands

Thiery, Jean-Paul
Laboratoire de Physiopathologie du Développement, Centre National de la Recherche Scientifique et Ecole Normale Supérieure, 46 rue d'Ulm, 75230 Paris, Cédex 05, France

Willcox, Nicholas
Neurosciences Group, Institute for Molecular Medicine, John Radcliffe Hospital, Headington, Oxford, OX3 9DU, UK

Zegers, Ben J. M.
Department of Immunology, University Children's Hospital "Het Wilhelmina Kinderziekenhuis", Nieuwe Gracht 137, 3512 LK Utrecht, The Netherlands

PART 1: T LYMPHOCYTE DIFFERENTIATION IN THE HUMAN THYMUS

CHAPTER 1

PROPERTIES OF PRE-T CELLS AND THEIR CHEMOTACTIC MIGRATION TO THE THYMUS

BEAT A. IMHOF, MARIE-ANGE DEUGNIER, BRIGITTE BAUVOIS, DOMINIQUE DUNON, AND JEAN-PAUL THIERY

Laboratoire de Physiopathologie du Développement, Centre National de la Recherche Scientifique et Ecole Normale Supérieure, 46 rue d'Ulm, 75230 Paris Cédex 05, France

ABSTRACT

The thymic rudiment in vertebrate embryos appears as an outpocketing of the endoderm at the level of the branchial pouches. Hemopoietic cells colonize the thymic epithelium early during embryogenesis to give rise to T cells and to accessory cells. Hemopoietic precursors originating from discrete intra-embryonic foci, the fetal liver, and the bone marrow are recruited by the thymic epithelium via a chemotactic mechanism. In this report we show that the thymic epithelium produces a chemotactic peptide named thymotaxin which induces directed migration of bone marrow hemopoietic precursors. Rat cells that migrate towards thymotaxin in Boyden chambers are of lymphoid type but are devoid of all available differentiation markers. In functional assays such migrated hemopoietic precursors can be differentiated to express T-cell markers, either *in vitro* by coculture with embryonic thymus, or *in vivo* by intrathymic injection into thy-1 congenic animals. Chemotactic molecules have been found to be secreted by avian and rat thymic epithelia. Thymotaxin from rat has been found to be active also on mouse and sheep hemopoietic cells. Therefore we suggest that chemotaxis is a general mechanism involved in homing of hemopoietic precursors to the thymus operating in all vertebrates including humans.

By the early 1960s it had already been found that removal of the mouse thymus on the day of birth resulted in a deterioration in allograft rejection (Miller, 1961). It turned out that T (thymus derived) lymphocytes were responsible for this immune reaction and that these cells developed in the thymus which acted as the major differentiation site (Metcalf, 1966). Based on transplantation experiments and on ultrastructure of embryonic thymic rudiments, Auerbach (1960) and subsequently Ackerman and Knouff (1965) hypothesized that lymphocytes were derived from thymic epithelial cells. However two years later Moore and Owen (1967) found that lymphocytes in the thymus were of extrinsic origin, though, the thymus has to be colonized by hemopoietic precursors.

I SITES OF ORIGIN AND CELL DYNAMICS OF HEMOPOIETIC PRECURSORS IN THE EMBRYO AND IN THE ADULT

During embryonal life the first hemopoietic precursors can be found in the mesodermal part of the visceral yolk sac before intraembryonic angiogenesis starts. *In vitro* experiments showed that hemopoietic foci from the yolk sac of mammals have the potential to give rise to every lineage of blood cells (Moore and Owen, 1967; Metcalf and Moore, 1971; Jotereau and Houssaint, 1977). However at least in birds hemopoietic cells originating from the yolk sac participate only during a transitory time period in the blood cell pool (Martin *et al.*, 1980) and the first lymphoid cells that arrive in the thymus are of intraembryonic origin, presumably generated from para-aortic foci (LeDouarin *et al.*, 1984).

In mammals intraembryonic foci are constituted in the liver. For example, the mouse hepatic rudiment contains the first hemopoietic cells at day 9.5 of gestation and a final shift from yolk sac to liver occurs after 12 days of development. Grafting techniques of mouse fetal liver into anaemic recipients provided proof for the totipotency and self-renewal of hemopoietic stem cells (Fleischman *et al.*, 1982). However liver hemopoietic foci disappear during the first week after birth and are replaced by the main site of hemopoiesis, the bone marrow (Nagel *et al.*, 1981).

In the adult, the bone marrow is the most complex hemopoietic organ. All blood cell lineages, aside from maturing T-cells, are found at different levels of maturation (Till and McCulloch, 1961; Abramson *et al.*, 1977; Jotereau and Le Douarin, 1978). Hemopoietic stem cells appear as early as 14 days in the bone marrow of mouse embryos and before 10 days in the quail (Jotereau and Le Douarin, 1978). In 11 day-old quail embryos hemopoiesis is fully active and lymphoid precursors are present, able to colonize the thymus and differentiate into T lymphocytes (Ben Slimane *et al.*, 1983; Jotereau *et al.*, 1980; Champion *et al.*, 1986).

By making chick-quail thymic chimeras Le Douarin and Jotereau (1980) established the exact time when colonization of the thymic rudiment by hemopoietic precursors starts. After grafting of quail thymic rudiments into two successive hosts two embryonic periods of thymus colonization have been characterized. In quail embryos the first wave of colonization starts as early as 5 days of incubation and the second at 11 days, each of them lasting only 24 hours. Finally a third generation of hemopoietic cells invades the thymus one day after hatching.

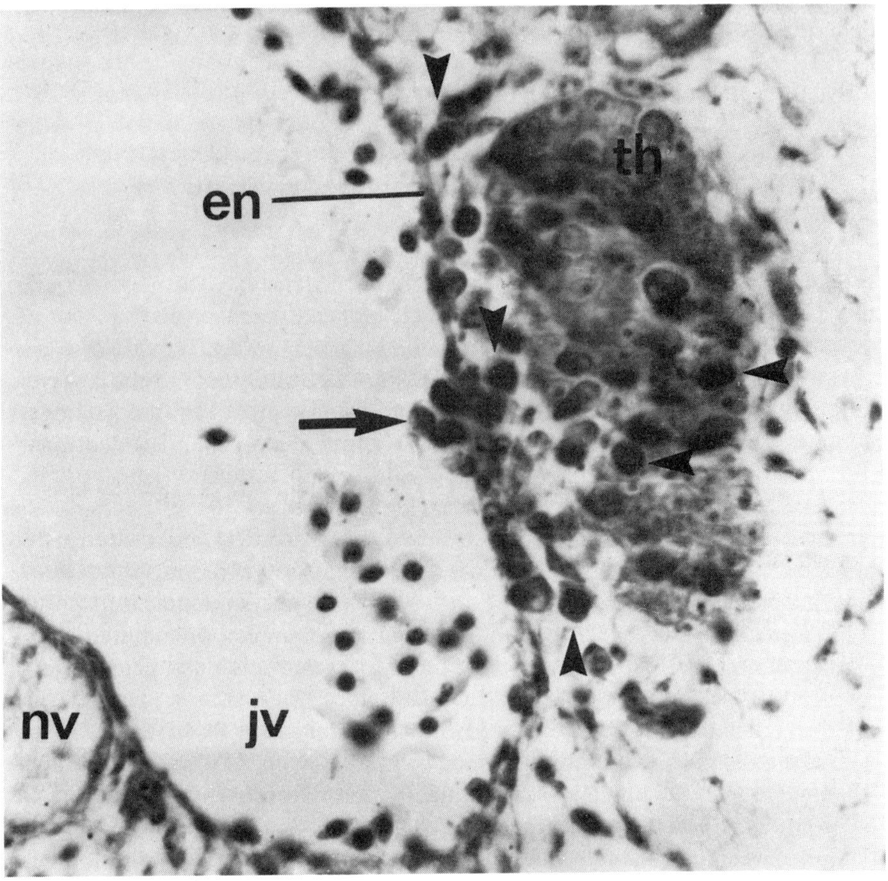

Figure 1. Morphology of a quail embryonic thymus during colonization by hemopoietic precursors. Transverse section through the cervical region of a 6 day-old quail embryo which was fixed and embedded in paraffin. The thymic rudiment lies along the jugular vein in the vicinity of the vagus nerve. Basophilic hemopoietic precursors are present in the jugular vein adhering to the endothelium (arrow). They can also be found in the mesenchyme (vertical arrowheads) and within the thymus (horizontal arrowheads). en, endothelium; jv, jugular vein; nv, vagus nerve; th, thymus.

Recently Jotereau *et al.* (1987) described the colonization of the mouse thymus by two different generations of hemopoietic precursors. They grafted an embryonic thymus into successive hosts that was allogenic for the thy-1 antigen. Taking thy-1 as a T-cell marker extensive thymus colonization by hemopoietic precursors and their rapid differentiation could be observed during 10 and 12.5 days of embryogenesis. A second generation of hemopoietic cells invades the thymus after day 13. However these cells stay in the thymus for one week before they differentiate into T-cells. Intravenously injected bone marrow cells into non-irradiated mouse babies colonized the thymus and differentiated into T-cells (Kyewski, 1987). Influx of hemopoietic precursors from the bone marrow to the thymus might then continue during adult life but at a very low level (Scollay *et al.*, 1986).

II CHEMOTACTIC MIGRATION OF HEMOPOIETIC PRECURSORS TO THE THYMUS

In the late 1970s Le Douarin suggested that the cyclic renewal of hemopoietic cells that colonize the thymus is governed by a chemotactic mechanism (Le Douarin, 1978). This idea was supported by the finding that hemopoietic precursors from yolk sac, fetal liver, thymus or bone marrow colonized a 10-day-old mouse thymic rudiment in a transfilter assay (Fontaine-Pérus *et al.*, 1981). When these colonized thymuses were cultured subsequently on a chick chorioallantoic membrane, hemopoietic cells differentiated into thy-1 positive cells.

Direct evidence for chemotaxis of avian hemopoietic precursors had been obtained initially by using Zigmond chambers (Zigmond, 1977) where migration of individual cells could be simultaneously observed (Figure 2; Ben Slimane *et al.*, 1983). Bone marrow cells from 11.5 day-old quail embryos migrated towards a soluble gradient created by the secretory activity of an embryonic thymus explanted during the first embryonic period of its colonization and placed in one reservoir of the Zigmond chamber. Locomotion of single cells was found to be oriented (chemotactic) and their speed of locomotion was as fast as 25µ m/min in comparison to only 12µm/min when the cells were exposed to control medium. Chemotactic migration was lost when the thymic rudiment originating from an older embryo placed into the Zigmond chamber was in a refractory state with respect to its colonization *in vivo*. However in this case the speed of locomotion of randomly migrating bone marrow cells was also increased to 25µm/min. Thus it turned out that two thymic effects interfered in hemopoietic cell migration: Chemotaxis, induced by secreted products from embryonic thymuses in the attractive phase, and chemokinesis effected by secretory material from thymuses at any stage of embryogenesis.

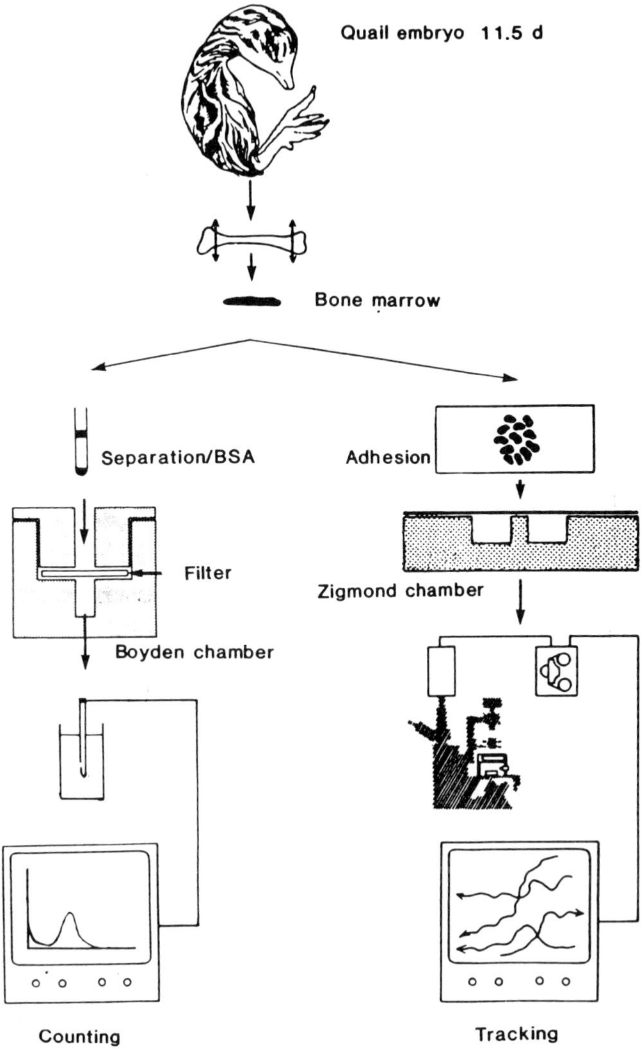

Figure 2. Outline of the two procedures used to test chemotactic migration *in vitro*.
Bone marrow cells from 11.5 day-old quail embryos or from young rats and mice were harvested from the tibia and femur. For Boyden chamber assays hemopoietic precursors were separated from total bone marrow by centrifugation on a BSA density gradient. Subsequently, responding cells migrated through a nuclepore filter (5μm pore size) towards a soluble gradient of material tested. Migrated cells were then counted with a Coulter Counter (Imhof *et al.*, 1988b). For Zigmond chamber assays embryonic bone marrow cells were first attached to glass cover slips and were subsequently mounted upside down onto the gradient chamber. Migration of hemopoietic precursors was observed simultaneously by a videomicroscope equiped with a time-lapse recorder.

In an attempt to identify the cell type that secretes substances involved in hemopoietic precursor homing to the thymus we prepared various embryonic cells in tissue culture (Champion *et al.*, 1986). Pure thymic epithelial cell cultures, thymic fibroblasts, macrophages, or dendritic cells could be obtained by sequential attachment of dissociated thymus. Kidney epithelial cells, heart fibroblasts and liver cell cultures were initiated by dissociation of the corresponding embryonic tissues by trypsin digestion. To probe the production of attractive factors, migration of hemopoietic precursor cells was analyzed in Boyden chambers (Figure 2).

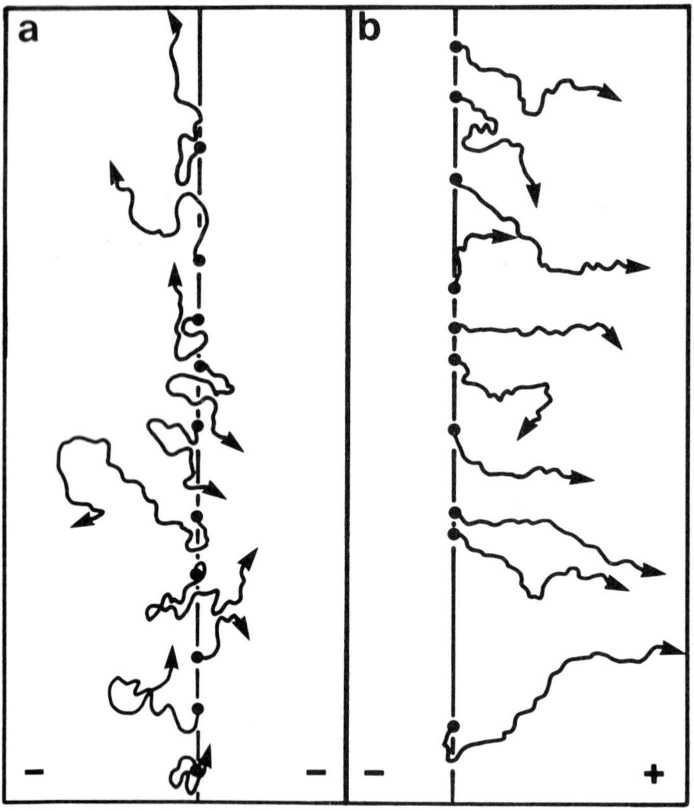

Figure 3. Migration of hemopoietic cells in a Zigmond chamber.
Conditioned medium from quail primary thymic epithelial cultures was tested for chemotactic attraction of hemopoietic precursors. (a) Both gradient reservoirs (indicated as -/-) contained control medium resulting in random migration. (b) Reservoir on the right was filled with conditioned medium and the one on the left contained control medium (indicated as -/+) to establish a soluble gradient. Cells migrated in an oriented way.

Among all supernatants obtained by cultured cells under serum free conditions, solely thymic epithelium conditioned medium stimulated migration of hemopoietic precursors. Both chemotactic and chemokinetic elements were found when thymic epithelial cell derived products were distinguished by checkerboard analysis (Zigmond and Hirsch, 1973). Migration of individual hemopoietic cells towards media conditioned by thymic epithelia was also followed in a Zigmond chamber observed by a video microscope connected to a time-lapse recorder. All locomoting cells displayed oriented migration directly towards the soluble gradient created by thymic epithelial cell conditioned medium. These experiments suggested that thymic epithelial cells which developed from the pharyngeal endoderm of the third and fourth branchial pouches during embryogenesis, secrete chemotactic and chemokinetic products attracting hemopoietic precursor cells. It was possible to separate these two different activities by fractionating the conditioned medium on a Sephadex G-25 column. Substances with a molecular weight of 4 kDa displayed almost exclusively a chemotactic activity whereas products with a molecular weight higher than 5 kDa were chemotactic and chemokinetic.

Chemotaxis involved in colonization of the thymus by hemopoietic precursors had also been suggested to operate in mammals (Pyke and Bach, 1981; Haar and Loor, 1981; Imaizumi *et al.*, 1987). Recently it became possible to establish a rat thymic cell line (IT-45 R1; Itoh *et al.*, 1982) that shows all the morphological criteria of epithelial cells (Figure 4). In addition, these cells fulfilled various functional aspects attributed to the thymic epithelium such as secretion of thymulin and thymopoietin (Savino *et al.*, 1986). Most importantly, rat bone marrow cells migrate towards a soluble gradient of IT-45 conditioned medium in Boyden chambers. Such *in vitro* migrated bone marrow cells were of immature phenotype with respect to the absence of differentiation markers on their cell surface (Imhof *et al.*, 1988a).

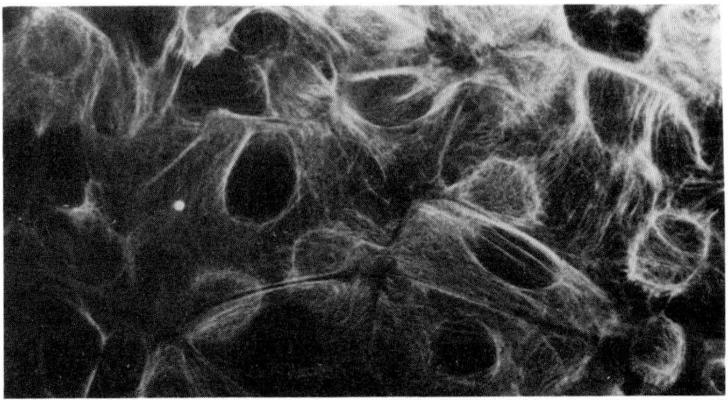

Figure 4. IT-45 R1, a rat thymic epithelial cell line.
The photograph shows an immunofluorescence staining for keratin intermediate filaments using an anti-pan-keratin (LU-5) antibody. These cells secreted peptides chemotactic for bone marrow cells from rat or mice.

In comparison to primary embryonic cultures of thymic epithelium which secreted factors during only a limited time period, working with the immortal thymic epithelial cell line IT-45 opened new perspectives. Several liters of serum-free IT-45 conditioned medium were loaded onto hydrophobic Sep-Pak C18 cartridges. Chemotactic activity could be eluted with the organic solvent acetonitrile. This procedure allowed a purification factor of 40 and a very convenient reduction in volume by a factor of 1000. The concentrate was then fractionated by Sephadex G-50 gel chromatography and the samples tested for chemotactic activity on hemopoietic precursors derived from bone marrow. These cells migrated in Boyden chambers towards molecules that eluted with a molecular weight of 11kDa and 4kDa (Figure 5). The 11kDa material was analyzed on SDS-gels and identified as a single band that we named thymotaxin (Imhof *et al.*, 1988b).

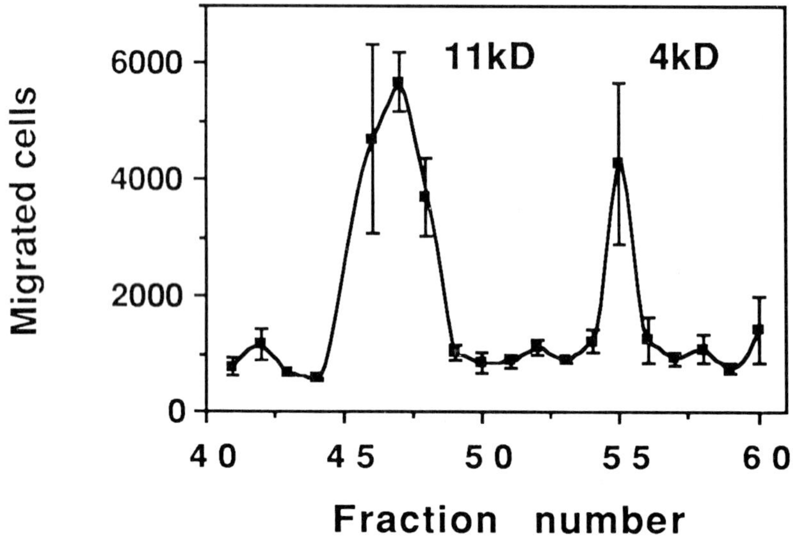

Figure 5. Purification of thymotaxin produced by thymic epithelial cells
Conditioned medium from IT-45 R1 cells was concentrated on Sep-Pak cartridges. This material was fractionated on a Sephadex G-50 column and fractions were analyzed for chemotactic migration of hemopoietic precursors from rat bone marrow in Boyden chambers. Material corresponding to a molecular weight of 11kDa and 4 kDa attracted preferentially bone marrow derived cells of immature lymphoid type. The 11kDa peptide was named thymotaxin.

TABLE I. Behaviour of hemopoietic cells after migration towards thymotaxin

	Cells migrated towards thymotaxin	Cells migrated towards control medium
a) ^{3}H-thymidine incorporation	-	+
Survival on thymic stroma of rat embryos	+	+
b) Differentiation into T-cell-antigen positive cells	+	-
c) GM-CFC potentialities	-	+

Thymotaxin responding cells are in a resting state and require a rat embryonic thymic stroma to survive and to differentiate into cells bearing T-cell antigens. They do not give rise to myeloid progeny in semi-solid cultures. These functional properties are not shared by bone marrow cells which migrate unspecifically towards control medium.
a) ^{3}H-thymidine (1mCi/ml) was incubated with cells for 3 hours. Cytocentrifuged cells were fixed, dipped with photo-emulsion and developed one week later. Cells were stained by the May-Grunwald-Giemsa technique.
b) Cells were cocultured on a monolayer of embryonic thymic stroma under steroid-free conditions.
c) Semi-solid cultures in methyl-cellulose contained Interleukin 1, Interleukin 3 and Erythropoietin.

III CHARACTERIZATION OF HEMOPOIETIC CELLS THAT MIGRATED TOWARDS THYMOTAXIN

Hemopoietic cells responding to thymotaxin account for 0.1% of the total nucleated bone marrow cells of juvenile rat. A ten-fold enrichment in responding cells could be routinely obtained by separating the total bone marrow cells on a 28% bovine serum albumin gradient and by using the low-density fraction in migration assays. This fraction was partially depleted in myeloid and erythroid cells (Fig.6a) and was enriched in hemopoietic precursor cells as judged by the high incidence of thy-1$^+$ cells (approximately 50%) and the low frequency of cells expressing T-cell and B-cell differentiation markers (<2%). It should be pointed out that in the rat bone marrow, thy-1 is expressed by immature cells, T-cell precursors and B-cell precursors (Hunt , 1979; Opstelten *et al.*, 1986; Hale *et al.*, 1987). Owing to large preparative migration chambers, it became possible to collect substantial number of cells responding to thymotaxin. This allowed the characterization of responding cells in terms of phenotype, cycling ability and differentiation capacity in microculture systems. We found that low-density cells selected by chemotactic migration towards thymotaxin have a lymphoid morphology. The responding subpopulation mainly contained small lymphocytes and medium-sized lymphoid cells

(Fig. 6b). These cells were in a resting state since they did not incorporate ^{3}H-thymidine (Table 1). On the other hand, immunofluorescence studies revealed that lymphoid cells responding to thymotaxin displayed an immature thy-$1^{+}T^{-}$ B^{-} phenotype as evidenced by the lack of CD4, CD5 and CD8 expression and the absence of cytoplasmic and surface IgM expression. Thus the small subset of cells responding to thymotaxin is composed of phenotypically immature lymphoid cells which are predominantly in a resting state.

Cells that migrated towards thymotaxin died in a xenogeneic coculture sytem which has been reported to promote the maintenance of immature rat lymphoid cells probably committed to the B lineage (Hayashi et al., 1984; Opstelten *et al.*, 1987). In contrast, a lymphoid-free stroma derived from embryonic rat thymic tissue provided a suitable microenvironment for the survival of lymphoid cells responding to thymotaxin (Fig. 7). To establish such lymphoid-free stroma, thymic lobes were taken from rat embryos at day 14-15 of gestation, that is before the period of extensive colonization of the thymus. After 3 days in vitro in a culture medium designed for the selective growth of epithelial cells (Small *et al.*, 1984), the spread thymic explants were surrounded by a monolayer of epithelial cells and by growing macrophage-like cells (Deugnier *et al.*, submitted for publication).

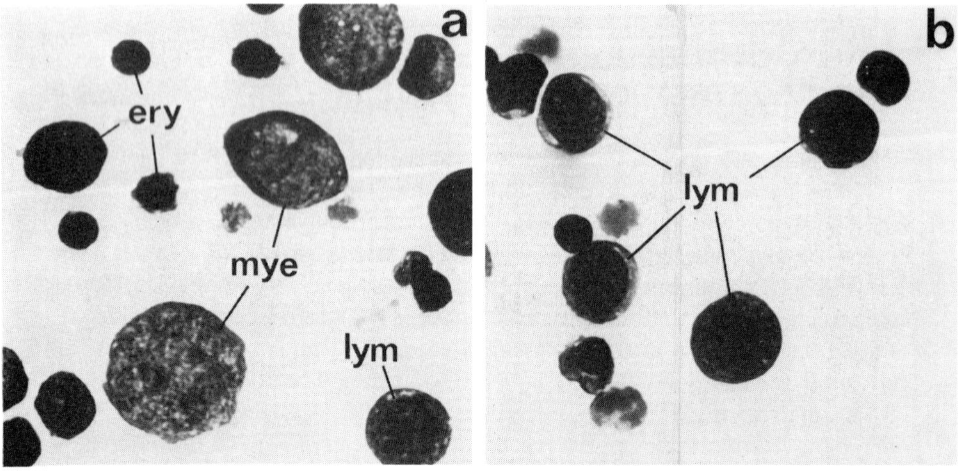

Figure 6. Phenotype of bone marrow derived hemopoietic cells that migrate towards thymotaxin. Rat bone marrow cells were enriched for hemopoietic precursors on a BSA gradient. These cells were exposed to a gradient of thymotaxin in Boyden chambers. For analysis cells were mounted onto a slide by a cytocentrifuge and stained by the May-Grunwald-Giemsa technique. (a) bone marrow precursors before chemotactic migration. (b) hemopoietic precursors after migration towards thymotaxin. ery, erythroid progenitors; lym, lymphoid cells; mye, myeloid progenitors.

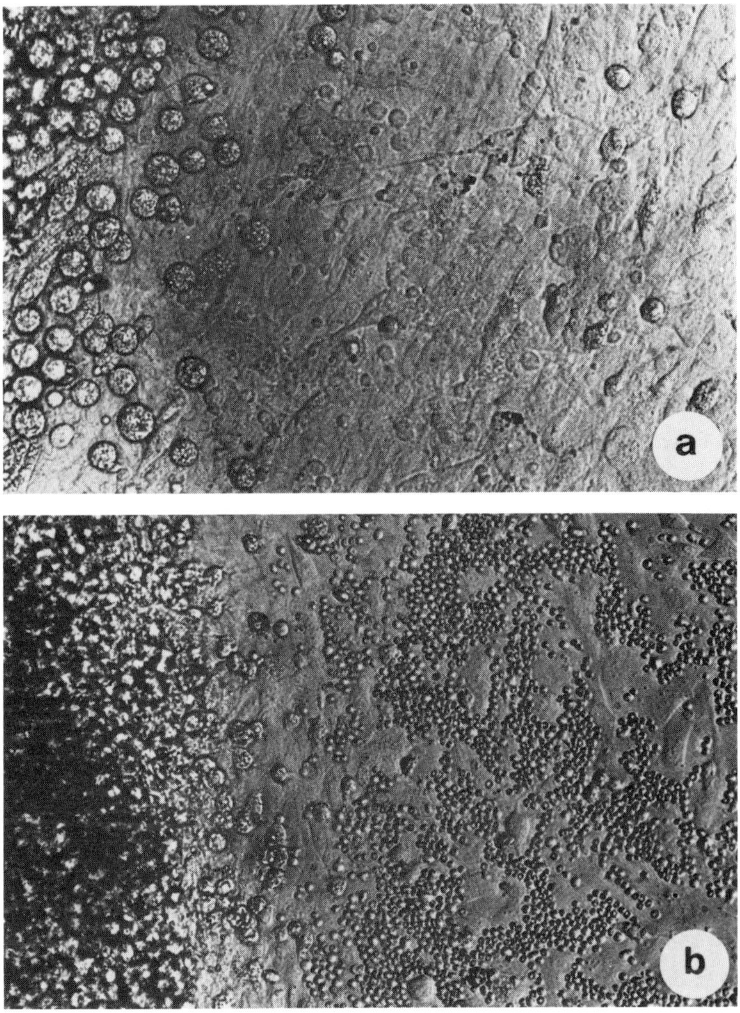

Figure 7. Hemopoietic precursors differentiated into T lineages by coculture with thymic stroma cells.
Embryonic thymuses of 14 to 15 day-old rat were explanted and cultured in a medium that favours growth of epithelial cells (Champion *et al.*, 1986). (a) Cells grew out of the thymic tissue as an epithelial monolayer after three days. (b) Culture medium was changed into conditions that favour maintenance of lymphoid cells (steroid-free serum). Hemopoietic cells were then cocultured on this thymic epithelial monolayer for an additional three days. Lymphoid cells adhere to thymic epithelial cells on the apical surface.

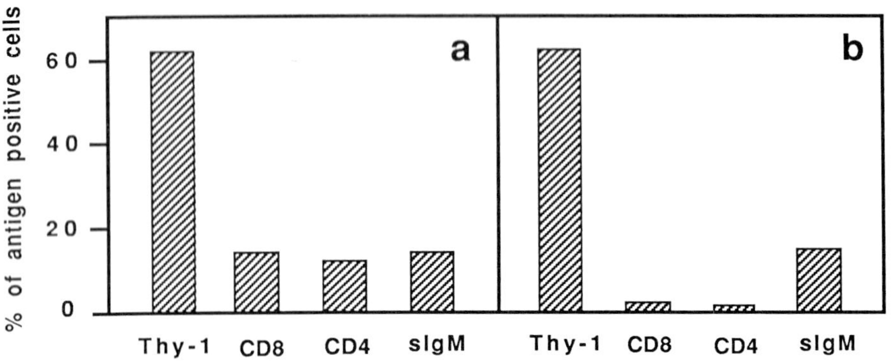

Figure 8. Hemopoietic precursors differentiated into T lineages by coculture with thymic stroma cells.
Analysis of differentiation antigens on cocultured hemopoietic cells by monoclonal antibodies. Loosely adherent cells were harvested from the cocultures and stained for immunofluorescence. Counting of positive cells was restricted to the small lymphoid cells. (a) Cocultured cells that migrated chemotactically towards thymotaxin. (b) Cocultured cells that migrated unspecifically towards control medium. Bone marrow derived thy-1$^+$, T$^-$, B$^-$ lymphoid cells that respond to thymotaxin acquire T-cell differentiation molecules after in vitro coculture on embryonic thymic stroma.

Thymotaxin-responding cells derived from bone marrow were loaded onto the thymic tissue and cultured for an additional period of 3 days under steroid-free conditions. At day 3 in coculture a significant fraction of lymphoid cells expressed T-cell differentiation markers (Fig. 8a). Phenotypic maturation of thy-1$^+$ T$^-$ B$^-$ cells responding to thymotaxin accounted for the presence of cells bearing T-cell differentiation antigens. Indeed lymphoid cells were not generated by the pre-established thymic stroma as indicated by phenotype analysis of embryonic thymic tissue cultures . On the other hand, selection and amplification of putative mature T-cells attracted by thymotaxin and grown in coculture are unprobable events considering the very low incidence of CD4$^+$ and CD8$^+$ cells (0% and 0.3%, respectively) among the responding cell subset at day 0 and the lack of significant lymphoid proliferation in these cocultures. Finally it should be noted that B-cell differentiation also appeared to occur in our *in vitro* system. Nevertheless expression of surface IgM was not restricted to thymotaxin-responding cells as shown by control experiments performed with hemopoietic cells which had migrated unspecifically towards a Dulbecco culture medium (Fig.8b). In contrast, acquisition of T-cell differentiation antigens was specific to thymotaxin-responding cells. Based on these *in vitro* experiments we suggest that the cell population responding to thymotaxin is enriched in T-cell precursors. To try to support these *in vitro* experiments we recently started *in vivo* assays of T-cell differentiation.

Since rat thymic peptides cross-react with mouse bone marrow cells, *in vivo* differentiation of migrated cells have been performed using C57 Bl / Ka thy-1.1, thy-1.2 congenic mice (Fig. 9). Bone marrow cells from thy-1.2 mice were taken after their chemotactic migration in preparative Boyden chambers and 10^5 cells were injected intrathymically into thy-1.1 mice which had been irradiated sublethally before injection. Thymocytes were analyzed twenty days later for expression of host (thy-1.1) and donor (thy-1.2) type molecule. Cells that migrated towards thymic peptides repopulated the thymus with 29% thy-1.2^+ cells in respect to all thymocytes. In contrast bone marrow cells that migrated towards control medium contributed only to 11% of all thymocytes and total bone marrow cells from donor mice achieved 39%. These *in vivo* results confirmed our data in the in vitro system which described the potential for T-cell differentiation of bone marrow cells which migrated towards thymic peptides.

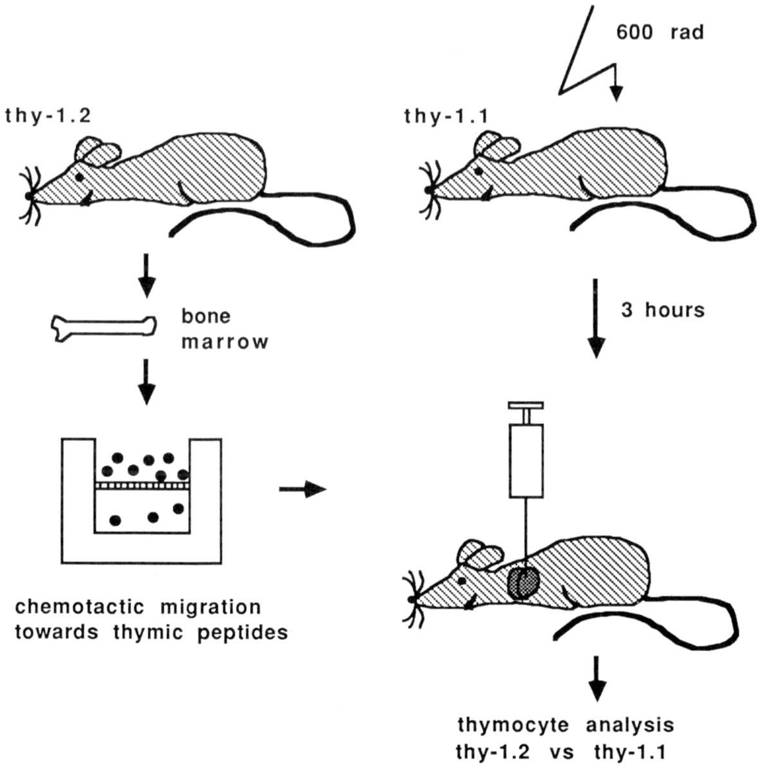

Figure 9. Scheme of the *in vivo* T-cell differentiation assay using intrathymic injection of hemopoietic cells into thy -1 congenic mice.

The question arose whether migrating cells are committed to the T-cell lineage or if these cells are multipotent. Therefore an *in vitro* system of semi-solid cultures was applied driving hemopoietic cells to myeloid or erythroid lineages. Thymotaxin-responding cells did not give rise to erythroid or myeloid progeny under Interleukin 1, Interleukin 3 and Erythropoietin stimulation (Table 1). Interestingly, the hemopoietic cell subset which had migrated unspecifically towards Dulbecco medium behave differently from the chemoattracted one. This cell subset represented less than 0.05% of the total nucleated bone marrow and contained early myeloid progenitors at a significant frequency of 2/1000.

In order to come to a conclusion concerning the state of differentiation of hemopoietic cells which migrated chemotactically towards thymotaxin we have put information of the above experiments together. First of all, bone marrow cells that respond to thymotaxin have the potential for T-cell differentiation. Pre-B cells, mature B and T-cells as well as myeloid and erythroid progenitors have lost the ability to migrate towards thymic chemotactic peptides. Interestingly thymic T-cell progenitors rapidly become unresponsive to thymotaxin since $CD^{-}4^{-}$, $CD^{-}8^{-}$, double negative thymocytes contained less than 0.5% migrating cells (manuscript in preparation). Hemopoietic precursors responding to thymic chemotactic peptides were predominantly resting cells that do not proliferate under Interleukin 1 and Interleukin 3 stimulation. Thus chemoattracted cells could be of putative lymphoid restricted lineage or pro-thymocytes already committed to the T-lineage. However up to now appropriate *in vitro* assays for the detection of very immature cells (stem-cells) are lacking. Therefore we cannot exclude the possibility that migrating cells belong to the resting subset of stem cells.

IV THYMOTAXIN: PERSPECTIVES IN HUMAN THYMUS RESEARCH

Thymotaxin from rat thymic epithelial cells has been found to be chemotactic on rat, mouse and sheep hemopoietic cells from the bone marrow. A similar chemotactic molecule was characterized from embryonic thymus of birds. This suggests that chemotactic migration of hemopoietic cells to the thymus is a general mechanism operating in all vertebrates probably including humans. It could open a new avenue of research which could lead to the possibility of sorting bone marrow derived precursor cells able to reconstitute patients with severe acquired immune deficiencies. Such cells in contrast to total bone marrow may repopulate the thymus without eliciting graft versus host diseases. This procedure could also be applied in the case of malignant thymic lymphomas after irradiation. Most importantly the thymotaxin-selected bone marrow precursors will allow lineage studies in order to specify the degree of commitment of these cells to become T-cells and/or accessory cells. It will also better define the very early events in differentiation into double negative thymocytes. Hemopoietic precursors sorted by chemotactic migration might be the cells of choice for studies of cell-cell interactions with thymic epithelium, macrophages, nurse cells and dendritic cells operating intrathymically during the cascade of T-cell differentiation.

ACKNOWLEDGMENTS

We are especially grateful to Monique Denoyelle and Jeanne-Marie Girault for excellent technical assistance. We thank Dr Sophie Ezine for intrathymic injection and helpful discussions and Dr Jelena Gavrilovic for critical reading of the manuscript.

This work was supported by grants from the Ligue Nationale Française contre le Cancer, CNRS, and INSERM CRE 86-4015.

REFERENCES

ACKERMAN, G.A., and R.A., KNOUFF. (1965). The epithelial origin of the lymphocytes in the thymus of the embryonic hamster. *Anat. Rec.* **152**: 35-54.

AUERBACH, R. (1960). Morphogenetic interactions in the development of the mouse thymus gland. *Dev. Biol.* **2**: 271-284.

BEN SLIMANE, S., F., HOULLIER, G.C., TUCKER, and J.P., THIERY. (1983). *In vitro* migration of avian hemopoietic cells to the thymus: Preliminary characterization of a chemotactic mechanism. *Cell Diff.* **13**: 1-24.

CHAMPION, S., B., IMHOF, P., SAVAGNER, and J.P., THIERY. (1986). The embryonic thymus produces chemotactic peptides involved in the homing of hemopoietic precursors. *Cell* **44**: 781-790.

FLEISCHMAN, R.A., R.P., CUSTER, and B., MINTZ. (1982). Totipotent hematopoietic stem cells: Normal self-renewal and differentiation after transplantation between mouse fetuses. *Cell* **30**: 351-359.

FONTAINE-PERUS, J.C., F.M., CALMAN, C., KAPLAN, and N.M., LE DOUARIN. (1981). Seeding of the 10-day mouse embryo thymic rudiment by lymphocyte precursors *in vitro*. *J. Immunol.* **126**: 2310-2315.

HAAR, J.L., and F., LOOR. (1981). Selective migration of "null" cells towards a thymus factor *in vitro*. *Thymus* **3**: 187-194.

HALE, M.L., D.L., GREINER, and K.F., MCCARTHY. (1987). Characterization of rat prothymocyte with monoclonal antibodies recognizing rat lymphocyte membrane antigenic determinants. *Cell Immunol.* **107**: 188-200.

HAYASHI, J., E.S., MEDLOCK, and I., GOLDSCHNEIDER. (1984). A selective culture system for generating terminal deoxynucleotidyl transferase-positive (TdT^+) lymphoid precursors cells *in vitro*. I. Description of the culture system. *J. Exp. Med.* **160**: 1622-1639.

HUNT, S.V. (1979). The presence of Thy-1 on the surface of rat lymphoid stem cells and colony-forming units. *Eur. J. Immunol.* **9**: 853-859.

IMAIZUMI, A., M., TORISU, and T., YOSHIDA. (1987). A chemotactic factor for rat thymocytes may regulate T-lymphocyte migration toward the thymic microenvironment. *Cell Immunol.* **108**: 53-63.

IMHOF, B.A., M.A., DEUGNIER, T., ITOH, and J.P., THIERY. (1988a). T-lymphocyte precursors migrate towards chemotactic peptides secreted by embryonic or juvenile thymus. In Proceeding of the 9th International Conference on Lymphatic Tisssues and Germinal Centres in Immune Reactions. *Adv. Exp. Med. Biol.* in press.

IMHOF, B.A., M.A., DEUGNIER, J.M., GIRAULT, S., CHAMPION, C., DAMAIS, T., ITOH, and J.P., THIERY. (1988b). Thymotaxin: A thymic epithelial peptide chemotactic for T-cell precursors. *Proc. Natl. Acad. Sci. USA. In press.*

ITOH, T., S., AIZU, S., KASAHARA, and T., MORI. (1982). Establishment of a functioning epithelial cell line from the rat thymus: A cell line that induces the differentiation of rat bone marrow cells into T cell lineages. *Biomed. Res.* **2**: 11-19.

JOTEREAU, F., F., HEUZE, V., SALOMON-VIE, and H., GASCAN. (1987). Cell kinetics in the fetal mouse thymus: Precursor cell input, proliferation and emigration. *J. Immunol.* **138**: 1026-1030.

JOTEREAU, F.V., and E., HOUSSAINT. (1977). Experimental studies on the migration and differentiation of primary lymphoid stem cells in the avian embryo. In *Developmental Immunobiology*. J.B. Saloman and J.B. Morton eds. Elsevier North Holland Biomedical Press, Amsterdam. 123-130.

JOTEREAU, F.V., and N.M., LE DOUARIN. (1978). The developmental relationship between osteocytes and osteoclasts: A study using the quail-chick nuclear marker in endochondral ossification. *Dev. Biol.* **63**: 253-265.

JOTEREAU, F.V., and N.M., LE DOUARIN. (1982). Demonstration of a cyclic renewal of the lymphocyte precursor cells in the quail thymus during embryonic and perinatal life. *J. Immunol.* **129**: 1869-1877.

KYEWSKI, B.A. (1987). Seeding of thymic microenvironments defined by distinct thymocyte-stromal interactions is developmentally controlled. *J. Exp. Med.* **166**: 520-538.

LE DOUARIN, N.M. (1978). Ontogeny of hematopoietic organs studied in avain embryo interspecific chimeras. In *Differentiation of normal and neoplastic hematopoietic cells.*. Cold Spring Harbor, New York. 5-31.

LE DOUARIN, N.M., and F.V., JOTEREAU. (1980). Homing of lymphoid stem cells to the thymus and the bursa of Fabricius studied in avian embryo chimaeras. In *Immunology* **80**. Fougereau, M. and Dausset, J. eds. Academic Press, London. 285-302.

LE DOUARIN, N.M., F., DIETERLEN-LIÈVRE, and P.D., OLIVER. (1984). Ontogeny of primary lymphoid organs and lymphoid stem cells. *Am. J. Anat.* **170**: 261-299.

LEPAULT, F., R.L., COFFMAN, and I.L., WEISSMAN. (1983). Characteristics of thymus-homing bone marrow cells. *J. Immunol.* **131**: 64-69.

MARTIN, C., D., BEAUPAIN, and F., DIETERLEN-LIÈVRE. (1980). A study of the development of the hemopoietic system using quail-chick chimeras obtained by blastoderm recombination. *Dev. Biol.* **75**: 303-314.

METCALF, D. (1966). Experimental and clinical studies. In *The Thymus*. Wolstenholme G.E.W. and Porter R., eds. London, Churchill Livingstone. 242-262.

METCALF, D., AND M.A.S., MOORE. (1971). In *The Haemopoietic cells*. North Holland Publ. Co. Amsterdam.

MILLER, J.F.A.P. (1961). Immunological functions of the thymus. *Lancet* **2**: 748-749.

MOORE, M.A.S., and J.J.T., OWEN. (1967). Chromosome marker studies in the irradiated chick embryo. *Nature* **215**: 1081-1082.

MOORE, M.A.S., and J.J.T., OWEN. (1967). Experimental studies on the development of the thymus. *J. Exp. Med.* **126**: 715-726.

NAGEL, M.D., J., NAGEL, and R., JACQUOT. (1981). Early erythropoiesis in foetal rat bone marrow: Evidence for a liver-to-bone marrow relay. *J. Embryol. Exp. Morph.* **64**: 275-293.

OPSTELLEN, D., G.J., DEENEN, J., ROZING, and S.V., HUNT. (1986). B lymphocyte-associated antigens on terminal deoxynucleotydul transferase-positive cells and pre-B cells in bone marrow of the rat. *J. Immunol.* **137**: 76-84.

PYKE, K., and J.-F., Bach. (1981). *In vitro* migration of potential hemopoietic precursors from the murine foetus. *Thymus* **3**: 1-7.

SAVINO, W., T., ITOH, B.A., IMHOF, and M., DARDENNE. (1986). Immunohistochemical studies on the phenotype of murine and human thymic stromal cell lines. *Thymus* **8**: 245-256.

SCOLLAY, R., J., SMITH, and V., STAUFFER. (1986). Dynamics of early T cells: Prothymocyte migration and proliferation in the adult mouse thymus. *Immunol. Rev.* **91**: 129-157.

SMALL, M., L., BARR-NEA, and M., ARONSON. (1984). Culture of thymic epithelial cells from mice and age-related studies on the growing cells. *Eur. J. Immunol.* **14**: 936-942.

TILL, J.E., and E.A., MCCULLOCH. (1961). A direct measure of the radiation sensitivity of normal mouse bone marrow cell. *Radiat. Res.* **14**: 213-222

ZIGMOND, S.H., and J.G., HIRSCH. (1973). Leukocyte locomotion and chemotaxis: New methods for evaluation and demonstration of a cell-derived chemotactic factor. *J. Exp. Med.* **137**: 387-410.

ZIGMOND, S.H. (1977). The ability of polymorphonuclear leukocytes to orient in gradients of chemotactic factors. *J. Cell Biol.* **75**: 606-616.

CHAPTER 2

REARRANGEMENT AND EXPRESSION OF T CELL RECEPTOR GENES AND THE SELECTION OF THYMOCYTES

ROBERT I. LECHLER

Department of Immunology, Royal Postgraduate Medical School, Hammersmith Hospital, Du Cane Road, London W12.ONN.

INTRODUCTION

The capacity of the immune system to distinguish self from altered-self has fascinated cellular immunologists for many years, and continues to fuel active investigation. A key mechanism that contributes to this self *versus* altered self discrimination is the selection of the mature T lymphocyte repertoire in the thymus. It is generally thought that both positive and negative selective influences are imposed on differentiating thymocytes. The consequence of thymic selection is a mature T cell repertoire with the capability of using self (thymic) MHC molecules as restriction elements for the recognition of a wide array of foreign antigens while having a limited capability of mounting auto-aggressive T cell immunity. Advances made during the past three years have provided tools for the detailed examination of the generation and selection of the repertoire of T lymphocyte specificities. These advances have included the cloning and sequencing of T cell receptor (Tcr) genes, and the generation of monoclonal antibodies (mAb) against monomorphic and variable portions of CD3/Tcr complexes. The application of these molecular and serological tools has provided some important answers, but has also thrown up some complex questions. In this review, the current state of knowledge about Tcr genetics and thymocyte selection will be outlined, drawing on murine and human data, and the unanswered questions will be set out.

GENOMIC ORGANIZATION AND REARRANGEMENT OF T CELL RECEPTOR GENES

Four rearranging Tcr genes have been identified, α (Chien *et al.*, 1984; Saito *et al.*, 1984b), β (Yanagi *et al.*, 1984; Hedrick *et al.* 1984), γ (Saito *et al.*, 1984a) and δ (Chien *et al.*, 1987), each encoding a polypeptide of the same name. These gene products are expressed at the cell surface with the CD3 complex as two separate heterodimers, αβ, referred to as Ti, (Allison, McIntyre and Bloch, 1982; Meuer *et al.*, 1983) and γδ (Brenner *et al.*, 1986; Bank *et al.*, 1986; and Lew *et al.*, 1986). These four genes are located on three chromosomes as outlined in Table I for man and for mouse.

TABLE I Chromosomal locations of T cell receptor genes.

Gene	Man		Mouse	
α	14	(Caccia *et al.*, 1985)	14	(Dembic *et al.*, 1985)
β	7	(Caccia *et al.*, 1984)	6	(Caccia *et al.*, 1984)
γ	7	(Rabbitts *et al.*, 1985)	13	(Kranz *et al.*, 1985)
δ	14	(Loh *et al.*, 1987)	14	(Chien *et al.*, 1987a)

The germline organization and nucleotide sequence of these genes bear many parallels with the immunoglobulin genes with which the T cell receptor genes are thought to share a common ancestor. Schematic diagrams of the germline arrangement of the four human genes is shown in Fig.1. The Tcr β gene is the best-defined, and the Tcr δ gene the least defined. There are a number of notable features of each of these genes. The Tcr α gene is a large gene and appears to be spread out over approximately 1000 kilobases (kb), comprising approximately 50 V, 100 J, and a single C gene segment in man (Klein *et al.*, 1987). The estimated number of Tcr Vβ gene segments in man is >70 (Kimura *et al.*, 1987) while in the mouse it appears that only 20-25 Vβ genes exist (Barth *et al.*, 1985). These Vβ genes are dispersed through mostly small, 1-5 member, families. The Vβ segments can rearrange to one of the D segments following the juxtaposition of a Dβ to one of 12 (mouse) (Malissen *et al.*, 1984) or 13 (man) (Toyonaga *et al.*, 1985) functional Jβ segments. The duplicated C segments in the β locus, each with a cluster of J and one D segment is an unique structural feature amongst rearranging genes. The Tcr β locus has one other unique feature, namely that the two D segments have three open reading frames, creating six possible Dβ gene products from two genes (Barth *et al.*, 1985; Toyonaga *et al.*, 1985).

The Tcr γ gene provides less scope for the generation of diversity, there being only 10 identified Vγ segments in man (Lefranc *et al.*, 1986), and 5 in the mouse (Garman *et al.*, 1986; Iwamoto *et al.*, 1986), a smaller number of Jγ segments and two (man) (Lefranc *et al.*, 1986) or three (mouse) (Iwamoto *et al.*, 1986) functional Cγ segments. The location of the δ gene locus within the α gene is unusual. In the mouse, and in man, two Dδ, two Jδ and a single Cδ gene are located a short distance 5' to the most 5' known Jα segment, and approximately 75kb 5' to Cα (Chien *et al.*, 1987; Baer *et al.*, 1988). The number of Vδ elements is not known, but appears to be very limited. One cDNA gene clone has been isolated in which a V gene that would appear to belong to a Vα family has rearranged to a Dδ segment (Chien *et al.*, 1987b). One obvious consequence of the location of the δ gene is that the rearrangement of a Vα to a Jα gene leads to the deletion of the Tcr δ gene. Indeed in the majority of peripheral Tiαβ-expressing T cells, the δ gene is absent (Lindsten *et al.*, 1987). The major difference from other 'three element' rearranging genes, IgH and Tcrβ, is that Vδ Dδ joining can occur prior to DδJδ rearrangement. This was noted in a thymocyte genomic library, and by Southern blot analysis of thymocyte DNA (Chien *et al.*,

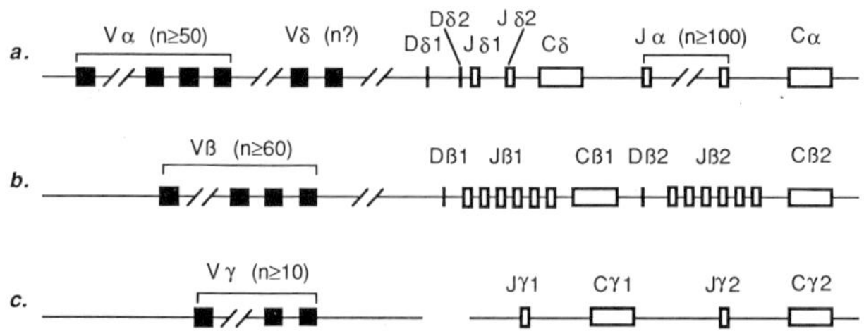

Figure 1. The genomic organization of human Tcr α and δ (a), β (b), and γ (c) Tcr genes are schematically represented. The drawing is not to scale. The region containing Vδ genes has not yet been linked to the Dδ, Jδ, Cδ region.

1987b).

The mechanisms of rearrangement of the Tcr genes closely resemble those of the immunoglobulin genes, utilizing the conserved heptamer and nonamer sequences, and following the 12 base pair (bp) 23 bp, spacer rule (Chien *et al.*,1984). One distinctive feature of the Tcr β and δ genes is that the D segments are spanned by a 12 bp spacer at the 5' and a 23 bp spacer sequence at the 3' end (Siu *et al.*, 1984b). This feature of Tcr Dβ and Dδ genes creates the possibility of several different rearrangement options, as portrayed in Fig.2. Though theoretically possible, options *b* and *c* have not been seen in thymocyte or mature T cell complementary DNA (cDNA) sequences of Tcr β genes, though a murine cDNA clone with a DδDδ rearrangement has been reported (Chien *et al.*, 1987b). Allelic exclusion applies to the rearrangement and expression of

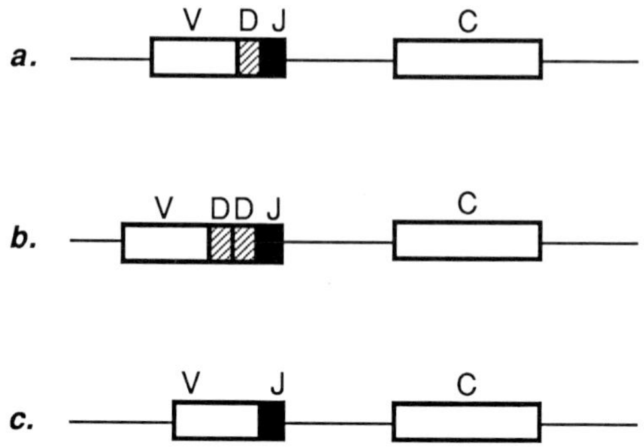

Figure 2. The possible rearrangements of the Tcrβ locus are shown. One of the Dβ segments can join to any 3' J segment (a); alternatively Dβ1 can ,in theory, join to Dβ2 which in turn can rearrange to a Jβ2 segment (b); a Vβ segment subsequently rearranges to a DJβ segment, or can, in theory, join directly to an unrearranged Jβ (c).

Tcr genes as for Ig genes in B cells. Thus the successful, productive rearrangement of any one of these genes terminates further recombination within that locus.

DIVERSITY GENERATION BY T CELL RECEPTOR GENES

The complicated germline organization of the T cell receptor genes exists in order to furnish the host with a mature T cell repertoire that is capable of recognizing the world of extrinsic antigens. All but one of the mechanisms employed by the immunoglobulin genes are used by the Tcr genes in order to generate diversity of the T cell repertoire.

The shared mechanisms are first, recombinatorial diversity, due to the stochastic rearrangement of individual V,D,J combinations (Siu *et al.*, 1984a). A second mechanism is junctional diversity due to the deletion of small numbers of nucleotides from the joining ends of V, D, and J germline segments, and the addition of non-germline-encoded (N) nucleotides by the enzyme, terminal deoxytidyltransferase (TdT). One of the possible consequences of imprecise joining of individual gene segments is a frame shift downstream of the junction, leading to an out of frame transcript. Two structural features of the Tcr β gene, one of which is shared by the δ gene, serve to minimize thymocyte wastage due to out of frame rearrangements. The existence of two D,J,C clusters in the β gene is an unique feature amongst rearranging genes, and affords thymocytes with four attempts at making an in frame rearrangement by using the two C segments on two chromosomes. The second feature, shared by β and δ genes, is the possession by the Dβ and Dδ segments, of three open reading frames. Not only does this reduce the likelihood of thymocyte wastage, but it effectively, trebles the number of D genes. Furthermore the occurrence of DδDδ joining creates three recombinational joining events in rearranged δ genes (V to D, D to D, and D to J), thereby creating enormous potential for junctional diversity. The third source of diversity is the association of individual combinations of α and β, or γ and δ, chains. The major difference between immunoglobulin and Tcr genes is the absence of somatic mutation of Tcr genes. No direct, nucleotide sequence data has been reported in evidence of somatic mutation of rearranged Tcr genes. There is a very plausible reason for this major difference from Ig genes, as discussed below.

For the human αβ receptor heterodimer the potential diversity can be crudely calculated as follows:-

α diversity 5 0(V) x 100 (J) x > 10 (N) = > 5×10^4

β diversity 70 (V) x "6"(D) x 13 (J) x > 100 (N) = > 5.5×10^5

αβ diversity due to random α and β chain pairing = > 2.75×10^{10}

The figures of 10 and 100 for N region diversity are an estimate that includes the contribution of junctional diversity, and almost certainly represents an underestimate. The higher figure for the β gene reflects the two sites of recombination compared to one site for the α genes. This figure is similar to that arrived at for Ig genes and their products using the same calculations. For Ig molecules there is no reason why this potential diversity should not be realized, for Tcr molecules, however, there appear to be constraints imposed on developing T cells within the thymus which limit the expressed repertoire considerably. These constraints are discussed below.

CHRONOLOGY OF T CELL RECEPTOR GENE REARRANGEMENT AND EXPRESSION INTHE THYMUS

It has long been known that the thymus is the environment in which T cells mature, and it is within the thymus that an ordered programme of genetic rearrangement occurs. This genetic programme is much more amenable to investigation in the mouse, and the data that are included here are from mouse experiments. Immigration into the thymus by T cell precursors from the foetal liver commences on the 10th and 11th day of gestation as judged by the appearance of non-stromal cells (Moore and Owen, 1967; Owen and Raff, 1970). These large basophilic cells undergo rapid proliferation and differentiation and give rise to cells with the surface phenotype of T cells, the pan-T cell marker Thy-1 appearing on day 15 of gestation (van Ewijk, Jenkinson and Owen, 1982). Cells expressing Ti αβ dimers at the cell surface can first be detected on day 17 of gestation (Roehm *et al.*, 1984). The chronological sequence and timing of Tcr gene expression has been studied using in *situ* hybridization by Pardoll *et al.* (1987). Based on these, and other results, a model of early thymocyte ontogeny is outlined in Fig.3. Transcripts of Tcr messenger RNA (mRNA) were detectable at the earliest time point studied namely day 12 of gestation. Expression of the β gene was first noted on day 12.5, and α transcripts were first seen on day 16. No phenotypic marker of "prothymocytes", ie. bone marrow derived cells that are committed to the T cell lineage but have not yet entered the thymus, has yet been identified. It is difficult, therefore, to determine whether Tcr γ-gene rearrangement and transcription occurs pre-thymically. One group has detected Tcr γ mRNA in murine fetal liver (Haars *et al.*, 1986), a primitive haemopoietic organ, and γδ-expressing CD3 positive cells are found in athymic Nude mice (Yoshikai *et al.*, 1986; Pardoll *et al.*, 1988), which suggests that rearrangement of these genes can occur extrathymically. Whether or not Tcr γ and δ gene rearrangement and expression occur pre-thymically, it is clear that they are the first loci to be expressed.

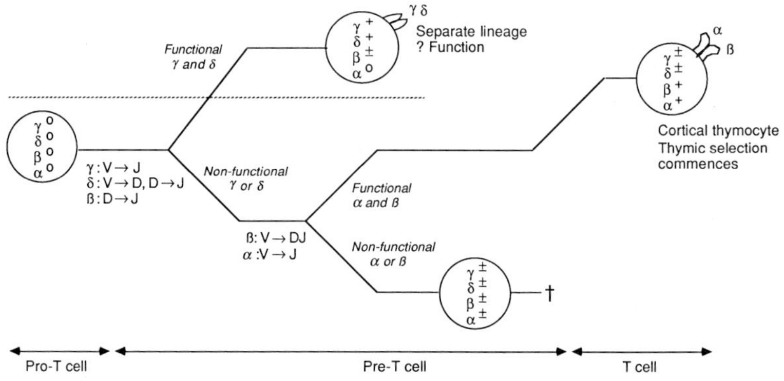

Figure 3. A model of the early stages of thymocyte differentiation is shown; the probable relationship between Tcr gene rearrangements and branch points in the differentiation pathway are indicated. Unrearranged genes are shown as *o*, genes that are partially rearranged or that have produced out of frame rearrangements are shown as ±, productive rearrangements are shown as +.

It is known from previous studies of immunoglobulin gene expression that gene transcription can occur prior to rearrangement (Nelson, Kelley and Perry; 1985). Seemingly, when the chromatin "opens" prior to recombination, some transcription of germline sequences occurs. Because the data on early thymocyte Tcr gene expression are based on in *situ* hybridization which gives no information about transcript size, it is very possible that the earliest detection of Tcr-specific RNA for each gene locus reflects germline transcription. In the same *in situ* hybridization study, a Vβ3-specific probe was used, and hybridization was not seen until day 14, 1.5 days after the earliest appearance of constant region transcripts. This further suggests that the β mRNA detected on days 12 and 13 is transcribed from germline, or incompletely rearranged DJC genes. Another approach to this question was taken by Born *et al.* (1985). They immortalized thymocytes on different days of gestation by fusion to a tumour cell, and examined gene rearrangements in the resulting hybridomas. From these experiments it appeared that β gene rearrangement first occurred on day 15. However, the caveat must be borne in mind that the thymocytes which underwent fusion may have been an unrepresentative sample, and that rearrangement could have been occurring somewhat closer to the time at which gene expression was first detected. Though the precise timing of Tcr gene rearrangement is difficult to establish, it is clear that the γ and δ genes are the first to be rearranged and expressed co-incident with, or possibly prior to, the entry of cells into the thymus. The rearrangement of β and α genes follows after a short time interval.

The products of these genes give rise to two distinct heterodimeric receptors. The large majority of human and murine T cells express a clonotypic αβ (Ti) dimeric receptor (Allison *et al.*, 1982; Meuer *et al.*, 1983), immature thymocytes and a small fraction of peripheral T cells express a receptor composed of a γδ dimer (Brenner *et al.*, 1986; Bank *et al.*, 1986 and Lew *et al.*, 1986). Both of these heterodimers are co-expressed with the invariant, transmembrane-signalling, CD3 complex. It appears that αβ-expressing and γδ-expressing cells represent phenotypically and functionally discrete populations, no cell expressing both types of receptor complex has ever been identified. Given the ordered sequence of Tcr gene expression described above, the question arises as to whether the early thymocytes that productively rearrange and express γδ receptors differentiate into αβ-expressing cells, or whether they are terminally differentiated cells. Two lines of evidence suggest that intra-thymic γδ-expressing cells are terminally differentiated. First, it was noted that the sequences of murine Tcr γ cDNA clones isolated from peripheral αβ-expressing T cells were, without exception, out of frame (Reilly *et al.*, 1986). If the peripheral T cells studied in these experiments were the descendants of γδ-expressing precursors, it would have to be postulated that a mutation occurred leading to a frame-shift in the productively rearranged γ gene. Not only does this seem unlikely, but somatic mutation has not been documented in Tcr genes. A further piece of evidence for intra-thymic γδ cells being end stage cells is that no expression of the non-glycosylated Vγ1.2-Cγ2 polypeptide is detectable on γδ-bearing thymocytes. In contrast, Vγ1.2-Cγ2 mRNA accounts for the majority of transcripts in peripheral αβ-expressing T cells (Lew *et al.*, 1986). It also appears that distinct populations of γδ-expressing cells may exist. In the mouse it has been noted that thymic and peripheral γδ-cells have markedly different frequencies of usage of Cγ regions. The vast majority of γ-chains in the thymic population contain Cγ1, while in the spleen there is equal usage of Cγ1 and Cγ4 (Pardoll *et al.*, 1988). The pattern of Tcr γ rearrangement and expression in man is similar (Lanier *et al.*, 1987), and the available evidence is consistent with there being two lineages of human γδ-expressing cells. Cells with the phenotype $CD3^+$, 4^- 8^- isolated from the peripheral blood of healthy individuals express a disulphide-linked γδ heterodimer. The γδ receptor expressed by thymocytes and peripheral cells from immunodeficiency patients has no disulphide linkage (Lanier *et al.*,1987). These two forms of receptor utilize Cγ1 and Cγ2 genes respectively. The possibility exists that peripheral γδ-expressing cells represent a totally independent lineage from their thymic counterparts, and that they arise extrathymically. However, the reduced numbers of splenic γδ cells in Nude mice, and the restoration of numbers following thymus grafting these animals provides some evidence against this suggestion (Pardoll *et al.*, 1988). One cell type that warrants further investigation in this regard is the $CD3^+$, $\gamma\delta^+$, Thy-1^+ dendritic epidermal cell (Koning *et al.*, 1987). It is conceivable that the skin epithelium provides an alternative site for the differentiation of these γδ-expressing cells. While a number of unanswered questions remain, it seems clear that the intrathymic Tcr γδ-expressing cells are terminally differentiated cells, and do not give rise to the cells that rearrange and express Tcr α and β genes.

Limited data exist regarding the functional properties of Tcr γδ-bearing cells. These cells are able to secrete interleukin-2 (IL-2) in response to immobilized anti-CD3 monoclonal antibody (mAb) (Bank *et al.*, 1986; Bluestone *et al.*, 1987), and in response to stimulation by the T cell mitogen concanavalin A (Bluestone *et al.*, 1987). Furthermore it has recently been reported that, when activated by anti-CD3 mAb, γδ-expressing cells are capable of MHC-unrestricted cytolysis (Brenner *et al.*, 1987), however, the receptor specificity and function of thymocytes and T cells remains obscure. Given the capacity to generate diversity of γ and δ variable gene products, it is tempting to speculate that they interact with antigen and an MHC product in a comparable manner to Ti αβ-dimers. One set of experiments reported by Owen *et al.* (1988), suggests that intrathymic γδ-cells may play a role in supporting the full rearrangement of Tcrβ genes. Using an *in vitro*, murine, thymic organ culture system, they noted that 12d fetal thymic organ cultures performed in the presence of anti-CD3 mAb failed to generate any thymocytes with full VDJ Tcrβ rearrangements. However, cells with incomplete DJβ, and full VJα rearrangements were seen. One interpretation of this result is that γδ-bearing thymocytes, upon interaction with their ligand produce factors that support the rearrangement of Vβ to DJβ partially rearranged complexes. Thus the γδ-expressing thymocytes might represent a specialized subpopulation of cells facilitating the differentiation of other thymocytes into mature T cells.

INTRATHYMIC CONSTRAINTS ON THE T CELL REPERTOIRE

(I) Negative Selection.

It has long been known that a high percentage of thymocytes dies within the thymus. Estimates of between 80% (Kampinga, personal communication) and 98% (Scollay *et al.*, 1980) have been made by different workers using different means to address the issue. One probable reason for thymocyte death is the failure to generate in-frame transcripts of rearranged Tcrα and β genes (Kronenberg *et al.*, 1986). It is clear, however, that a more active process of selection is also involved. Although mature T cells mount strong proliferative and cytotoxic responses if co-cultured with MHC incompatible stimulator cells, much weaker proliferation and no cytotoxicity is generated by autologous cells. Furthermore T cell immunity is readily induced against xenogeneic proteins, but under the same conditions usually cannot be induced against the equivalent autologous protein. It is assumed that one of the major mechanisms of self-non-reactivity is the deletion in the thymus of auto-reactive cells, with specificity for self MHC and fragments of self antigens.

Experimental evidence has recently been reported that substantiates this theory of intrathymic clonal abortion, in some cases at least. Studying the expression of the Tcr Vβ segment, Vβ17a, Kappler *et al.*, (1987b) noticed a negative correlation between I-E expression and the expression of Vβ17a on mature mouse T cells. In I-E$^-$ mice, ie. those with H-2^b or H-2^s haplotypes, approximately 10% of peripheral T cells are Vβ17a positive. In contrast in I-E$^+$ mice less than 1% peripheral T cells use this Vβ gene. When Vβ17a positive T cell hybridomas were made from I-E$^-$ mice a remarkably high frequency of anti-I-E alloreactivity was detected; 70% of the hybridomas responded to I-E^k and 45% to I-E^d allogeneic cells (Kappler *et al.*, 1987a). Thus it appears that the Vβ17a product confers high affinity for I-E molecules, and thymocytes expressing β chains with this variable portion are deleted in the thymus. When thymocyte subpopulations from I-E$^+$ and I-E$^-$ mice were compared, the numbers of Vβ17a positive immature cortical thymocytes was very similar, but amongst mature medullary thymocytes Vβ17a cells were almost totally depleted in the I-E$^+$ strains. Immature cortical thymocytes go through a stage of expressing two cell surface molecules, CD4 and CD8 (Scollay *et al.*, 1984; Mathieson and Fowlkes, 1984), which interact with monomorphic determinants on MHC class II (Doyle and Strominger, 1987) and class I (Swain 1983) molecules respectively. Mature, medullary thymocytes and peripheral T cells express either CD4 or CD8 according to the class of MHC product to which they are restricted. It is interesting to note that in I-E$^+$ mice there are no CD8$^+$ CD4$^-$ Vβ17a-bearing T cells (Kappler *et al.*, 1987b). This may be interpreted in two ways. Either clonal deletion occurs during the double positive CD4$^+$ CD8$^+$ stage of cortical differentiation, or if clonal deletion occurs in the medulla the affinity of Vβ17a for I-E is so high that even a CD8$^+$ CD4$^-$Vβ17a medullary thymocyte would be aborted. The results of two recent sets of experiments favours the former possibility. First, high doses of anti-CD4 monoclonal antibody were given to T cell-depleted mice during the reconstitution phase following bone marrow stem cell injection. As a result of this treatment substantial numbers of CD8$^+$ Vβ17a$^+$ T cells were detected in the periphery (Fowlkes *et al.*, in press). Unless CD4$^+$ cells are somehow normally involved in the deletion of CD8$^+$ Vβ17a$^+$ cells, this result suggests that the deletion of Vβ17a$^+$ thymocytes occurs at the CD4$^+$8$^+$ stage of thymocyte differentiation. The second piece of experimental evidence for clonal deletion of double-positive thymocytes derives from the study of Tcr transgenic mice reported by Kisielow *et al.* (1988). The α and β genes from an H-Y-specific, H-2Dd-restricted, CD8$^+$ T cell clone were introduced into the germline of H-2d-expressing mice. Thymocytes expressing the transgene-encoded receptor were deleted in male mice at the CD4$^+$8$^+$ stage of differentiation.

A very similar set of observations has been documented in mice carrying the minor lymphocyte stimulating (Mls) *a* allele. T cells expressing Vβ6 (MacDonald *et al.*, 1988) and Vβ8.1 (Kappler *et al.*, 1988) products appear to be deleted in Mls *a*$^+$ mice while they are well represented in the periphery of Mls *a*-mice. As for Vβ17a, both CD4$^+$ and CD8$^+$ T cells with either of these Vβ products are depleted, although Mls *a*-reactive T cells are class II dependent. It may be argued that these results add a new dimension to the observation of clonal abortion of Vβ17a positive thymocytes described above because the specificity of the "forbidden receptors" in this last case is not primarily for an MHC molecule, but rather for a self molecule, presumably seen in the context of self MHC. This begs the question, however, as to the nature of Mls determinants. One hypothesis proposes that the difference between Mls *a* and *b* resides in the MHC class II molecules themselves. Though the MHC genes may be identical in Mls *a* and *b* congenic strains of mice, it is possible that the phenotypic difference may result from post-translational modification of the α and/or β glycoprotein due to the action of an enzyme which is present in some and not other mouse strains. Although this is a theoretical possibility, there is no supporting biochemical data. It is more likely that Mls determinants are created by a high affinity interaction between an endogenous peptide and a class II MHC molecule (Kappler *et al.*, 1988).

These examples of almost complete depletion of T cells expressing particular Vβ segments, irrespective of the other variable portions of the β chain, and of the partner α chain, may be exceptional, but they lend strong support to the hypothesis of intrathymic negative selection of self-reactive cells. It seems likely that in most instances self-reactivity to binary complexes of self MHC and peptides of self molecules will be created by contributions from VDJβ and VJα polypeptides, rather then by a single Vα or Vβ segment alone. Indeed in a detailed examination of the receptors of a series of T cell clones reactive with cytochrome-C and I-E, but with differing fine specificities, correlations were noted between junctional differences surrounding the Dβ segments and fine antigen specificity (Fink *et al.*, 1986). Wholesale deletions of Vα or Vβ segments may therefore often not be seen, and more subtle approaches may be necessary in order to detect further examples of clonal deletion secondary to self-reactivity conferred by D, J, N and junctional sequences. As mentioned earlier, somatic mutation of Ti genes has not been documented, and appears not to be used as a mechanism of diversity and increased affinity for antigen, as is seen in Ig genes and their products. Given the intrathymic deletion of autoreactive thymocytes, were somatic mutation to occur in peripheral T cells, autoreactive specificities might well result. It is presumably in order to avoid this risk that somatic mutation of rearranged genes is avoided by T cells.

(ii) Positive Selection.

Whether or not thymocytes with affinity for self-MHC alleles are positively selected in the thymus remains a contentious issue (Zinkernagel *et al.*, 1978; Klein and Nagy, 1982). The answer depends, in part, upon whether or not the germline repertoire of Tcr genes is skewed towards the MHC products of the same species. If there is no germline MHC bias, then a great deal of selection is required, since all peripheral T cells have a requirement to co-recognize antigen with an MHC molecule. On the other hand if the germline genes have evolved to have intrinsic affinity for MHC products, as seems likely, and as is suggested by the Vβ17a observations described above, then it could be argued that little further selection for self-MHC recognition may be necessary. However, the results of a large series of murine chimaera and/or thymus graft experiments suggest that thymocytes are positively selected for recognition of thymic MHC specificites. The general design of these experiments involved the injection of MHC heterozygous (AxB) bone marrow stem cells into leucocyte depleted A or B strain recipient mice, or the transplantation of an A or B thymus into an athymic (A x B)F_1 recipient. The T cell repertoire of these animals following differentiation in an A or a B strain thymus appears to be strongly biased toward the thymic MHC type (Bevan and Fink, 1978; Zinkernagal *et al.*, 1978; Lo and Sprent, 1986). The recently reported results of Marrack *et al.*, (1988) supplement these chimaera models. They observed that thymocytes from H-2*b* x *k* mice treated from birth with either anti-I-A^b or anti-I-A^k monoclonal antibody, when transferred into irradiated *b* x *k* recipients with antigen, gave rise to antigen specific cells strongly biased towards the untreated I-A allele. These studies avoid the complexities of chimaeras and lend support to the concept that MHC alleles expressed in the thymus are responsible for selection of thymocyte receptors with specificity for thymic MHC. In addition, the extreme rarity of antigen-specific T cells from I-Ab mice which are able to recognize antigen with I-A*bm*12 antigen presenting cells, and *vice-versa* (Kanamori *et al.*, 1984), implies that T cells are selected for self-MHC recognition. The I-A *bm*12 mutant differs from I-Ab at only three amino acid residues, 67, 70, and 71, in the Aβ1 domain (McIntyre and Seidman, 1984). By modelling MHC class II structure on the recently published HLA-A2 three-dimensional structure (Bjorkman *et al.*, 1987), residues 67 and 71 are predicted to contribute to the antigen- binding site of Aβ, and residue 70 is predicted to point upwards towards the T cell receptor. It seems unlikely that residues 67 and 71 impose constraints on the binding of all antigens. Rather, the discrimination by T cells between I-A^b and I-A*bm*12 may be taken to reflect, at least in part, the intrathymic selection of receptors with affinity for one or other Aβ polypeptide, a process in which residue 70 would be envisaged as playing a prominent role.

The arguments against positive selection include the recent observation that, once cortical thymocytes express Tiαβ dimers, cell division appears to cease. This was deduced from correlating the incorporation by thymocytes of tritiated thymidine with the appearance of cell surface Ti molecules (Parkin, Owen and Jenkinson, 1988). A negative correlation was noted. This does not rule out further selection however; it could be that positive signals are required in order for Ti-expressing thymocytes to survive and to migrate into the medulla. Such positive signals could be derived from epithelial cells with which thymocytes with self-MHC affinity interact. In the absence of positive signalling the Ti-bearing thymocytes may be programmed to die. A further argument against positive selection is the absence of any documented correlations between the usage of particular Vα or Vβ gene usage by mature T cells, and the expression of individual MHC alleles. Detection of correlations of this kind may require a more detailed screening of the use of D and J segments and even of junctional sequences. Nonetheless, if the need for positive selection does impose major constraints on the T cell repertoire, some MHC:Ti complementarity should be detectable.

(iii) Sites of Selection.

The thymic cells responsible for positive and negative selection, and the stages of thymocyte differentiation at which these events occur has long been a contentious issue (Longo and Schwartz, 1980; Lo and Sprent, 1986). Until recently, in so far as a consensus view existed, it was that positive selection resulted from a "favourable" interaction between thymic epithelial cells and cortical thymocytes, and that tolerance or clonal deletion was determined by the interaction of medullary dendritic cells with mature thymocytes. The clonal deletion of Vβ17a-expressing (Kappler, Roehm and Marrack, 1987b) and Vβ6/Vβ8.1 -expressing (MacDonald *et al.*, 1988; Kappler *et al.*, 1988) thymocytes apparently at the $CD4^+8^+$ cortical stage of differentiation (Fowlkes *et al.*, in press) taken together with the deletion of anti-H-Y Ti-αβ transgenic thymocytes at the double-positive stage, strongly suggests that tolerance and negative selection can occur in the cortex. A set of I-Eα-transgenic, H-2b, mice has recently been investigated which express I-E molecules with a wild-type distribution, or only on thymic epithelial cells (van Ewijk *et al.*, 1988). The results of these studies provide further evidence suggestive that the deletion of at least some cortical thymocytes is due to their interaction with thymic epithelium. In the transgenic mice in which I-E expression was confined to the thymic epithelium, the numbers of peripheral Vβ17a-T cells was markedly reduced, and reactivity to the thymic I-E type was eliminated. Assuming that the presence of small numbers of $I\text{-}E^+$ dendritic cells was not overlooked, it appears that thymic epithelial cells play a role in negative selection. The apparent inconsistency of these data may be resolved in the following way: the interaction between thymic epithelial cells and $CD4^+8^+$ thymocytes (with low levels of Ti expression) may be primarily responsible for the selection of cells with an intermediate affinity for self-MHC. However, thymocytes with very high affinity for self MHC (such as $V\beta17a^+$ cells) may be deleted due to the receipt of a qualitatively or quantitatively different signal. The majority of self reactive cells are not "exposed" until they reach the medulla, raise their level of Ti expression, and encounter dendritic cells with very high levels of MHC expression. Until the nature of positive and negative signalling is more clearly understood, such hypotheses will remain highly speculative.

Future Directions

As is clear from this review, although the last few years have seen major advances in our understanding of T cell recognition, many fundamental questions remain unanswered. Three such questions are (i) what is the molecular basis of intrathymic selection? The signals (and their induction) that lead to thymocyte differentiation on the one hand, and death on the other are unknown. (ii) Does tolerance to tissue-specific antigens occur in the thymus? The presentation of peptides derived from ubiquitous and/or circulating auto-antigens can easily be evisaged, but self molecules that are sequestered in specific tissues, such as myelin basic protein, do not have ready access to the thymus. (iii) What is the specificity and function of $\gamma\delta$-expressing thymocytes? The evidence suggests that these cells constitute a distinct intrathymic lineage, but the ligands with which they interact, and their possible function in supporting the differentiation of other thymocytes need to be defined.

The application of transgenic technology to these problems is likely to continue to generate interesting and sometimes interpretable results. Rearranged Tcr genes have already been introduced into the mouse germline (Kisielow *et al.*, 1988), and doubtless more such experiments are in train. This approach should provide further information about allelic exclusion, tolerance, and the respective contributions of the α and β chains of the receptor. *In vitro* formation of the tri-molecular complex of MHC, antigen, and Ti using soluble components should pave the way for crystals of this complex to be grown and analyzed. Definition of the three-dimensional structure of such a complex would provide a major breakthrough in determining the physical interactions underlying thymic selection. Given the complexity of events, and of the microenvironment in which they take place, this area of cellular immunology should keep immunologists in business for some years to come.

REFERENCES

ALLISON, J.P., McINTYRE, B.W. and BLOCH, D. (1982) Tumour-specific antigen of murine T-lymphoma defined with monoclonal antibodies. *Journal of Immunology*, **129**, 2293-2300.

BAER, R., BOEHM, T., YSSEL, H., SPITS, H. and RABBITTS, T.H. (1988) Complex rearrangements within the human Jδ-Cδ/Jα-Cα locus and aberrant recombination between J segments. *EMBO Journal* **7**, 1661-1668.

BANK, I., DePINHO, R.A., BRENNER, M.B., CASSIMERIS, J., ALT, F.W. and CHESS, L.A. (1986) A functional T3 molecule associated with a novel heterodimer on the surface of immature human thymocytes. *Nature*, **322**, 179-181.

BARTH, R., KIM, B., LAN, N., HUNKAPILLER, T., SOBIECK, N., WINOTO, A., *et al.* (1985) The murine T-cell receptor uses a limited repertoire of expressed Vβ gene segments. *Nature*, **316**, 517-523.

BEVAN, M.J., and FINK, P.J. (1978) The influence of thymus H-2 antigens on the specificity of maturing killer and helper cells. *Immunological Reviews*, **42**, 3-19.

BJORKMAN, P.J., SAPER, M.A., SAMRAOUI, B., BENNETT, W.S., STROMINGER, J.L. and WILEY, D.C. (1987) Structure of the human class I histocompatibility antigen, HLA-A2. *Nature*, **329**, 506-512.

BLUESTONE, J.A., PARDOLL, D.M., SHARROW, S.O. and FOWLKES, B.J. (1987) Ontogeny phenotype and functional characterization of murine thymocytes possessing T-cell associated T cell receptor structures. *Nature*, **326**, 82-84.

BORN, W.J., YAGUE, E., PALMER, J., KAPPLER, J. and MARRACK, P. (1985) Rearrangement of T-cell receptor β-chain genes during T-cell development. *Proceedings of the National Academy of Science, USA*, **82**, 2925-2929.

BRENNER,M.B., McLEAN,J., DIALYNAS,D.P., STROMINGER,J.L., SMITH,J.A., OWEN,F.L., *et al.* (1986) Identification of a putative second T-cell receptor. *Nature*, **322**, 145-149.

BRENNER, M.B., McLEAN, J., SCHEFT, H., RIBERDY, J., SIEW-LAN, A., SEIDMAN, J.G., *et al.* (1987) Two forms of the T-cell receptor protein found on peripheral blood cytotoxic lymphocytes. *Nature*, **325**, 689-694.

CACCIA, N., BRUNS, G.A.P., KIRSCH, I.R., HOLLIS, G.F., BERTNESS, V. and MAK, T.W.(1985) T-cell receptor α chain genes are located on chromosome 14 q 12 in humans. *Journal of Experimental Medicine*, **151**, 1255-1260.

CACCIA, N., KRONENBERG, M., SAXE, D., HAARS, R., BRUNS, G., GOVERMAN, J., *et al.*, (1984) The T-cell receptor β chain genes are located on chromosome 6 in mice and chromosome 7 in humans. *Cell*, **37**, 1091-1099.

CHIEN, Y., BECKER, D., TINDSTEN, T., OKAMURA, M., COHEN, D. and DAVIS, M. (1984a) A third type of murine T-cell receptor gene. *Nature*, **312**, 31-35.

CHIEN, Y., GASCOIGNE, N., KAVALER, J., LEE, N., and DAVIS, M. (1984b) Somatic recombination in a murine T-cell receptor gene. *Nature*, **309**, 322-326.

CHIEN, Y., IWASHIMA, M., KAPLAN, K.B., ELLIOTT, J.F. and DAVIS, M.M. (1987a) A new T-cell receptor gene located within the alpha locus and expressed early in T-cell differentiation. *Nature*, **327**, 677-682.

CHIEN, Y., IWASHIMA, M., WETTSTEIN, D.A., KAPLAN, K.B., ELLIOTT, J.F., BORN, W. and DAVIS, M.M. (1987b) T-cell receptor δ gene rearrangements in early thymocytes. *Nature*, **330**, 722-727.

DEMBIC, Z., BANNWARTH, W., TAYLOR, B.A. and STEINMETZ, M. (1985) The gene encoding the T cell receptor α-chain maps close to the Np-2 locus on mouse chromosome 14. *Nature*, **314**, 271-273.

DOYLE, C. and STROMINGER, J.L. (1987) Interaction between CD4 and class II MHC molecules mediates cell adhesion. *Nature*, **330**, 256-259.

FINK, P.J., MATIS, L.A., McELLIGOTT, D.L., BOOKMAN, M. and HEDRICK, S.M. (1986) Correlations between T-cell specificity and the structure of the antigen receptor. *Nature*, **321**, 219-225.

FOWLKES, B.J., SCHWARTZ, R.H. and PARDOLL, D.M. (in press) Deletion of self-reactive thymocytes occurs at a $CD4^+8^+$ precursor stage. *Nature*.

GARMAN, R.D., DOHERTY, P.J. and RAULET, D.H. (1986) Diversity, rearrangement and expression of murine T cell gamma genes. *Cell*, **45**, 733-742.

HAARS, R., KRONENBERG, M., GALLATIN, W.M., WEISSMAN, I.L., OWEN, F.L. and HOOD, L. (1986) Rearrangement and expression of T-cell antigen receptor and gamma genes during thymic development. *Journal of Experimental Medicine*, **164**, 1-24.

HEDRICK, S.M., COHEN, D.I., NIELSON, E.A. and DAVIS, M.M. (1984) Isolation of cDNA clones encoding T cell-specific membrane-associated proteins. *Nature*, **308**, 149-153.

IWAMOTO, A., RUPP, F., OHASHI, P.S., WALK, C.C., PIRCHER, H., JOHO, R., HENGARTNER, H. and MAK, T.W. (1986) T-cell-specific γ genes in C57 BL/10 mice: sequence and expression of new constant and variable genes. *Journal of Experimental Medicine*, **163**, 1203-1212.

KANAMORI, S., WALSH, W.D., HANSEN, T.H. and TSE, H.Y. (1984) Assessment of antigen-specific restriction sites on Ia molecules as defined by the bm12 mutation. *Journal of Immunology*, **133**, 2811-2814.

KAPPLER, J., WADE, T., WHITE, J., KUSHNIR, E., BLACKMAN, M., BILL, J., *et al.* (1987a) A T cell receptor Vβ segment that imparts reactivity to a class II major histocompatibility complex product. *Cell*, **49**, 263-272.

KAPPLER, J., ROEHM, N. and MARRACK, P. (1987b) T cell tolerance by clonal elimination in the thymus. *Cell*, **49**, 273-280.

KAPPLER, J.W., STAERZ, U., WHITE, J. and MARRACK, P. (1988) Self-tolerance eliminates T cells specific for Mls-modified products of the major histocompatibility complex. *Nature*, **332**, 35-40.

KIMURA, N., TOYONAGA, B., YOSHIKAI, Y., DU, R. and MAK, T. (1987) Sequences and repertoire of the human T-cell receptor α and β chain variable region genes in thymocytes. *European Journal of Immunology*, **17**, 375-383.

KISIELOW, P., BLUTHMANN, H., STAERZ, U., STEINMETZ, M. and von BOEHMER, H. (1988) Tolerance in T-cell-receptor transgenic mice involves deletion of nonmature $CD4^+8^+$ thymocytes. *Nature*, **333**, 742-746.

KLEIN, J. and NAGY, Z.A. (1982) MHC restriction and Ir genes. *Advances in Cancer Research*, **37**, 234-317.

KLEIN, M., CONCANNON, P., EVERETT, M., KIM, L., HUNKAPILLER, T. and HOOD, L. (1987) Diversity and structure of human T-cell receptor α-chain variable region genes. *Proceedings of the National Academy of Science USA*, **84**, 6884-6888.

KONING, F., STINGL, G., YOKOYAMA, W.M., YAMADA, H., MALOY, W.L., TSCHACHLER, E., SHEVACH, E.M. and COLIGAN, J.E. (1987) Identification of a T3-associated γ/δ T cell receptor on Thy-1^+ dendritic epidermal cell lines. *Science*, **236**, 834-837.

KRONENBERG, M., SIU, G., HOOD, L.E. and SHASTRI, N. (1986) The molecular genetics of the T-cell antigen receptor and T-cell antigen recognition. *Annual Reviews of Immunology*, **4**, 529-591.

LANIER, L.L., FEDERSPIEL, N.A., RUITENBERG, J.J., PHILLIPS, J.H., ALLISON, J.P., LITTMAN, D., *et al.* (1987) The T-cell antigen receptor complex expressed on normal peripheral blood $CD4^-$, $CD8^-$ T lymphocytes: A CD3-associated disulfide-linked γ chain heterodimer. *Journal of Experimental Medicine*, **165**, 1076-1094.

LEFRANC, M.-P., FORSTER, A., BAER, R., STINSON, M.A. and RABBITTS, T.H. (1986) Diversity and rearrangement of the human T-cell rearranging γ genes: Nine germ-line variable genes belonging to two subgroups. *Cell*, **45**, 237-246.

LEW, A.M., PARDOLL, D.M., MALOY, W.L., FOWLKES, B.J., KRUISBEEK, A., CHENG, S-F., *et al.* (1986) Characterization of T-cell receptor gamma chain expression in a subset of murine thymocytes. *Science*, **234**, 1401-1405.

LINDSTEN, T., FOWLKES, B.J., SAMELSON, L.E., DAVIS, M.M. and CHIEN, Y. (1987) Transient rearrangements of the T-cell antigen receptor δ locus in early thymocytes. *Journal of Experimental Medicine*, **166**, 761-775.

LO, D. and SPRENT, J. (1986) Identity of cells that imprint H-2-restricted T-cell specificity in the thymus. *Nature*, **319**, 672-675.

LONGO, D.L. and SCHWARTZ, R.H. (1980) T cell specificity for H-2 and Ir gene phenotype correlates with the phenotype of antigen-presenting cells. *Nature*, **287**, 44-46.

MacDONALD, H.R., SCHNEIDER, R., LEES, R.K., HOWE, R.C., ACHA-ORBEA, H., FESTENSTEIN, H., *et al.* (1988) T-cell receptor Vβ us predict reactivit an toleranc t Mls*a*-encoded antigens. *Nature*, **332**, 40-45.

MALISSEN, M., MINARD, K., MJOLSNESS, S., KRONENBERG, M., GOVERMAN, J., HUNKAPILLER, T., *et al.* (1984) Mouse T-cell antigen receptor: Structure and organization of constant and joining gene segments encoding the β polypeptide. *Cell*, **37**, 1101-1110.

MARRACK, P., KUSHNIR, E., BORN, W., McDUFFIE, M. and KAPPLER, J. (1988) The development of helper T cell precursors in mouse thymus. *Journal of Immunology*, **140**, 2508-2514.

MATHIESON, B.J. and FOWLKES, B.J. (1984) Cell surface antigen expression on thymocytes: Development and phenotypic differentiation of intrathymic subsets. *Immunological Reviews*, **82**, 141-173.

McINTYRE, K.R. and SEIDMAN, J. (1984) Nucleotide sequence of mutant I-Aβ*bm*12 gene is evidence for genetic exchange between mouse immune response genes. *Nature*, **308**, 551-553.

MEUER, S.C., FITZGERALD, K.A., HUSSEY, R.E., HODGDON, J.C., SCHLOSSMAN, J.F. and REINHERZ, E.L. (1983) Clonotypic structures involved in antigen-specific human T cell function. *Journal of Experimental Medicine*, **157**, 705-719.

MOORE, M.A.S. and OWEN, J.J.T. (1967) Experimental studies on the development of the thymus. *Journal of Experimental Medicine*, **126**, 715-725.

NELSON, K.J., KELLEY, D.E. and PERRY, R.P. (1985) Inducible transcription of the unrearranged *k* constant region locus is a common feature of pre-B cells and does not require DNA or protein synthesis. *Proceedings of the National Academy of Science, USA*, **82**, 5305-5309.

OWEN, J.J.T. and RAFF, M.C. (1970) Studies on the differentiation of thymus-derived lymphocytes. *Journal of Experimental Medicine*, **132**, 1216-1232.

OWEN, J.T.T., OWEN, M.J., WILLIAMS, G.T., KINGSTON, R., and JENKINSON, E.J. (1988) The effects of anti-CD3 antibodies on the development of the T-cell receptor $\alpha\beta^+$ lymphocytes in embryonic thymus organ cultures. *Immunology*, **63**, 639-642.

PARDOLL, D.M., FOWLKES, B.J., LECHLER, R.I., GERMAIN, R.N. and SCHWARTZ, R.H. (1987) Early genetic events in T cell development analyzed by *in situ* hybridization. *Journal of Experimental Medicine*, **165**, 1624-1638.

PARDOLL, D.M., FOWLKES, B.J., LEW, A.M., MALOY, W.L., WESTON, M.A., BLUESTONE, J.A., *et al.* (1988) Thymus-dependent and thymus-independent developmental pathways for peripheral T cell receptor-γδ-bearing lymphocytes. *Journal of Immunology*, **140**, 4091-4096.

PARKIN, J.G., OWEN, J.J.T. and JENKINSON, E.J. (1988) Proliferation of thymocytes in relation to T-cell receptor β-chain expression. *Immunology*, **64**, 97-99.

RABBITTS, T.H., LEFRANC, M.P., STINSON, M.A., SIMS, J.E., SCHROEDER, J., STEINMETZ, M., *et al.* (1985) The chromosomal location of T-cell receptor genes and a T-cell rearranging gene: Possible correlation with specific translocations in human T-cell leukaemia. *EMBO Journal*, **4**, 1461-1465.

REILLY, E.B., KRANZ, D.M., TONEGAWA, S. and EISEN, H.N. (1986) A functional γ gene formed from known γ gene segments is not necessary for antigen-specific responses of murine cytotoxic T-lymphocytes. *Nature*, **321**, 878-880.

ROEHM, N.L., HERRON, J., CAMBIER, D., DIGUSTO, K., HASKINS, J., KAPPLER, J. and MARRACK, P. (1984) The major histocompatibility complex-restricted antigen receptor on T cells: distribution on thymus and peripheral T cells. *Cell*, **38**, 577-584.

SAITO, H., KRANZ, D., TAKAGAKI, Y., HAYDAY, A., EISEN, H. and TONEGAWA, S. (1984a) Complete primary structure of a heterodimeric T cell receptor deduced from cDNA sequences. *Nature*, **309**, 757-762.

SAITO, H., KRANZ, D., TAKAGAKI, Y., HAYDAY, A., EISEN, H. and TONEGAWA, S. (1984b) A third rearranged and expressed gene in a clone of cytotoxic T lymphocytes. *Nature*, **312**, 36-40.

SCOLLAY, R.G, BARTLETT, P. and SHORTMAN, K. (1984) T cell development in the adult murine thymus: Changes in the expression of the surface antigens Ly2, L3T4 and B2A2 during development from early precursor cells to emigrants. *Immunological Reviews*, **82**, 79-103.

SCOLLAY, R.G., BUTCHER, E.C. and WEISSMANN, I.L. (1980) Thymus cell migration. Quantitative aspects of cellular traffic from the thymus to the periphery in mice. *European Journal of Immunology*, **10**, 210-218.

SIU, G., CLARK, S.P., YOSHIKAI, Y., MALISSEN, M., YANAGI, Y., STRAUSS, E., MAK, T.W. and HOOD, L. (1984a) The human T-cell antigen receptor is encoded by variable, diversity, and joining gene segments that rearrange to generate a complete V gene. *Cell*, **37**, 393-401.

SIU, G., KRONENBERG, M., STRAUSS, E., HAARS, R., MAK, T.W. and HOOD, L. (1984b) The structure, rearrangement and expression of Dβ gene segments of the murine T-cell antigen receptor. *Nature*, **311**, 344-350.

SWAIN, S.L. (1983) T cell subsets and the recognition of MHC class. *Immunological Reviews*, **74**, 129-142.

TOYONAGA, B., YOSHIKAI, Y., VADASZ, V., CHIN, B. and MAK, T.W. (1985) Organization and sequences of the diversity, joining and constant region genes of the human T-cell receptor β chain. *Proceedings of the National Academy of Science USA*, **82**, 8624-8628.

Van EWIJK, W.E., JENKINSON, E.J. and OWEN, J.J.T. (1982) Detection of Thy-1, T-200, Lyt-1 and Lyt-2-bearing cells in the developing lymphoid organs of the mouse embryo *in vivo* and *in vitro*. *European Journal of Immunology*, **12**, 262-271.

van EWIJK, W., RON, Y., MONACO, J., KAPPLER, J., MARRACK, P., LE MEUR, M., *et al.* (1988) Compartmentalization of MHC class II gene expression in transgenic mice. *Cell*, **53**, 357-370.

YANAGI, Y., YOSHIKAI, Y., LEGGETT, K., CLARK, S.P., ALEKSANDER, I. and MAK, T.W. (1984) A human T cell-specific cDNA clone encodes a protein having extensive homology to immunoglobulin chains. *Nature*, **308**, 145-149.

YOSHIKAI, Y., REIS, D.M. and MAK, T.W. (1986) Athymic mice express a high level of β- and γ-chain T-cell receptor messages. *Nature*, **324**, 482-485.

ZINKERNAGEL, R.M., CALLAHAN, G.N., KLEIN, J. and DENNERT, G. (1978) Cytotoxic T cells learn specificity for self H-2 during differentiation in the thymus. *Nature*, **271**, 251-3.

CHAPTER 3

ONTOGENY OF THE HUMAN T CELL RECEPTORS - SINGLE CELL STUDIES

GEORGE JANOSSY AND DARIO CAMPANA

Department of Immunology, Royal Free Hospital School of Medicine, Hampstead, London NW3 2QG

SUMMARY

During T lymphoid development there is an orderly acquisition of cytoplasmic CD3 → cytoplasmic TCRβ → and TCRα followed by the insertion of the functional CD3-TCR into the maturing T cell's membrane. This stepwise development has been studied in the central lymphoid organs with newly standardized monoclonal antibodies (MAbs) to the various proteins, such as CD3ε, TCRβ, and assembled TCRαβ, and to TCRδ. The investigations reviewed represent multiparameter analyses where the expression of other T cell differentiation antigens (CD7, CD1 and CD4 plus CD8), the nuclear marker of terminal deoxynucleotidyl transferase (TdT) as well as the signs of proliferative activity during the cell cycle (the incorporation of bromo-deoxyuridine and labelling with the Ki-67 antibody). First, proliferating cytoplasmic $CD3^+$ ($cCD3^+$) cells (TdT^-, $CD1^-$) are seen in the fetal liver at 7th week gestational age. When the fetal thymus is populated, the proliferating cortical thymocytes studied from the 10th week onward are $cCD3^+$, TdT^-, $CD1^-$ and $cCD3^+$, TdT^-, $CD1^+$, and their majority develop into TCRαβ bearing cells. In man the TCRγδ lineage is apparently a very small minority (%) which includes a few $CD1^+$ cortical thymocytes. The dominant TdT^+, $CD1^+$, TCRαβ committed lineage appears around the 19th week of fetal development and continues in the postnatal thymus. Within the latter lineage, the display of leucocyte common antigen molecules (LCA) also changes during thymic development: the typical $UCHL1^+$ cortical thymocytes give rise to $CD45R^{+}$ ($2H4^{+}$) 'virgin' T lymphocytes through a "double negative" ($UCHL1^{-}$, $CD45R^{-}$) stage. It will be interesting to pay attention to the infant thymocytes while they are at their recently identified intermediate stage of development: already TdT^{-} but still $CD1^{+}$, doubly $CD4^{+}$, $CD8^{+}$ with membrane CD3 display and probably a low expression of both UCHL1 and CD45R. These cells might be the candidates for thymic selection on the basis of their newly displayed TCR specificity.

INTRODUCTION

In previous studies it has been established that the human thymus contains three different main populations of lymphoid cells, the large thymic blasts, typical cortical thymocytes and medullary thymic lymphocytes (reviewed by Janossy *et al.* 1986). These cell types not only exhibit different phenotypic features (Reinherz *et al.* 1980; Bradstock *et al.* 1980; Tax *et al.* 1984) but also show a markedly different distribution of terminal deoxynucleotidyl transferase (Bradstock *et al.* 1980), a nuclear enzyme regarded as a 'somatic mutagen' (Desiderio *et al.* 1984) as well as that of the purine degradative and synthetic enzymes such as 5'nucleotidase and purine nucleoside phosphorylase (Ma *et al.* 1982). In addition, in the developing and infant thymus these three major populations are primarily segregated into the areas of subcapsular cortex, cortex and medulla, respectively (Bhan *et al.* 1980; Janossy *et al.* 1980; Ritter, Sauvage and Cotmore, 1981). Nevertheless, this is only a loose association, and signs of cell traffic such as the arrival of cells with cortical thymic phenotype into the medullary regions can also be decoded from the frozen landscapes of the immunohistological preparations (Janossy *et al.* 1986). With the standardization of monoclonal antibodies against the different types of epithelial cells in the human thymus (Haynes, 1984; Bofill *et al.* 1985; De Maagd *et al.* 1985; Lobach, Scearce and Haynes, 1985) and with the identification of various epithelial cell types at ultrastructural level (Kendall *et al.* 1985), the pathology of thymic tumours and inflammatory diseases such as myasthenia gravis can be interpreted at a cell population level (Chilosi *et al.* 1986; Willcox *et al.* 1987).

Although the observations in animal experiments have indicated that there is a unidirectional flow between the three cellular compartments discussed above (cf. Weissman, 1973; Fowlkes *et al.* 1985; Penit, 1986), a number of questions have remained unresolved, including the possibilities of cortical thymocytes representing a metabolically non-viable group of 'end-stage' cells (Shortman and Jackson, 1974; Ma *et al.* 1982) and the existence of a separate medullary loop of thymic T lymphoid differentiation (Goldschneider *et al.*

1982). In this sense, the illustration of the phenotypic changes in the human thymus along a single lineage pathway has always remained an over-simplification or a calculated guess rather than a set of scientifically confirmed facts (Reinherz *et al.* 1980; Janossy *et al.* 1986).

The technique of organ cultures for thymic development represents a particularly exciting experimental approach in the mouse (Jenkinson, Kingston and Owen, 1987; Born *et al.* 1987) but is not practical in humans where there is no regular supply of samples. Thus the phenotypic analysis of human thymic lymphoid populations is still an important approach, which during the last couple of years has become richer with a more meaningful dissecting power for the following reasons. It has gradually emerged that in the early precursor cells of both the immature B and T cell lineages the antigen specific receptor complex is first synthesized in the endoplasmic reticulum. This phenomenon was originally documented as the synthesis of cytoplasmic IgM (μ heavy chain) in the pre-B cells of the fetal murine liver (Raff *et al.* 1976) and more recently as the cytoplasmic expression of a human B cell specific antigen (135 kd) detected by CD22 antibodies in bone marrow borne immature B cells (Campana *et al.* 1985). The important issues here are two-fold. Firstly, these moieties are inserted into the membrane during cellular differentiation and for this reason the shift of cytoplasmic - membrane expression more firmly signifies lineage development than the previously applied markers; and secondly, multi-chain complexes of prime functional importance such as the Ig receptor together with CD22 antigen in the B lineage and the T cell receptor (TCR) in the T lineage are the best examples of this phenomenon.

The application of the study of TCR related proteins for studying thymic development has been fostered in four stages. Firstly, a great progress has been achieved by characterizing the two sets of TCR heterodimers, TCRαβ and TCRγδ, which are found on separate subsets of T cells. The three major types of relevant monoclonal antibodies (MAbs) are (i) against the non-polymorphic parts of the TCRβ chain (Brenner *et al.* 1987), (ii) against a configurational epitope formed by the fully assembled TCRαβ receptor (Spits *et al.* 1985) and (iii) against the non-polymorphic parts of TCRδ chain (Band *et al.* 1987). The standardization of MAbs of such high quality (Table I) is important because it has opened up the possibilities for the sensitive analysis of single cells which express TCR associated proteins, and this area of interest ideally complements the investigations of TCR gene rearrangements and transcription studied at cell population level. Unlike the mouse where antibodies against the TCRβ react only with 20-25% of T cells (McDuffie *et al.* 1986), the MAbs to human TCR proteins shown in Table I appear to cover the majority if not all of the relevant human thymocyte/T cell populations. With the three types of reagents it has been possible to confirm the major categories of thymocyte differentiation, but further reagents will still be required to investigate fine details, such as the gradual assembly of TCR on the immature T cells' surface (see below).

TABLE I Monoclonal antibodies and heterologous antisera used for analysing T cell ontogeny in humans

	CD	Class	Source	ref
Monoclonal antibodies				
βF1to detect TCRβ	-	IgG1	TCS Cat No. TA1051	1
TCRδ-1 to detect TCRδ	-	IgG1	Dr.M.B.Brenner, Boston, U.S.A	2
WT31 to detect TCRαβ	-	IgG1	BD Cat No. 7770	3
UCHT1 to detect CD3ε	3	IgG1	Dr.P.Beverley, London, UK	4
NA1/34	1	IgG2a	Prof.A.McMichael, Oxford, UK	
RFT2	7	IgG2	RFH	
RFT4	4	IgG1	"	
RFT8	8	IgM	"	
Ki67	-	IgG1	Dakopatts, Denmark, Cat No.M722	5
anti-BrdU	-	IgG1	Becton-Dickinson	6
Heterologous antisera:				
Rabbit anti-TdT			SL Cat No. CT-004	7
G-anti-R IgG FITC			SL Cat No. CT-008	

Abbreviations: TCS: T Cell Sciences Inc., Cambridge, MA; BD: Becton Dickinson Mountain View, CA; RFH: Royal Free Hosp. London; SL: Sera-Lab Ltd.,Sussex, UK.
References: 1: Brenner *et al.* 1985; 2: Band *et al.* 1987; 3: Spits *et al.* 1985; 4: Beverley, 1987; Campana *et al.* 1987; 5: Gerdes *et al.* 1984; 6: Gratzner, 1982; 6: Janossy *et al.* 1980.

Secondly, co-modulation and immunoprecipitation experiments have revealed that both TCRαβ and TCRγδ are non-covalently associated with a non-polymorphic group of molecules recognized by CD3 MAbs (Brenner, Trowbridge and Strominger, 1985). Of the five CD3 chains two are homologous glycoproteins (δγ) and three are non-glycosylated (ε, ζ and ω; Clevers *et al.* 1988). The assembly of the CD3/TCR complex in the endoplasmic reticulum (ER) and its insertion to the membrane has been documented (Alarcon *et al.* 1988; Berkhout, Alarcon and Terhorst 1988). The crucial issue here is that a strict T lineage commitment can be assigned to the immature cells which synthesize ER-bound 'cytoplasmic' CD3 antigen (cCD3) (Campana *et al.* 1987), including the blast cells of T-acute lymphoblastic leukaemia (T-ALL), the malignant variants of thymic precursor cells (Furley *et al.* 1986; Campana *et al.* 1987). At the same time no other non-T precursor cells or non-T leukaemias express cCD3 antigen (reviewed by Janossy and Campana, 1988). Among CD3 MAbs those reactive with the CD3ε chain, such as UCHT1, appear to be the most reliable reagents for cCD3 labelling (Campana *et al.* 1987; Van Dongen *et al.* 1988). Thirdly, methods for double and triple labelling have been standardized in order to study the concomittant or exclusive expression of various nuclear, cytoplasmic and membrane antigens on T lineage cells (Campana, Coustan-Smith and Janossy, 1988). Fourthly, an extension of this analysis has been the study of cells in the proliferative cell cycle with two recently standardized MAbs. The MAb Ki67 reacts with cells in the G_1, S, M and G_2 phase (Gerdes *et al.* 1984) and MAbs to bromo-deoxyuridine (BrdU) identify DNA synthetic cells following pre-incubation with BrdU (Gratzner, 1982) used in various combinations with each other and with other reagents (Fig.1).

All these methods provide a framework for a multiparameter analysis of func-

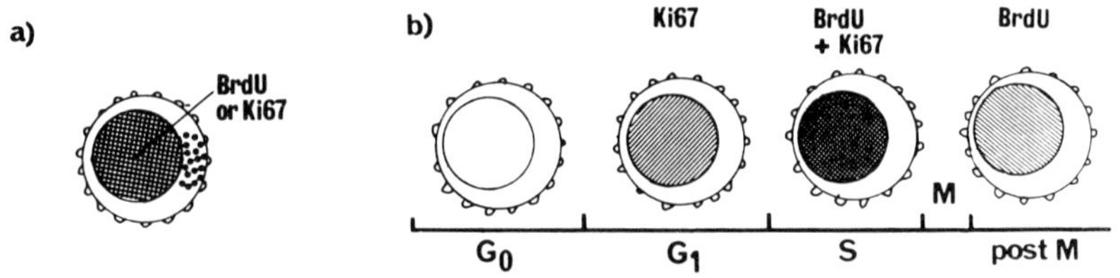

Figure 1. Two variants of triple marker analysis applied for investigating proliferative activity. (a) The precise identification of cells with two markers can be performed in combination with anti-BrdU or Ki67 Abs. These triple combinations are frequently required because two-colour combinations on single cells are often necessary for the precise definition of the cell's phenotype. Clear examples of these are the study of immature T lineage cells ($cCD3^+$, TdT^+ or TdT^-) or the virgin and memory T cell populations within both the $CD4^+$ and the $CD8^+$ subsets of T cells. (b) Alternatively, cells identified with one phenotypic marker can be positioned in the cell cycle using two proliferation associated markers such as anti-BrdU and Ki67 Abs together (from Campana, Coustain-Smith and Janossy, 1988; with the permission of *Journal of Immunological Methods*).

tionally important cellular proteins during thymocyte ontogeny. These observations in the human sphere have been paralleled and complemented by the investigations of thymic differentiation in rodents (Mathieson and Fowlkes, 1984; Owen, Kingston and Jenkinson, 1986; Farr *et al.* 1985; Penit, 1986; Born *et al.* 1987; Penit and Vasseur, 1988) and in the chicken where the equivalents of major T cell differentiation antigens (Chan *et al.* 1988) as well as the TCRγδ (referred to as TCR1) and TCRαβ (TCR2) have also been found (Chen *et al.* 1988). Thus the observations made in the human thymus can be (cross-)related to the observations obtained in experimental animal studies.

In this review we have four aims: (i) the documentation of early T lineage development in the fetal liver; (ii) the summary of the quantitative observations about the emergence of human lymphoid cells belonging to the two T-cell lineages TCRαβ and TCRγδ, in comparison with similar studies performed in the chicken; (iii) the analysis of the majority populations TCRαβ in the infant thymus in terms of their proliferative capacity, and (iv) the speculations about the cell type which is the target to tolerance induction in the human thymus, and the possible significance of common leucocyte antigen display on some thymic cells (Beverley, 1987; Sanders *et al.* 1988; Akbar *et al.* 1988).

T LYMPHOID DEVELOPMENT IN THE FETAL LIVER

The features of human prothymocytes are still controversial, and the potential importance of these cell types in clinical bone marrow transplantation is still ignored. In the human fetal liver around the 7th gestational week (gest. wk) cells with lymphoid morphology express CD7 (p40) antigen (Lobach *et al.* 1985). Similar $CD7^+$ cells also gather around the thymic rudiments. A more recent observation is that $cCD3^+$ cells representing about 10-15% of cells with lymphoid morphology can be observed in the fetal liver of 7-9th gest. wks of age; 90% of these $cCD3^+$ cells carry the proliferation associated Ki67 nuclear antigen (Fig.2) but do not express TCRβ and mCD3 as yet. On these cells the only other T-cell associated marker is the CD7 antigen (p40) which is not a fully T cell specific marker because some additional precursor cells and a few acute myeloblastic leukaemias also express it (Van Dongen *et al.* 1985; Vodinelich *et al.* 1983). Neither Class II nor nuclear TdT are detectable, and other markers such as CD2, CD5, CD4 and CD8 are also missing at this early stage (Campana *et al.* 1988). Thus these forms are rapidly cycling immature T-lineage cells. The hypothesis that $CD7^+$, $cCD3^+$ cells include prothymocytes is testable. Experiments in which such cells are seeded onto bone marrow stroma and thymic epithelial cell lines (Mizutani *et al.* 1987) might provide valuable information about these cells' options and T lineage restriction. We feel that such studies should be performed in comparison with populations selected on the basis of their CD34 positivity perhaps even before more patients are given 'pure stem cell populations' positively selected by CD34 for rescue after ablative therapy (Berenson *et al.* 1988). This is because the CD34 antigen is apparently absent on $CD7^+$ putative early T lineage cells and the thymi in these patients may not reconstitute.

Around the 10-11th gest. wk cells expressing TCRβ chains and mCD3 appear in the liver (Fig. 2), and the proportions of $TCR\beta^+$ and $mCD3^+$ cells continue to increase up to the 20th gest. wk. Double marker experiments clearly demonstrate that the maturing $mCD3^+$ cells have stopped proliferating ($Ki67^-$, $BrdU^-$) but throughout the same developmental period a few $cCD3^+$, $TCR\beta^+$, $mCD3^-$ cells still remain detectable and these peculiar cells are in the cell cycle (Ki67$^+$). From the recent experiments of colonization of mouse thymic epithelium *in vitro* (Jenkinson, Kingston and Owen, 1987) and from the observations of avian ontogeny (Sowder *et al.* 1988) as well as from previous observations on nude animals one would expect that the thymic microenvironment is of prime importance for initiating the synthesis of TCR chains. According to this prevailing view, the mCD3$^+$ cells in the liver should be a sample of recent thymic emigrants which return to the liver. On the other hand the presence of proliferating TCRβ$^+$, mCD3$^-$ cells seen in the liver suggests to us that at least a minimal T-cell differentiation may take place inside the fetal liver *in situ*, for it would be unusual to such proliferating cells of intermediate features to leave the thymus. Finally, it should be pointed out that the CD7$^+$, cCD3$^+$, mCD3$^-$ cells might be similar to the precursor cells which form T cell colonies *in vitro* (Mossalayi *et al.* 1988). In these latter studies it has been demonstrated that after extensive depletion of mCD3$^+$ mature T lymphocytes from the bone marrow a few CD7$^+$ cells remain which generate T colonies without passing through a CD1$^+$, TdT$^+$ cortical thymocytic phase of development.

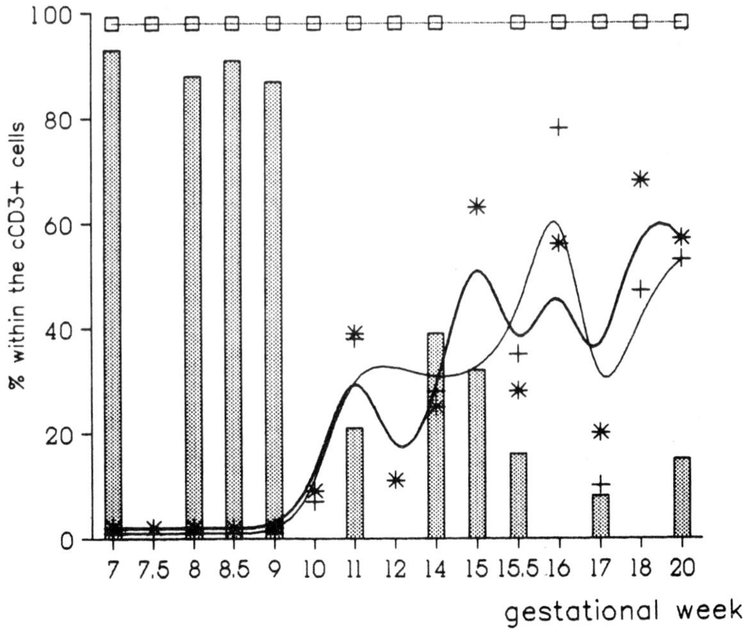

Figure 2. The characterization of cCD3$^+$ cells in the human fetal liver during the 7-20th gestational week. All cCD3$^+$ cells are also CD7$^+$ (□). Cells expressing Ki67, representing cells in the proliferation cycle, are shown by bars (▩). The ascending numbers of cells positive for TCRβ (+, detecting both cytoplasmic and membrane antigens) and for mCD3 (*) are also depicted (from Campana *et al.* 1988).

THE EMERGENCE OF TCRγδ CARRYING THYMOCYTES IN THE FETAL THYMUS

Although the potential significance of the murine TCRγδ lineage has been introduced with flare and inspiration (Janeway, Jones and Hayway, 1988) much less is known about the human TCRγδ lineage, and to our surprise we were unable to find $TCR\delta^+$ cells in the fetal liver at any stage of development between the 8th to the 20th gest. wks.

It has previously been suggested that $TCR\gamma\delta^+$ cells might be generated in the human thymus. Nevertheless the proportions of such cells are consistently lower in the thymus than in the circulating blood, and therefore thymic $TCR\gamma\delta^+$ cells might be contaminating blood lymphocytes (Borst *et al.* 1988). The putative evidence for the rearrangement of the TCRγδ genes in immature thymocytes mainly derives from the investigation of their configuration in selected $CD4^-$, $CD8^-$ ('double negative') cells, regarded as T cell precursors (Bank *et al.* 1986; Crispe *et al.* 1987). Since most peripheral and circulating $TCR\gamma\delta^+$ cells are also 'double negative' and these may contaminate the thymus, the finding of rearranged TCRγδ loci in such populations is not entirely unexpected, and the suggestion that these clones are derivatives of immature cells is not convincing. For this reason the $TCR\gamma\delta^+$ cells have been investigated in the fetal and infant thymus using double and triple staining combinations for cortical thymocyte markers such as CD1 and TdT (Fig. 3). The proportions of thymic $TCR\delta^+$ cells in the earliest samples tested between the 10-13th gest. wks were extremely low (0.02%), and 30-40% of the $TCR\delta^+$ cells seen at 12-13th gest. wk were doubly labelled for CD1 but TdT negative. In the samples taken at the 14-20th gest. wks the proportions of $TCR\delta^+$ cells averaged between 0.04-0.7%, and again 20-40% of these cells expressed CD1 but no TdT. An identical low range of positivity (mean O.4%) was seen in the infant thymus and the higher limit of CD1 positivity within this subset was 45% without concomittant TdT positivity. When tested in cytocentrifuge preparations, $TCR\delta^+$ cells carried the CD3 antigen, although with this method it has not been established whether this was cCD3 or mCD3. Circulating $TCR\gamma\delta^+$ cells, on the other hand, are invariably $CD1^-$, and for this reason the observations above have been interpreted as a direct evidence in favour of the cortical thymic features of an admittedly minor $TCR\gamma\delta^+$ population which nevertheless lacks nuclear TdT (Campana *et al.* 1988).

These findings, when compared to the findings seen in other species point to certain differences. Firstly, it appears from the investigation of the avian immune system that the proportions of TCRγδ bearing cells in the thymus are higher when compared to humans (8-10% versus 0.02-1%). Clearly, the early wave of TCRγδ predominance over TCRαβ cells seen in the chicken embryos

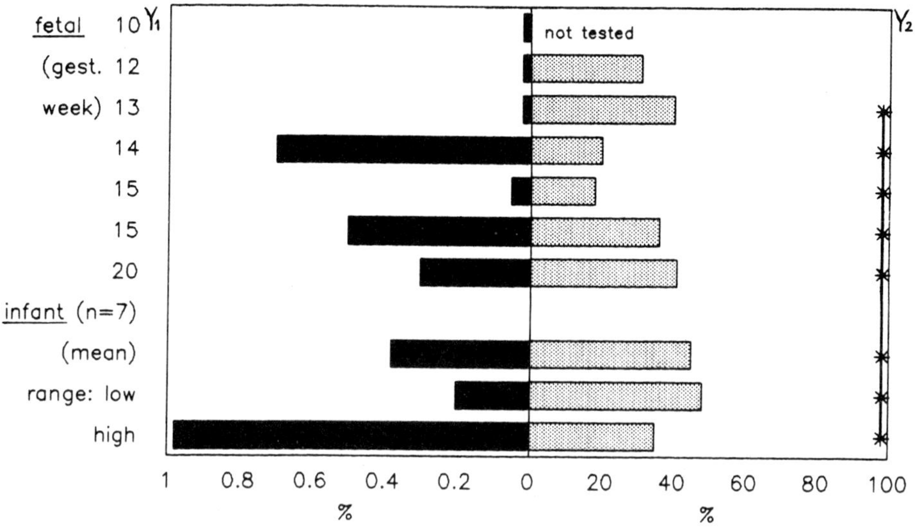

Figure 3. The proportions (Y1) and phenotypes (Y2) of TCRδ-positive cells in the fetal and infant thymus. The bars represent the values observed at the different ages as shown. On the left the percentages of TCRδ$^+$ cells are shown in the total cell suspension (■). On the right the proportions of CD1$^+$ cells are shown within the TCRδ population (□). In the same diagram the CD3$^+$ cells are also shown (*) including both cCD3$^+$ and mCD3$^+$ cells For the detailed observations see Campana *et al.* 1988.

(Chen *et al.* 1988) is not detectable in human fetal thymic samples studied. In the human thymus the proportions of mature TCRαβ bearing (WT31$^+$) cells always exceeds that of TCRγδ thymocytes (see above and Borst *et al.* 1988). The very low relative proportions of human thymic TCRγδ$^+$ cells is even more obvious when immature cells expressing cytoplasmic TCRβ are also counted. In chicken, the result of this particular test has not been shown, although one might expect that even at the time of the peak TCRγδ occurence in the 14-day embryonic thymus a substantial proportion of 70-75% TCRγδ$^-$ cells are immature members of the TCRαβ committed populations which may contain cytoplasmic TCRβ (Chen *et al.* 1988). Another apparent species difference is that in humans the few TCRγδ bearing cells are invariably CD4$^-$ and only a small minority (<15%) are CD8$^+$ in the various tissues such as the thymus, peripheral blood and the spleen; in the mouse and chicken the TCRγδ cells tend to acquire CD8 in higher proportions (50%; Janeway, Jones and Hayway, 1988).

TDT NEGATIVE TCRαβ POPULATIONS IN THE HUMAN THYMUS

In the fetal thymus samples taken between the 10-20th gest. wks the expression of TdT, cCD3, TCRβ, WT31 (detecting the assembled TCRαβ) and mCD3 has been studied (Fig.4), and previous observations demonstrating the distinctive features of cortical and medullary thymic populations including their different phenotypes (Asma, van den Bergh and Vossen, 1986; Lobach and Haynes, 1987) have been confirmed from the 12th gest. wk onwards. This was despite the absence of TdT until the 19th gest. wk (Bodger *et al.* 1983). The new findings are that the TdT$^-$ populations also carry TCRβ and the fully assembled TCRαβ as detected by WT31 (Campana *et al.* 1988). The thymocytes in the immature fetus therefore give rise to perfectly formed T lymphocyte populations, although these cells' poor reactivity to foreign antigens is known. Consequently, the possibility exists that the lack of full thymic diversification due to the lack of TdT might be related to the poor reactivity to foreign antigens. A possible analogy for such a phenomenon is the normally populated thymus with poor function in the 'bare lymphocyte' syndrome where thymic education suffers for another reason: due to HLA deficiency (Schuurman *et al.* 1985). The fact remains that T cells seeding from the early TdT$^-$ thymus 'calmly' establish the T dependent microenvironments throughout the body, and then are soon replaced by a new wave of T cells which descend from TdT$^+$ precursors (see below).

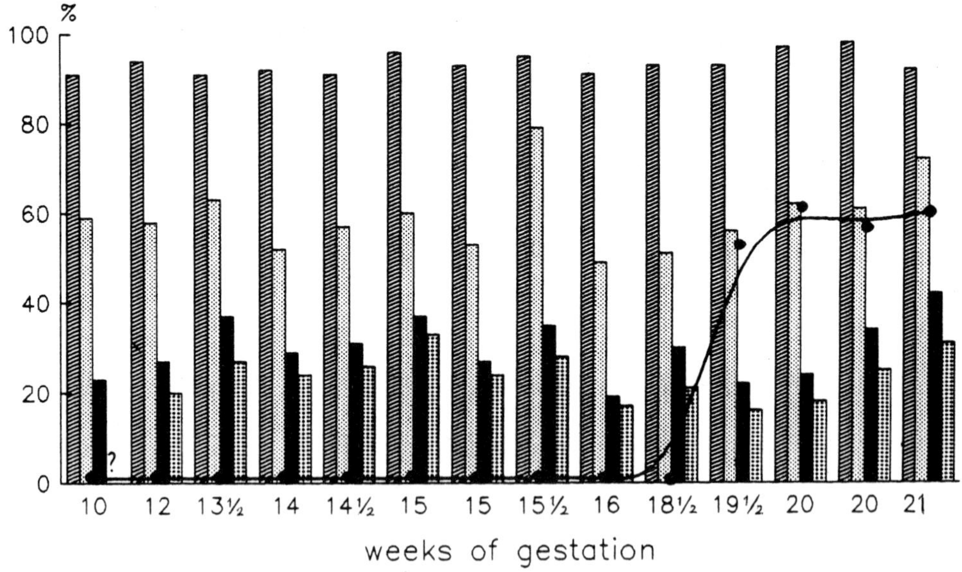

Figure 4. The proportions of fetal thymocytes labelled for cCD3 (▨), TCRβ (▒), mCD3 (■), TCRαβ (▦) and TdT (•–•) at the various gestational ages, as modified from Campana *et al.* 1988.

THE DEVELOPMENT OF MAJORITY TCRαβ POPULATIONS IN THE THYMUS

From the 19-20th gest. wks onward the mainstream of thymic development is the generation of the TdT^+ lineage. This population remains to be detectable in the thymus until old age, 71 years, and one regrets that Steinman (1986) has not investigated for TdT the cortical region seen in the thymus of the oldest individual at 106 years of age since we would have really liked to argue here that he or she survived so long because of having a thymus with TdT positivity. The phenotypic observations about the $TCR\alpha\beta^+$ infant thymocyte populations are shown in Fig. 5.

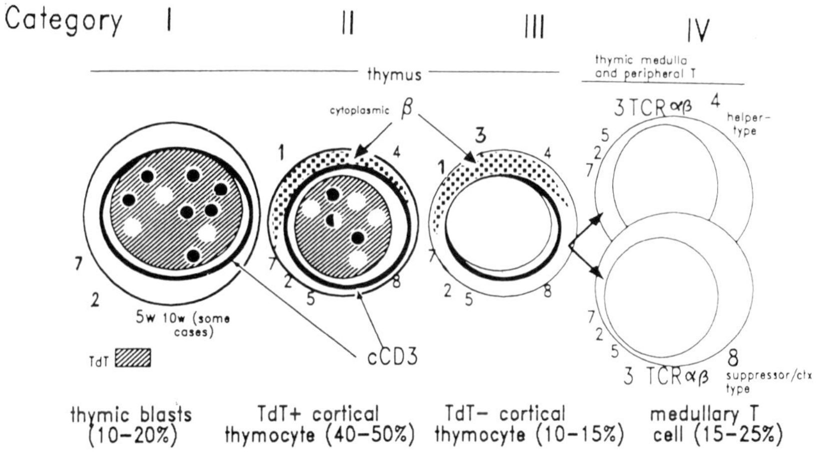

Figure 5. Cell kinetic observations and phenotypic analyses of thymocytes support a single lineage development of a dividing $cCD3^+$,TdT^+ precursor cells into a dichotomizing $CD4^+$,$CD8^-$ and $CD8^+$,$CD4^-$ medullary T lymphocyte population. The numbers refer to the CD designation of T cell antigens. In these phenotypic studies of human lymphoid populations the following components of the T cell receptor complex (TCR) are investigated: (i) cytoplasmic (cCD3) and membrane CD3 antigens (mCD3; using UCHT1 MAb), (ii) cytoplasmic and membrane TCRβ chain (using βF1 MAb) and (iii) the assembled TCRαβ chain expressed on the T cell membrane (using WT31 MAb). Each white circle drawn in the nucleus corresponds to 10% $Ki67^+$ cells, and the additional black circle corresponds to 10% $BrdU^+$ cells within a given population: virtually all large thymic blasts are $Ki67^+$ and 70% are $BrdU^+$, but no such cells are seen among the $mCD3^+$ cells (modified from Campana and Janossy, 1988).

Four complementary concepts are summarized:

(i) Membrane antigens associated with helper-type ($CD4^+$) and suppressor/cytotoxic type cells ($CD8^+$) identify maturation changes along the thymocyte differentiation pathway and define a basic dichotomy during T cell development (Reinherz *et al.* 1980; Tidman *et al.* 1981). The corresponding scheme obtained in the mouse with MAbs anti-L3T4 (CD4) and anti-Lyt-2 (CD8) is referred to as the double negative - double positive - single positive (DN-DP-SP) pathway (Mathieson and Fowlkes 1984; Scollay, Smith and Strauffer, 1986; Penit and Vasseur, 1988). In the human thymus these forms correspond to large thymic blasts ($CD4^-$, $CD8^-$), cortical thymocytes ($CD4^+$, $CD8^+$) and to medullary T cells which constitute a mixture of $CD4^+$, $CD8^-$ (majority) and $CD4^-$, $CD8^+$ cells (minority; Fig.5). The similarities among the different species are striking. One example is that from DN ($CD4^-$, $CD8^-$) populations DP ($CD4^+$, $CD8^+$) cells are generated through immature $CD4^-$, $CD8^+$ transit cells seen both in the mouse (Penit and Vasseur 1988) and in humans (Piantelli *et al.* 1986).

(ii) Nuclear TdT is a marker of lymphocyte immaturity (Janossy *et al.* 1980) except in the fetus where immature thymocytes are TdT^- (see above). At a population level the presence or absence of TdT in thymocytes roughly corresponds to the steroid-sensitivity and -resistance in thymocyte subsets, respectively (Weissman, 1973).

(iii) MAbs to TCR associated proteins define the functional stages of thymocyte/T cell development (see above and Campana *et al.* 1988).

(iv) These methods are combined with the analysis of cells in the proliferative cell cycle. The observations shown in Fig.5 indicate that during normal thymocyte development the TdT^+ cells include virtually all $Ki67^+$ and BrdU incorporating cells and no identifiable mature T cells ($WT31^+$) remain in the proliferative cycle (Campana and Janossy 1988), although these cells may obviously respond to antigens during an immune response and then they divide (see below). Similar observations have been made in the mouse (Penit and Vasseur 1988). It is also clear that the TdT^+ and $WT31^+$ (TCRαβ) cells represent two entirely different non-overlapping populations.

Thus the following categories of cell types are identifiable. In Category I the cells express cCD3 and TdT but no cytoplasmic TCRβ chain. Some of the cells are large blasts and the rest are intermediate sized cells. This is the most actively dividing population (Ki67 95%, anti-BrdU 60-70%) which are likely to rearrange their TCRβ genes at this stage and represent the 'target' cells for leukaemic transformation in T-ALL (Reinherz *et al.* 1980; Bradstock *et al.* 1980; Furley *et al.* 1986; Campana *et al.* 1987).

Cells in Category II are typical cortical thymocytes with TdT and cytoplasmic TCRβ chain; only 60-70% and 20-30% of these are $Ki67^+$ and $BrdU^+$, respectively. Thus, a proportion of these cells leave the cell cycle and become a resting population. As these cells already synthesize TCRβ, the TdT in these cells might be superfluous or may contribute to the TCRα gene rearrangement. The membrane phenotype is $CD1^+$, $CD8^+$ (not shown in Fig.5 but see Piantelli *et al.* 1986) and $CD1^+$, $CD8^+$, $CD4^+$ (majority: 60-70%).

Cells in Category III have been recently described in humans (Campana *et al.* 1988). These cells have no nuclear TdT but are negative with WT31, a MAb which detects the fully assembled TCRαβ. On the other hand, many of these cells carry mCD3. A careful re-investigation of the membrane expression of TCRβ and TCRα on these cells with additional MAbs is still warranted because this population may express these antigens either very weakly or in a form which is not appropriately recognized by WT31. These are the cells which are likely to be similar to murine thymocytes which form contacts with epithelial cells through TCRβ (Farr *et al.* 1985), and the lymphocyte selection on epithelial cells is likely to take place at this stage of development (cf. Kappler *et al.* 1988; Kisielow *et al.* 1988). The cells in Category III have low proliferative activity (<5%).

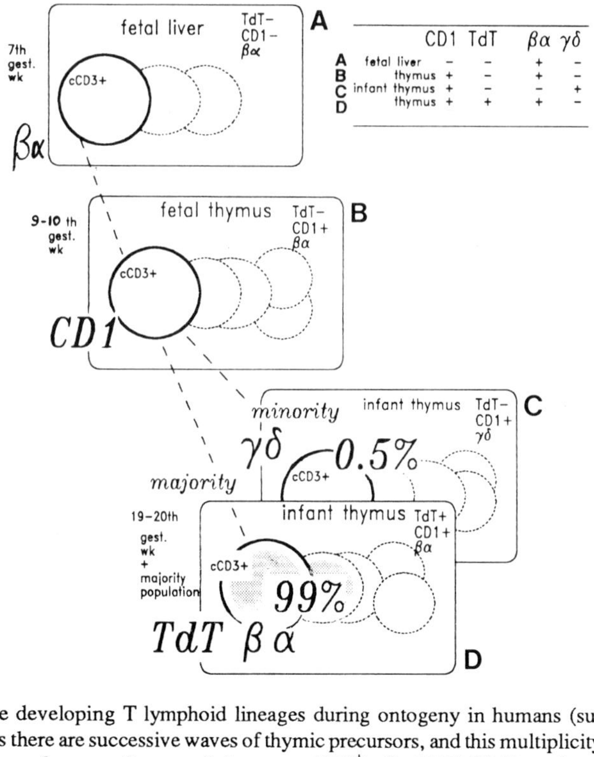

Figure 6. The developing T lymphoid lineages during ontogeny in humans (summary). In all animal species there are successive waves of thymic precursors, and this multiplicity is represented in the emergence of separate lineages. In humans, $cCD3^+$ cells (TdT^-,$CD1^-$) are first seen in the fetal liver (7th week gestational age; box A). When the thymus is populated around the 8-10th week the proliferating cortical thymocytes are TdT^-,$CD1^+$, their majority developing into TCRαβ populations (box B). In man the TCRγδ lineage is apparently a small minority from the 10th gestational week onward but includes a few $CD1^+$ cortical thymocytes (box C).The dominant TCRαβ,TdT^+,$CD1^+$ lineage appears around the 19th week of fetal development (box D). The cellular lineages in the individual boxes represent a series of cell types as illustrated for lineage D in Fig.5.

The phenotypically most mature populations (WT31$^+$, Category IV), correspond to the steroid resistant mature T cell populations in the thymic medulla and to mature T lymphocytes on the periphery. These 'single positive' cells are heterogeneous (CD4$^+$, CD8$^-$ and CD8$^+$, CD4$^-$) and non-proliferating.

The precursor-product relationships between the cells shown in Fig. 5 have recently been confirmed in animal experiments (Penit and Vasseur 1988). When mitotic inhibitors were given to mice followed by BrdU the regeneration started by the appearance of CD4$^-$, CD8$^-$ (DN) blasts, followed by CD4$^-$, CD8$^+$ and CD4$^+$, CD8$^+$ (DP) cells. It has also been demonstrated that the CD4$^+$, CD8$^-$ (SP) mature population of helper-type T cells are the descendants of DP cells (Penit and Vasseur 1988, Smith 1987).

The conclusion of the data so far reviewed is shown in Fig.6. There appears to be a development of cCD3$^+$ precursor cells which is depicted in the successive and dichotomizing 'vertical' progression of cells together with the indication of their first appearance. Each of these precursor cell types give rise to a 'horizontally' maturing cohort of T lymphoid cells shown in boxes A to D with their main characteristic features summarized.

THE IMPORTANCE OF LEUCOCYTE COMMON ANTIGENS (LCA)

In Fig. 7 a broader view of the T lymphoid development is shown (reviewed by Janossy, Campana and Akbar, 1988). The three salient points are as follow. Firstly, murine experiments have documented that from the large pool of cells which actively proliferate in the thymus, referred to as Cascade I, only mostly few leave this central lymphoid organ, and even these emigrating cells are mostly short-lived, when compared to the life span of lymph node T cells (Scollay, Smith and Strauffer, 1986; Wallis *et al.* 1979). Secondly, the recent thymic emigrants can be regarded as resting 'virgin' T cells, and the differentiation from a fragile short lived metabolic state to a long-lived physiology is likely to be linked with the exposure to a proliferative antigenic stimulus, referred to as Cascade II, which generates 'memory' T cells. Thirdly, in both mice and humans phenotypic alterations have been observed in the expression of LCAs during the activation of T cells by antigens or mitogens (Akbar *et al.* 1988, Sanders *et al.* 1988; Powrie and Mason, 1988) which seems to mark the stages of such transition: virgin T cells express CD45R in both the mouse and humans, while during antigenic activation CD45R is lost and the cells initiate the expression of UCHL1, another molecule governed by the LCA gene (Terry *et al.* 1988; Beverley, 1987).

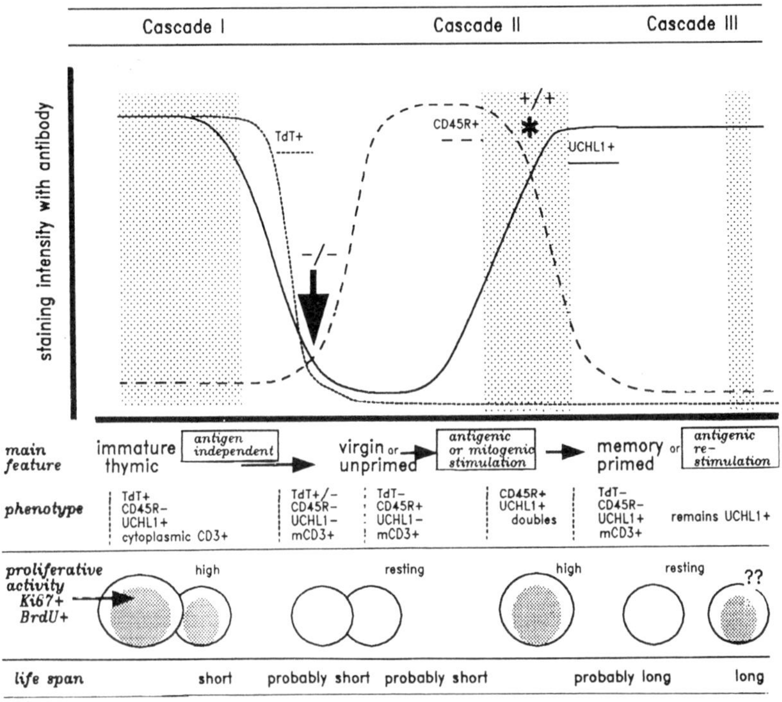

Figure 7. Triple cascade of T cell development in man. In the thymus, immature TdT^+ cells divide and give rise to a virgin population of T lymphocytes (Cascade I). After antigenic stimulation there is another wave of changes and memory T cells are formed (Cascade II). Memory cells actively recirculate and are exposed to antigen on the periphery but may come back to the lymphoid tissues for their induced proliferative cycle (Cascade III). Membrane moieties of the common leucocyte antigen group (CD45) such as CD45R and UCHL1 are differentially expressed during these phases of T cell development. The expression of UCHL1 is biphasic. During the late stages of thymic development UCHL1 is lost prior to the appearance of CD45R (double -/- cells; arrow). During activation of virgin T cells by antigen UCHL1 positivity is generated before the loss of CD45R (double +/+ cells; asterisk). From Janossy *et al.* 1988 with permission of *Immunology*.

From the point of view of the thymic physiology, it is important to note that UCHL1 shows a bi-phasic expression, and behaves like a typical 'jumping' differentiation antigen alternating in expression depending upon the the cells' developmental stage (Smith *et al.* 1986). In the thymus typical cortical thymocytes are $UCHL1^+$, $CD45R^-$, and these develop into 'virgin' T cells with $CD45R^+$, $UCHL1^-$ phenotype through a 'double negative' stage ($CD45R^-$, $UCHL1^-$; see large arrow in Fig.7). Nevertheless, UCHL1 is not an activation marker, because once the memory T cells have acquired UCHL1, they do not seem to lose it (Beverley, 1987; Janossy *et al.* 1988).

It is encouraging that the expression of LCA moieties might give us more insight into the life story of the T lymphoid lineage, and that the different animal species and man show similar features helping the design of experiments which then contribute to the interpretation of human diseases. From the point of view of further understanding, the identification of a transitional cell type (Category III) with the concomittant expression of CD1, CD4 and CD8 molecules together with the display of CD3 antigens and perhaps TCRαβ chains in a low or unstable configuration is interesting because the analysis of these cells will help explain how tolerance is induced. As was pointed out, similar cells in the mouse have been shown to form close contacts with epithelial cells (Farr *et al.* 1985) as a possible mechanism for such selection process. It is also likely that the same cells are weak expressors of the LCAs, and it will be fascinating to see whether this molecular group or the lack of it may also contribute in such a selection process.

REFERENCES

Akbar, A.N., Terry, L., Timms, A., Beverley, P.C.L. and Janossy, G. (1988) Loss of CD45R and gain of UCHL1 reactivity is a feature of primed T cells. *J Immunol* **140**, 217-226.

Alarcon, B., Berkhout, B., Breitmeyer, J. and Terhorst, C. (1988) Assembly of the human T cell receptor-CD3 complex takes place in the endoplasmic reticulum and involves intermediary complexes between the CD3-gamma, delta, epsilon core and single T cell receptor alpha or beta chains. *J Biol Chem* **263**, 2953-2961.

Asma, G.E.M., van den Bergh, R.L. and Vossen, J.M. (1986) Characterization of early lymphoid precursor cells using monoclonal antibodies and anti-terminal deoxynucleotidyl transferase in the human fetus. *Clin exp Immunol* **64**, 356-365.

Band, H., Hochstenbach, F., McLean, J., Hata, S., Krangel, M.S. and Brenner, M.B. (1987) Immunochemical proof that a novel rearranging gene encodes the T cell receptor δ subunit. *Science* **238**, 682-687.

Bank, I., De Pinho, R.A., Brenner, M.B., Cassimeris, J., Alt, F.W. and Chess, L. (1986) A functional T3 molecule associated with a novel heterodimer on the surface of immature human thymocytes. *Nature* **322**, 179-181.

Berenson, R.J., Andrews, R.G., Bensinger, W.I., Kalamasz, D., Knitter, G., Bernstein, I. and Buckner, C.D. (1988) Antigen $CD3^+$ marrow cells engraft lethally irradiated baboons *J. Clin. Invest.* **81**, 951-955.

Berkhout, B., Alarcon, B. and Terhorst, C. (1988) Transfection of genes encoding the T cell receptor-associated CD3 complex into COS cells results in assembly of the macromolecular stucture. J Biol Chem **263**, 8528-8536.

Beverley, P.C.L. (1987) Minireview: human T cell subsets. *Immunol Letters* **14**, 263-267.

Bhan, A.K., Reinherz, E.L., Poppema, S., McClusky, R.T. and Schlossman, S.F. (1980) Location of T cell and major histocompatibility complex antigens in the human fetus. *J exp Med* **152**, 771-782.

Bodger, M.P., Janossy, G., Bollum, F.J., Burford, G.D. and Hoffbrand, A.V. (1983) The ontogeny of terminal deoxynucleotidyl transferase cells in the human fetus. *Blood* **61**, 1125-1131.

Bofill, M., Janossy, G., Willcox, N., Chilosi, M., Trejdosiewicz, L.K., Bagott, M. and Newsom-Davis, J. (1985) Microenvironments in the normal and myasthenia gravis thymus. *Am J Pathol* **119**, 462-473.

Born, W., McDuffie, M., Roehm, N., Kushnir, E., White, J., Thorpe, D., Stefano, J., Kappler, J. and Marrack, P. (1987) Expression and role of the T cell receptor in early thymocyte differentiation *in vitro*. *J Immunol* **138**, 999-1008.

Borst, J., Van Dongen, J.J.M., Bolhuis, R.L.H., Peters, P., Haller, D.A., De Vries, E. and Van de Griend, R.J. (1988) Distinct molecular forms of human T cell receptor γδ detected on viable T cells by a monoclonal antibody. *J Exp Med*, (in press).

Bradstock, K.F., Janossy, G., Pizzolo, G., Hoffbrand, A.V., McMichael, A., Pilch, J.R., Milstein, C., Beverley, P. and Bollum, F.J. (1980) Subpopulations of normal and leukemic human thymocytes: an analysis using monoclonal antibodies. *J Natl Canc Inst* **65**, 33-42.

Brenner, M.B., McLean, J., Schaft, H., Warnke, R.A., Jones, N. and Strominger, J.I. (1987) Characterization and expression of the human α/β T cell receptor by using a framework monoclonal antibody. *J Immunol* **138**, 1502-1514.

Brenner, M.B., Trowbridge, I.S. and Strominger, J.L. (1985) Cross linking of human T cell receptor proteins: association between the T cell idiotype subunit and the T3 glycoprotein heavy subunit. *Cell* **40**, 183-190.

Campana, D., Coustan-Smith, E. and Janossy, G. (1988) Double and triple staining methods for studying the proliferative activity of human B and T lymphoid cells. *J Imm Meth* **107**, 79-88.

Campana, D. and Janossy, G. (1988) Proliferation of normal and malignant immature lymphoid cells. *Blood* **71**, 1201-1211.

Campana, D., Janossy, G., Bofill, M., Trejdosiewicz, L.K., Ma, D., Hoffbrand, A.V., Mason, D.Y., Lebacq, A.M. and Forster, H.K. (1985) Human B cell development. I. Phenotypic differences of B lymphocytes in the bone marrow and peripheral lymphoid tissue. *J Immunol* **134**, 1524-1529.

Campana, D., Janossy, G., Coustan-Smith, E., Tian, W.T., Ip, S., Wong, L. and Amlot, P.L. (1988) The expression of T cell receptor-associated proteins during T cell ontogeny in man. *J Immunol*, (in press).

Campana, D., Thompson, J.S., Amlot, P., Brown, S. and Janossy, G. (1987) The cytoplasmic expression of CD3 antigens in normal and malignant cells of the T lymphoid lineage. *J Immunol* **138**, 648-655.

Chan, M.M., Chen, C.L., Ager, L.L. and Cooper, M.D. (1988) Identification of the avian homologues of mammalian CD4 and CD8 antigens. *J Immunol* **140**, 2133-2143.

Chen, C.H., Cihak, J., Losch, U. and Cooper, M.D. (1988) Differential expression of two T cell receptors, TcR1 and TcR2, on chicken lymphocytes. *Eur J Immunol* **18**, 539-543.

Chilosi, M., Iannucci, A., Fiore-Donati, L., Tridente, G., Pampanin, M., Pizzolo, G., Ritter, M., Bofill, M. and Janossy, G. (1986) Myasthenia gravis: immunohistological heterogeneity in microenvironmental organization of hyperplastic and neoplastic thymuses suggesting different mechanisms of tolerance breakdown. *J Neuroimmunol* **11**, 191-204.

Clevers, H., Alarcon, B., Wileman, T. and Terhorst, C. (1988) The T cell receptor/CD3 complex: a dynamic protein ensemble. *Ann Rev Immunol* **6**, 629-662.

Crispe, I.N., Moore, M.W., Husmann, L.A., Smith, L., Bevan, M.J. and Shimonkevitz, R.P. (1987) Differentiation potential of subsets of $CD4^-8^-$ thymocytes. *Nature* **329**, 336-339.

DeMaagd, R.A., MacKenzie, W.A., Schuurman, H.J., Ritter, M.A. and Price, K.M. (1985) The human thymus microenvironment: heterogeneity detected by monoclonal antiepithelial cell antibodies. *Immunology* **54**, 745-754.

Desiderio, S.V., Yancopoulos, G.D., Paskind, M., Thomas, E., Boss, M.A., Landau, N., Alt, F.W. and Baltimore, D. (1984) Insertion of N regions into heavy chain genes is correlated with expression of terminal deoxynucleotidyl transferase in B cells. *Nature* **311**, 752-755.

Farr, A.G., Anderson, S.K., Marrack, P. and Kappler, J. (1985) Expression of antigen specific major histocompatibility restricted receptors by cortical and medullary thymocytes *in situ*. *Cell* **43**, 543-553.

Fowlkes, B.J., Edison, L., Mathieson, B.J. and Chused, T.M. (1985) Early T lymphocytes. Differentiation *in vivo* of adult intrathymic precursor cells. *J exp Med* **162**, 802-822.

Furley, A.J., Mizutani, S., Weilbaecher, K., Dhalival, H.S., Ford, A.M., Chan, L.C., Molgaard, H.V., Toyonaga, B., Mak, T., van der Elsen, P., Gold, D., Terhorst, C. and Greaves, M.F. (1986) Developmentally regulated rearrangement and expression of genes encoding the T cell receptor-T3 complex. *Cell* **46**, 75-87.

Gerdes, J., Lemke, H., Baisch, H., Wacker, H.H., Schwab, U. and Stein, H. (1984) Cell cycle analysis of a cell proliferation-associated human nuclear antigen defined by the monoclonal antibody Ki-67. *J Immunol* **133**, 1710-1715.

Goldschneider, I., Shortman, K., McPhee, D., Linthicum, S., Mitchell, G., Battye, F. and Bollum, F.J. (1982) Identification of subsets of proliferating low Thy 1 cells in thymus cortex and medulla. *Cell Immunol* **36**, 59-69.

Gratzner, H.G. (1982) Monoclonal antibody to 5-bromo- and 5-iododeoxyuridine. *Science* **218**, 474-475.

Haynes, B.F. (1984) The human thymic microenvironment. *Adv Immunol* **36**, 87-142.

Janeway, C.A., Jones, B. and Hayday, A. (1988) Specificity and function of T cells bearing gd receptors. *Immunology Today* **9**, 73-76.

Janossy, G., Bofill, M., Rowe, D. and Beverley, P.C.L. (1988) Tissue distribution of human T cell subpopulations expressing CD45R and UCHL1 leukocyte common antigens. *Immunology*, (in press).

Janossy, G., Bofill, M., Trejdosiewicz, L.K., Willcox, H.N.A. and Chilosi, M. (1986) Cellular differentiation of lymphoid subpopulations and their microenvironments in the human thymus. In: *The Human Thymus: Current Topics in the Microenvironment, histophysiology and pathology*, edited by H.K. Müller-Hermelink, Heidelberg/New York/Tokyo, Springer-Verlag, 89-125.

Janossy, G. and Campana, D. (1988) The pathophysiological basis of immunodiagnosis in acute lymphoblastic leukaemia. *Cancer Rev*, (in press).

Janossy, G., Campana, D. and Akbar, A. (1988) Kinetics of T lymphocyte development. In: *Cell Kinetics of the Inflammatory Reactions*. Edited by O.H. Iversen, Berlin, Heidelberg, Springer Verlag.

Janossy, G., Thomas, J.A., Bollum, F.J., Granger, S., Pizzolo, G., Bradstock, K.F., Wong, L., Ganeshaguru, K. and Hoffbrand, A.V. (1980) The human thymic microenvironment: an immunohistologic study. *J Immunol* **125**, 202-212.

Jenkinson, E.J., Kingston, R. and Owen, J.T. (1987) Importance of IL-2 receptors in intra-thymic generation of cells expressing T cell receptors. *Nature* **329**, 160-162.

Kappler, J.W., Staez, U., White, J. and Marrack, P.C. (1988) Self-tolerance eliminates T cells specific for Mls modified products of the major histocompatibility complex. *Nature* **332**, 35-38.

Kendall, M.D., van den Wijngaert, F.P., Schuurman, H.J., Rademakers, L.H. and Kater, L. (1985) Heterogeneity of the human thymus epithelial microenvironment at the ultrastructural level. *Adv Exp Med Biol* **186**, 289-297.

Kisielow, P., Bluthman, H., Staez, U.D., Steinmetz, M. and Boehmer, H.V. (1988) Tolerance in T cell receptor transgenic mice involves deletion of non-mature CD4+, 8+ thymocytes. *Nature* **333**, 742-746.

Lobach, D.F. and Haynes, B.F. (1987) Ontogeny of the human thymus during fetal development. *J Clin Immunol* **7**, 81-97.

Lobach, D.F., Hensley, L.L., Ho, W. and Haynes, B. (1985) Human T cell antigen expression during early stages of fetal thymic maturation. *J Immunol* **135**, 1752-1759.

Lobach, D.F., Scearce, R.M. and Haynes, B.F. (1985) The human thymic microenvironment. Phenotypic characterization of Hassall's bodies with the use of monoclonal antibodies. *J Immunol* **134**, 250-257.

Ma, D.D.F., Sylwestrowicz, T.A., Granger, S., Massaia, M., Franks, G., Janossy, G. and Hoffbrand, A.V. (1982) Distribution of terminal deoxynucleotidyl transferase, purine degradative and synthetic enzymes in subpopulations of human thymocytes. *J Immunol* **129**, 1430-1435.

Mathieson, B.J. and Fowlkes, B.J. (1984) Cell surface antigen expression on thymocytes: development and phenotypic differentiation of intrathymic subsets. *Immunol Rev* **82**, 141-173.

McDuffie, M., Born, W., Marrack, P. and Kappler, J. (1986) The role of the T-cell receptor in thymocyte maturation: effects *in vivo* of anti-receptor antibody. *Proc Natl Acad Sci USA* **83**, 8728-8732.

Mizutani, S., Watt, S.M., Robertson, D., Hussein, S., Healy, L.E., Furley, A.J. and Greaves, M.F. (1987) Cloning of human thymic subcapsular cortex epithelial cells with T-lymphocyte binding sites and hemopoietic growth factor activity. *Proc Natl Acad Sci USA* **84**, 4999-5003.

Mossalayi, M.D., Lecron, J.C., Goube de Laforest, P., Janossy, G., Debre, P. and Tanzer, J. (1988) Characterization of prothymocytes with cloning capacity in human bone marrow. *Blood* **71**, 1281-1287. @REF TEXT = Owen, J.J., Kingston, R. and Jenkinson, E.J. (1986) Generation of cells expressing cytoplasmic and/or surface T-cell receptor beta chains during the development of mouse fetal thymus. *Immunology* **59**, 23-27.

Penit, C. (1986) *In vivo* thymocyte maturation. BUdR labeling of cycling thymocytes and phenotypic analysis of their progeny support the single lineage model. *J Immunol* **137**, 2115-2121.

Penit, C. and Vasseur, F. (1988) Sequential events in thymocyte differentiation and thymus regeneration revealed by a combination of bromodeoxyuridine DNA labeling and antimitotic drug treatment. *J Immunol* **140**, 3315-3323.

Piantelli, M., Larocca, L.M., Aiello, F.B., Maggiano, N., Carbone, A., Ranelletti, F.O. and Musiani, P. (1986) Proliferation of phenotypically immature human thymocytes with and without interleukin 2 receptors. *J Immunol* **136**, 3204-3210.

Powrie, F., and Mason, D. (1988) Phenotypic and functional heterogeneity of $CD4^+$ cells. *Immunology Today* **9**, 274-277.

Raff, M.C., Megson, M., Owen, J.J.T. and Cooper, M.D. (1976) The production of intracellular IgM by B lymphocyte precursors in mouse. *Nature* **259**, 224-227.

Reinherz, E.L., Kung, P.C., Goldstein, G., Levey, K.H. and Schlossman, S.F. (1980) Discrete stages of human intrathymic differentiation analysis of normal thymocytes and leukaemic lymphoblasts of T-cell lineage. *Proc Natl Acad Sci USA* **77**, 1588-1592.

Ritter, M.A., Sauvage, C.A. and Cotmore, S.F. (1981) The human thymic microenvironment: *in vivo* identification of thymic nurse cells and other antigenically distinct subpopulations of epithelial cells. *Immunology* **44**, 439-446.

Sanders, M.E., Makgoba, M.W., Sharrow, S.O., Stephany, D., Springer, T.A., Young, H.A. and Shaw, S. (1988) Human memory T lymphocytes express increased levels of three cell adhesion molecules (LFA-3, CD2 and LFA-1) and three other molecules (UCHL1, CDw29 and Pgp-1) and have enhanced IFN-gamma production. *J Immunol* **140**, 1401-1407.

Schuurman, H.J., van den Wijngaert, F.P., Huber, J., Zeger, B.J., Schuurman, R.K., Roord, J.J. and Kater, L. (1985) The thymus in "bare lymphocyte" syndrome. *Adv Exp Med Biol* **186**, 921-928.

Scollay, R., Smith, J. and Strauffer, V. (1986) Dynamics of early T cells: prothymocyte migration and proliferation in the adult mouse thymus. *Immunological Reviews* **91**, 129-157.

Shortman, K. and Jackson, H. (1974) The differentiation of T lymphocytes I. Proliferation kinetics and interrelationship of populations of mouse thymus cells. *Cell Immunol* **12**, 230-240.

Smith, L. (1987) CD4 positive murine T cells develops from CD8 positive precursors *in vivo*. *Nature* **326**, 798-800.

Smith, S.H., Brown, M.H., Rowe, D., Callard, R.E. and Beverley, P.C.L. (1986) Functional subsets of human helper-inducer cells defined by a new monoclonal antibody, UCHL1. *Immunology* **58**, 63-70.

Sowder, J.T., Chen, C.H., Ager, L.L., Chan, M.M. and Cooper, M.D. (1988) A large subpopulation of avian T cells express a homologue of the mammalian T gamma/delta receptor. *J Exp Med* **167**, 315-322.

Spits, H., Borst, J., Tax, W., Capel, P.J.A., Terhorst, C. and De Vries, J.E. (1985) Characteristics of a monoclonal antibody (WT-31) that recognizes a common epitope on the human T cell receptor for antigen. *J Immunol* **135**, 1922-1928.

Steinman, G.G. (1986) Changes in the human thymus during aging. In: *The Human Thymus*, Müller-Hermelink, H.K., Ed., Heidelberg-New York-Tokyo, Springer-Verlag, 43-88.

Tax, W.J.M., Tidman, N., Janossy, G., Trejdosiewicz, L.K., Willems, R.M., Leeuwenberg, J.F.M., De Witte, T., Capel, P.J.A. and Koene, R.A.P. (1984) Monoclonal antibody (WT1) directed against a T cell surface glycoprotein: characteristics and immunosuppressive activity. *Clin exp Immunol* **55**, 427-436.

Terry, L.A., Brown, M.H. and Beverley, P.C.L. (1988) The monoclonal antibody UCHL1 recognizeds a 180,000 MW component of the human leucocyte common antigen CD45. *Immunology* **64**, 331-336.

Tidman, N., Janossy, G., Bodger, M., Granger, S., Kung, P.C. and Goldstein, G. (1981) Delineation of human thymocyte differentiation pathways utilizing double staining techniques with monoclonal antibodies. *Clin exp Immunol* **45**, 457-467.

Van Dongen, J.J.M., Hooijkaas, H., Comans-Bitter, M., Hahlen, K., De Klein, A., Van Zanen, G.E., Van't Veer, M.B., Abels, J. and Benner, R. (1985) Human bone marrow cells positive for terminal deoxynucleotidyl transferase, HLA-DR, and a T cell marker may represent prothymocytes. *J Immunol* **135**, 3144-3150.

Van Dongen, J.J.M., Krissansen, G.W., Wolvers-Tettero, I.L.M., Comans-Bitter, W.M., Adriaansen, H.J., Hooijkas, H., Van Wering, E.R. and Terhorst, C. (1988) Cytoplasmic expression of the CD3 antigen as a diagnostic marker for immature T-cell malignancies. *Blood* **71**, 603-612.

Vodinelich, L., Tax, W., Bai, Y., Pegram, S., Capel, P. and Greaves, M. (1983) A monoclonal antibody WT1 for detecting leukaemias of T cell precursors. *Blood* **63**, 1108-1114.

Weissman, I.L. (1973) Thymus cell maturation. Studies on the origin of cortisone resistant thymic lymphocytes. *J Exp Med* **137**, 504-511.

Wallis, V.J., Leuchars, E., Chaudhuri, M. and Davies, A.J.S. (1979) Studies on hyperlymphoid mice. *Immunology* **38**, 163-171.

Willcox, N., Schluep, M., Ritter, M.A., Schuurman, H.J., Newsom-Davis, J. and Christensson, B. (1987) Myasthenic and nonmyasthenic thymoma. An expansion of a minor cortical epithelial cell subset? *Am J Pathol* **127**, 447-457.

CHAPTER 4

IL-4, ONCOGENESIS AND THE THYMUS

MARK LARCHÉ AND MARY A. RITTER

Department of Immunology, Royal Postgraduate Medical School, Hammersmith Hospital, Ducane Road, London W12 0NN.

INTRODUCTION

The monoclonal antibody 'revolution' which followed Köhler and Milstein's publication, in which they described their technique for producing continuously growing cells that secrete antibody of a predefined specificity, has accelerated progess in countless fields (Köhler and Milstein, 1975). Few areas of immunology have escaped this 'revolution', and the thymus is no exception. By the turn of the decade a large panel of monoclonal antibodies was available which detected epitopes on cell surface molecules of T cells and their intra-thymic precursors. With these reagents it was possible to delineate phenotypically and functionally distinct subpopulations of T cells and thymocytes and to investigate the role of many of these molecules in cell activation, growth and differentiation (McMichael *et al*, 1987).

It was not until the early 1980's that thymologists began to turn their attention in earnest to the thymic microenvironment itself and its role in the generation of functionally mature T cells. For some time it has been recognized that the thymic microenvironment plays a vital role in the differentiation of T lymphocytes, and that both direct contact with thymic stromal cells and interaction with the humoral factors that they secrete are essential to this T cell developmental process (Stutman *et al*, 1969; Gelfand, 1980). By generating monoclonal antibodies against these functionally important molecules, be they soluble or cell surface bound, thymologists hoped to gain valuable insight into the workings of the thymus. Several groups pursued this line of research, generating many antibodies against thymic components in several species. Whilst few of these antibodies cross-reacted between species, the specificities generated were broadly similar, dissecting the microenvironment into the familiar subcapsular and medullary regions, a distinct cortical region and a variety of other discrete antigenic compartments such as Hassall's corpuscles, macrophage and dendritic cell populations (Ritter and Haynes, 1987; *see also* Session III of Rolduc report in this volume of *Thymus Update).*

MONOCLONAL ANTIBODY MR6

In our own laboratory several monoclonal anti-human thymic epithelial cell antibodies were generated in order to study functionally relevant molecules in the human thymus (described in detail in DeMaagd *et al,* 1985). One antibody, MR6, was selected for its reactivity with the cortical thymic epithelium. Since this antibody also weakly labelled medullary macrophages/dendritic cells, the MR6 hybridoma was repeatedly subcloned by limiting dilution in an attempt to separate the two staining patterns; however, this proved to be impossible, indicating that the MR6 antibody recognises an epitope shared by these cortical and medullary stromal cells (Fig. 1a). Immunoelectron microscopy (performed in collaboration with Dr. B. von Gaudecker, Kiel, West Germany) using gold-labelled secondary antibodies shows that MR6 reacts strongly with a surface membrane molecule on the 'spider-like' cortical epithelial cells and that, in addition, it appears to label weakly the surrounding thymocytes (Fig.1b). Immunofluorescence staining of thymocytes and T cells in suspension has also revealed that the molecule detected by MR6 is present at low density on the surface of these cells (Larché *et al,* 1987). Flow cytometric analysis shows a unimodal distribution and suggests that essentially all thymocytes carry at least some molecules to which MR6 binds, but that for some the number is so low that the level of labelling is below the sensitivity of detection of the assay (Fig. 2).
Biochemical analysis of this molecule by Western blotting of lysates from thymocytes and T cell lines demonstrated that MR6 binds to a protein/glycoprotein with a relative molecular mass of approximately 145,000d (Mr 145Kd) under both reducing and non-reducing conditions (p145-MR6; Larché *et al,* 1988b) indicating that the polypeptide chain to which MR6 binds is not covalently linked to other chains, although non-covalent associations may occur. Preliminary immunoprecipitation experiments with MR6 followed by SDS-PAGE (performed in collaboration with Dr. A. Bradbury, Cambridge, U.K.) using different levels of protease inhibitors suggest that the the MR6 antigen is highly sensitive to proteolysis

and that its true molecular weight may be as high as 200Kd. Whether or not there is a second associated polypeptide chain is not yet clear, although by

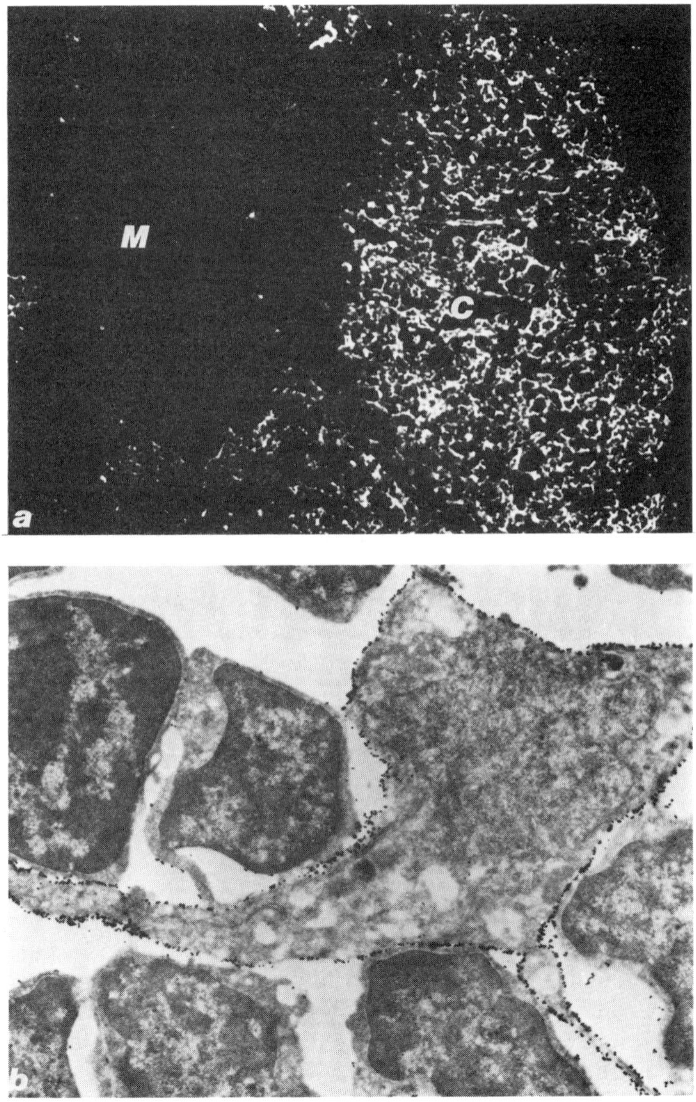

Figure 1.(a) Immunofluorescence analysis of frozen sections of paediatric human thymus using monoclonal antibody MR6. The cortical epithelium shows strong staining while the medullary macrophage/dendritic cells are only weakly labelled by MR6. Cortex (C) and medulla (M) are marked. Magnification 320x. From Larché *et al* 1988b with permission. (b) Immunogold electron microscopy with MR6 monoclonal antibody shows strong surface labelling of cortical epithelial cells; the surrounding lymphocytes may be very weakly MR6^{+}. Magnification 14,000x. Immunoelectron microscopy performed by Dr. Brita von Gaudecker, Kiel, West Germany.

analogy with the IL-2R it is possible that the receptor for IL-4 is heterodimeric in at least some situations (on activated lymphocytes).

A comparison of Western blotting data from lysates of thymocytes versus whole thymus (thymocytes plus stromal cells) shows the same p145-MR6 band, indicating that MR6 is detecting the same molecule on thymocytes, cortical epithelium and macrophage/dendritic cells. Furthermore, although our initial immunofluorescence studies of other haemopoietic lineages revealed no clear labelling with MR6, subsequent enhanced immunolabelling showed weak staining, and Western blotting analysis of cell lysates from these lines showed that at the protein level cells from the B cell, monocyte and granulocyte lineages do express the same 145Kd protein, but at levels too low to be detected by conventional immunofluorescence techniques.

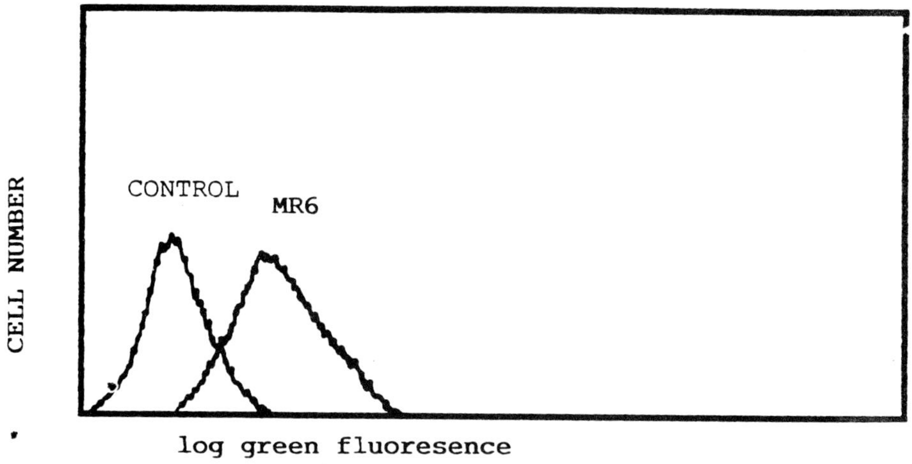

Figure 2. Flow cytometric analysis (EPICS Profile, Coulter, Hialeah, U.S.A.) of human paediatric thymocytes stained by indirect immunofluorescence using monoclonal antibody MR6. MR6 shows weak labelling with a unimodal distribution.

FUNCTIONAL STUDIES WITH MR6

Our observation that both thymocytes and mature activated T cells (T cell clones) were weakly positive for MR6 raised the possibility that the molecule detected might function in the periphery as well as within the thymus. In an attempt to assess the possible functional role for p145-MR6, the antibody was included in cultures of thymocytes or peripheral blood mononuclear cells (PBMC) responding to the mitogen phytohaemagglutinin (PHA). In these assays MR6 caused a dose dependent decrease in the proliferative response of these cell populations, as compared to isotype-matched control antibodies. Subsequently antigen- and interleukin-induced proliferative responses of cloned MHC class II-restricted helper T cells were used as assay systems. When MR6 was included in cultures of T cell clone HA1.7 (reactive with residues 306-319 of the carboxy-terminus of the influenza HA-1 molecule; Lamb *et al,* 1982) responding to either specific antigen or interleukin 2 (IL-2), the antibody induced a dose-dependent decrease in the cellular proliferative response reaching a level of 50-60% inhibition at an antibody concentration of 10mg/ml (Fig. 3a). This effect was also demonstrable in antigen-activated alloreactive T cell clones and in peripheral blood mononuclear cells (PBMC) responding to either Mycobacterium tuberculosis soluble extract (MTSE) or IL-2, demonstrating that the phenomenon was not merely a peculiarity of clone HA1.7. Interestingly, when interleukin 4 (IL-4) was used as the stimulating agent, MR6 was able to inhibit completely the proliferation of clone HA1.7 at a concentration of 10mg/ml (Fig. 3b). Collectively, these data suggest that MR6 is in some way interfering with proliferative signals that are generated during T lymphocyte activation. However the molecular weight and tissue distribution of p145-MR6 indicates that it is not one of the previously identified T cell surface activation molecules such as CD1, CD2, CD3, IL-2R or the T cell receptor for antigen.

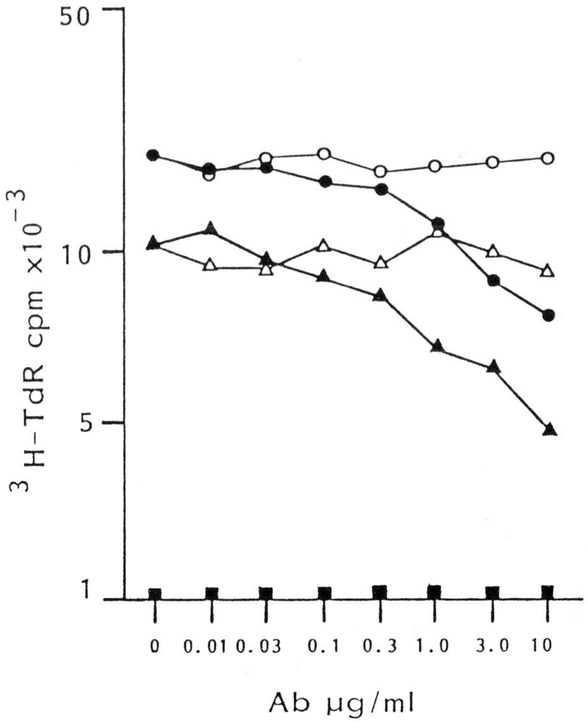

Figure 3a. Monoclonal antibody MR6 inhibits the proliferative response of HA1.7 cloned human T cells (reactive with residues 306-319 of the HA-1 molecule of influenza virus haemagglutinin) when stimulated with either antigen (●) or with IL-2 (▲). Control antibody, anti-beta 2 microglobulin, has no effect on either antigen (○) or IL-2 (△) induced proliferation. From Larché *et al* 1988a with permission.

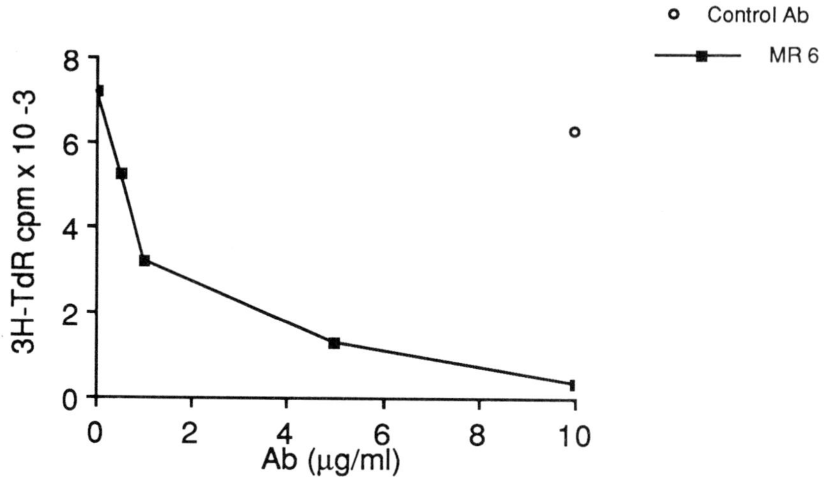

Figure 3b. MR6 monoclonal antibody (■) at 10{g/ml completely inhibits the proliferative response of HA1.7 to IL-4 (250 U/ml). Control antibody H17E2 (O) has no effect. From Larché *et al* 1988a with permission.

MR6 AND THE IL-4 RECEPTOR

Recently, Park *et al* (1987) reported the characterisation of the human IL-4 receptor (IL-4R). Using radiolabelled IL-4 and an anti-IL-4 antibody, the authors demonstrated that the human IL-4R has a relative molecular mass of approximately 139Kd and that this molecule is expressed at low surface density on cells from the T cell, B cell, monocyte, granuolocyte and fibroblast lineages. Since the molecule detected by MR6 appeared to be very similar in size, tissue distribution and density of expression to the molecule described by Park and colleagues, and since MR6 antibody completely inhibited the IL-4-dependent proliferation of T cell clone HA1.7, it was proposed that MR6 might bind to the IL-4R. Further evidence in support of this proposal came from experiments designed to measure IgE production by B cells from an atopic individual after stimulation with allergen in the presence of allergen-specific autologous T helper cell clones (O'Hehir *et al,* 1988; Larché *et al,* 1988a). In this system either one of two T cell clones were used. One clone, DE9 required exogenous IL-4 in order to drive IgE production by B cells. The other clone, DE26 was able to drive IgE production without exogenous IL-4 (this clone produces some endogenous IL-4), however antibody production in this system was enhanced by the addition of exogenous IL-4. MR6 was added to the assay on day 1 and day 5 at 5mg/ml. At this concentration MR6 completely blocked production of IgE driven by either of the T cell clones (Table I). The fact that MR6 blocks this IL-4 dependent function provides strong circumstantial evidence in favour of the view that this antibody recognises the IL-4R on human lymphocytes.

TABLE I. Inhibition of IL-4-dependent IgE Production by MR6 Antibody.

Exp.	Clone	E-PBMC	Allergen	IL-4	MR6	IgE (pg/ml)◆
1	DE26	+	Df*	-	-	309
	DE26	+	Df	+	-	439
	DE26	+	Df	+	+	0
	DE26	+	Df	+	neg	283
	-	-	Df	+	-	0
2	DE 9	+	Df	-	-	0
	DE 9	+	Df	+	-	150
	DE 9	+	Df	+	+	0
	DE 9	+	G^	-	-	0
	-	+	Df	-	-	0

T-helper cell clones DE9 and DE26 (2.5×10^5/ml) reactive with *D.farinae.* were cultured with autologous E^- peripheral blood mononuclear cells (2.5×10^5/ml) and allergen (5×10^3BU/ml) in the presence of MR6 or an isotype-matched negative control antibody (5.0μg/ml). Supernatant IgE was quantitated by ELISA after 10 days of culture.

Neg. negative control antibody (H17E2)
*Df Dermatophagoides farinae
^G mixed grass pollen (control allergen)
◆ background for the assay has been substracted.

(Table modified from Larché *et al* 1988a.)

MR6 AND EPITHELIAL CELLS

Analysis of the structure and function of the molecule to which MR6 binds on thymic cortical epithelial cells has posed a far greater problem since, in contrast to T lymphocytes, human thymic epithelial cells are unfortunately extremely difficult to grow as pure populations in tissue culture. Mixed primary cultures have been successfully grown by several groups, but only two well characterized transformed cell lines have been derived. These were produced by transfection of cultures with SV40 large-T and are both of subcapsular phenotype (Mitzutani *et al*, 1987). These lines probably reflect the fact that the epithelial cells obtained from primary cultures of human thymus are almost exclusively of the subcapsular/medullary phenotype. Cortical epithelium appears to be rare *in vitro*, although not absent (Haynes *et al*, 1984). However, MR6 positive cells are not found in thymus cultures (our unpublished observations; H.-J, Schuurman, Utrecht, The Netherlands, personal communication). This may reflect a requirement for thymocte/T cell derived factors such as IL-4 (thymocytes and T cells being absent from these cultures at this stage) for the expression of p145-MR6 by epithelial cells. Alternatively, this subpopulation of cells may simply not grow in culture. Thus we have been unable to isolate MR6$^+$ thymic epithelial cells free from contaminating thymic T cell populations; we have therefore been unable provide direct proof that the molecule expressed by epithelial cells is the same as the thymocyte/T cell p145-MR6 rather than simply a different molecule that shares an epitope recognised by MR6. However, the fact that we can only identify a single band from detergent lysates of whole thymus (thymocytes plus stromal cells) suggests that p145-MR6 is present on both epithelium and lymphocytes, although the possibility that we are only detecting lymphocyte antigen and that our solubilisation procedures do not extract the (different) epithelial antigen cannot be excluded.

In a previous study (Willcox *et al*, 1987) of human thymoma tissue from patients with myasthenia gravis, the neoplastic epithelium was found to be strongly MR6$^+$ and was apparantly derived by transformation and clonal expansion of a 'double positive' (MR6$^+$/MR19$^+$) subpopulation of normal cortical epithelial 'stem' cells (Lampert and Ritter, 1988). However, these tumours contain large numbers of thymic lymphocytes, presumably responding to the inductive microenvironment provided by the neoplastic epithelial cells, and so do not provide a 'clean' system in which to study the nature and origin of the epithelial antigen to which MR6 binds. We therefore screened a variety of other epithelial tumours that did not contain large numbers of proliferating/infiltrating lymphocytes. Of 20 carcinomas studied (from colon, rectum, lung, skin, ovary and breast) all were MR6$^+$, with those from the skin showing the greatest intensity of staining (Fig. 4; Al Jabaari *et al*, 1989). Interestingly, the normal tissue counterpart of these tumours were either negative or weakly MR6$^+$ (discussed below). Samples from some of these tumours were solubilised, separated by SDS-PAGE and analysed by Western blotting using antibody MR6 to determine the molecular weight of the epithelial cell antigen to which it bound. All carcinoma samples screened in this way were found to express the same 145Kd band as seen previously with T cell preparations indicating that the molecule recognised by MR6 on epithelial cell populations was indeed the same as that recognised on cells of haemopoietic origin (Al Jabaari *et al*, 1989).

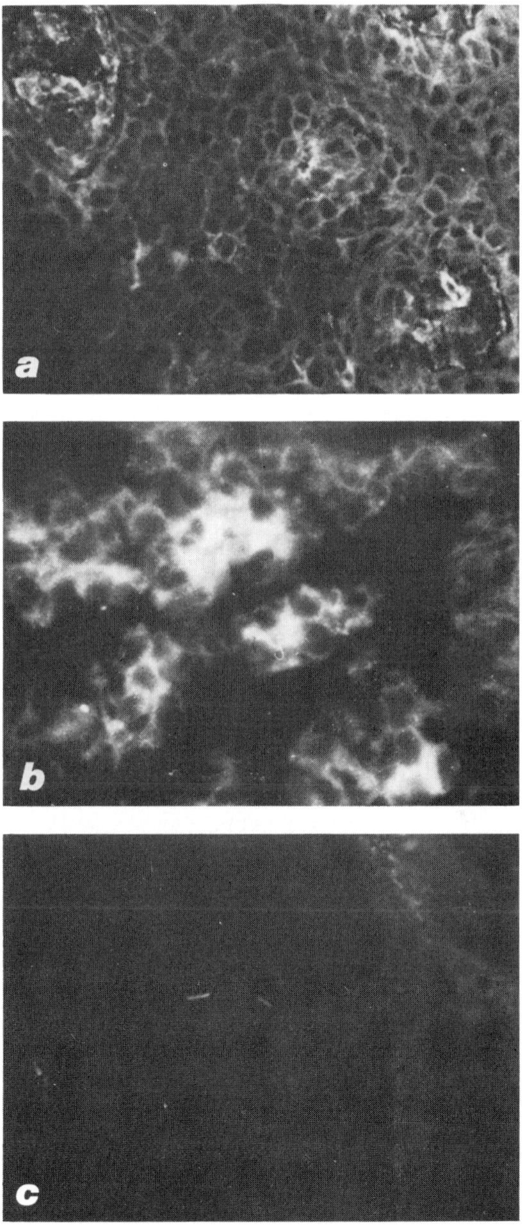

Figure 4. Immunofluorescence analysis of frozen sections of carcinoma tissue using monoclonal antibody MR6. Essentially all tumour cells in (a) squamous (skin) carcinoma and (b) ovarian carcinoma are strongly $MR6^+$. (c) Ovarian carcinoma. No staining is seen with the negative control antibody, Ox7. Magnification: (a) and (c) 320x; (b) 640x.

ONCOGENESIS

Elevated expression of normally occuring cell surface molecules in neoplastic conditions is a frequent event (Merlino *et al*, 1984; Harris *et al*; 1988). This aberrant expression may occur as either a primary or secondary event in relation to the process of transformation itself, an example of the latter being the increased expression of the transferrin receptor in many neoplastic tumours reflecting the increased requirement of these rapidly proliferating cells for iron. More frequently, upregulation of a surface molecule such as a growth factor receptor will play a primary role in oncogenesis (Hudziac *et al*; 1987). The association between oncogenes and neoplasia is well documented. More recently the protein products of a number of oncogenes have been identified, and several of these molecules appear to be receptors for growth factors (Adamson, 1987). When such receptor molecules are expressed in large numbers (amplification) or in a mutated form (such as truncation), cells may become permanently activated and grow at a much increased rate. However, the process of oncogenesis is thought to be a multistage one and the involvement of several oncogene products may be necessary to induce outright transformation of a cell (Land, Parada and Weinberg, 1983).

Our immunofluoresence data indicate that p145-MR6 is more highly expressed on carcinoma cells that on their normal counterparts, while the inhibitory effects of MR6 antibody in T cell functional assay systems suggest that p145-MR6 may be a receptor for the growth factor IL-4. Taken together, these two sets of observations raise the possibility that aberrant/amplified expression p145-MR6 may play a primary role in oncogenesis. Comparison of the molecular and cell distribution characteristics of p145-MR6 with the products of other known oncogenes suggests that this molecule may represent a novel oncogene product.

IL-4 RECEPTORS IN THE HUMAN THYMIC MICROENVIRONMENT

The expression of cellular growth factor receptors in an organ such as the thymus is in itself, an unremarkable event. The process of thymocyte differentiation and maturation requires many cellular interactions and growth factors. The role of growth factors such as IL-2 in thymic maturation has been the subject of intense investigation in recent years. More recently, thymocytes have also been shown to produce and respond to IL-4, with one T cell being able to produce enough IL-4 to stimulate thousands of other cells (Sideras *et al*, 1988; Mosmann, 1988). This lymphokine induces proliferation of double negative 'prothymocytes' and upregulates the level of IL-4R expression on these cells (Lowenthal *et al*, 1988). It might also be responsible for inducing the differentiation of some of these double negative cells into 'single positive' $CD8^+$ immature thymocytes, since it has recently been shown that IL-4 can mediate CD8 induction on $CD4^+$ T cell clones (Paliard *et al*, 1988). However, progress in this area is hampered by the lack of reagents directed against the IL-4R.

Ideas as to the function of IL-4 on thymic microenvironmental cells are considerably more speculative, owing to the lack of relevant data. Monoclonal antibody MR6 was originally generated by immunizing mice with homogenized thymic stroma and was shown to give strong staining of the cortical epithelial cells of the thymus and relatively weaker staining of medullary macrophage/dendritic cells (Fig. 1). Widespread expression of receptors for a growth factor such as IL-4 is well documented (Park *et al*, 1987), however the expression of large amounts of this protein by a relatively quiescent population of cells such as cortical epithelium presents something of a mystery. Mitotic figures are rarely seen in these cells and, apart from early in embrogenesis, they are not known to proliferate rapidly under normal conditions. Indeed the whole organ decreases in size with age. Why then, should thymic epithelial cells express such large amounts of this molecule? One clue may be found in the location of this molecule within the thymus. The cortical region of the thymus is known to be an important area for the expansion of immature thymocytes and may play a part in thymic selection of MHC-restricted cells, while medullary macrophage/dendritic cells, which are also $MR6^+$ are thought to be involved in deletion of autoreactive T cells (Lo and Sprent, 1986; Ready *et al*, 1984). A possible function for these large concentrations of receptor molecules could be sequestration of available IL-4 such that the remaining amounts of this factor available to thymocytes may be too small to allow full cell activation and thus these lymphokine starved autoreactive clones may be prevented from developing, via a mechanism comparable to that proposed for IL-2 in peripheral tolerance induction (Malkovsky and Medawar, 1984). Preliminary data (Larché *et al*, 1988a) suggest that peak expression of IL-4 receptors on antigen-activated T lymphocytes occurs before peak expression of the IL-2 receptor and may therefore play a central role in the early events of cell activation. Effective "starvation" of cells by sequestration of IL-4 may cause cell death or anergy and thus prevent functional maturation.

An alternative explanation for the expression of such large quantities of receptor may lie in the cell maintainance role of IL-4 proposed by de Vries and co-workers (Spits *et al*, 1988). IL-4 appears to be able to maintain populations of T cell clones without antigen or IL-2 for extended periods (up to 21 days) and thus expression of IL-4 receptors by thymic stromal cells may serve a similar function. If so, then the addition of IL-4 to *in vitro* cultures of thymic epithelium might be expected to enhance the survival and growth of the 'temperamental' cortical epithelial subpopulation. *In vivo*, IL-4 may be required for maintaining the fine network of epithelial cells that form a syncytium throughout the cortex (Kendall, 1986). IL-4 may also be partly responsible (with γinterferon) for the induction of MHC class II molecules on these epithelial cells, although under certain conditions these two factors are known to be antagonistic (Roehm *et al*, 1984; Snapper and Paul, 1987).

SUMMARY

Monoclonal antibody MR6, raised against human thymic stromal cells binds to a novel cell surface molecule on cortical thymic epithelium, thymocytes, mature T lymphocytes and cells of several other haemopoietic and epithelial lineages. This molecule has the physical and functional characteristics of the IL-4 receptor, however the possibility that MR6 recognises another molecule with similar characteristics cannot be excluded. Further experiments are required to clarify this issue. Aberrant/elevated expression of p145-MR6 on a variety of epithelial tumours suggests that this molecule may be involved in oncogenesis and could be the product of a novel oncogene. Thus in addition to its usefulness as a laboratory reagent, conjugation of MR6 with drugs or radionucleotides may provide an effective imaging and/or therapeutic reagent for patients with a variety of carcinomas. Finally, our original purpose in raising this antibody was to create a tool with which to analyse microenvironmental molecules involved in the induction of intra-thymic T lymphocyte differentiation. As yet, the role of p145-MR6 on cortical thymic epithelial cells remains a mystery, although it is possible that it is in some way involved in the induction of tolerance to self via clonal deletion of autoreactive cells within the thymus.

ACKNOWLEDGEMENTS

We would like to thank our many scientific colleagues for their collaboration and helpful discussions on the work presented in this chapter: R. Hargreaves, N. Imami-Shita, B. Al Jabaari, H. Ladyman, J. Lamb, R. Moqbel, R. O'Hehir, D. Quint, B. von Gaudecker, E. Zanders. This work was supported by the Medical Research Council.

REFERENCES

Adamson, E.D. (1987) Oncogenes in development. *Development* **99**, 449-471.

Al Jabaari, B., Ladyman, H.M., Larché, M., Sivolapenko, G.B., Epenetos, A.A. and Ritter, M.A. (1989) Elevated Expression of the Interleukin 4 Receptor in Carcinoma - A Target for Immunotherapy? *British Journal of Cancer*, in press.

De Maagd, R.A., MacKenzie, W.A., Schuurman, H-J, Ritter, M.A., Price, K.M., Broekhuizen, R. and Kater, L. (1985) The human thymus microenvironment: heterogeneity detected by monoclonal anti-epithelial cell antibodies. *Immunology* **54**, 745-754.

Gelfand, E.W., Dosch, H-M., Shore, A., Limbatibul, S. and Lee, J.W.W. (1980) The role of the thymus in human T cell differentiation. In: *Biological Basis of Immunodeficiency* (eds E.W. Gelfand and H-M Dosch), pp 39-65. Raven Press, New York.

Harris, A.L., Smith, K., Neal, D., Fennelly, J. and Hall, R.R. (1988) Epidermal growth factor receptor (EGFr) expression correlates with tumour recurrence, stage progression and overall survival in human bladder cancer. *Proceedings of the American Association of Cancer Research* **29**, 453.

Haynes, B.F., Grummett, B., Scearce, R.M., Wolf, L.S., Tuck, D. and Singer, K.H. (1987). Summary of Reactivity Patterns of the Thymic Epithelial Panel of Antibodies in Human Skin and Thymus In: *Leucocyte Typing III*. Edited by A.J. McMichael *et al.* Oxford University Press, 262-269.

Hudziac, R.M. Schlesinger, J. and Ullrich, A. (1987) Increased expression of the putative growth factor receptor p185 ^{HER2} causes transformation and tumorigenesis of NIH 3T3 cells. *Proceedings of the National Academy of Sciences USA* **84**, 7159-7163.

Kendall, M.D. (1986) The syncytial nature of epithelial cells in the thymic cortex. *Journal of Anatomy* **147**, 85-87.

King, A.G., Wierda, D and Landreth, K.S. (1988) Bone Marrow Stromal Cell Regulation of B-Lymphopoiesis (The Role of Macrophages, IL-1, and IL-4 in Pre-B Cell Maturation). *The Journal of Immunology* **141**, 2016-2026.

Kohler, G. and Milstein, C. (1975) Continuous cultures of fused cells secreting antibody of predefined specificity. *Nature* **256**, 495-497.

Lamb, J.R., Eckels, D.D., Lake, P., Woody, J.N. and Green, N. (1982) Human T cell clones recognise synthesised peptides of influenza haemagglutinin. *Nature (London)* **300**, 66-69.

Lampert, I.A. and Ritter, M.A. (1988) The origin of the diverse epithelial cells of the thymus: is there a common stem cell? *Thymus Update* **1**, 5-25.

Land, H., Parada, L.F., Weinberg, R.A. (1983) Tumorigenic conversion of primary embryo fibroblasts requires at least two cooperating oncogenes. *Nature* **304**, 596-602.

Larché, M., Ladyman, H. and Ritter, M.A. (1987) Human Thymic Epithelial Cells and Lymphocytes May Share Common Epitopes. In: *Leucocyte Typing III*. Edited by A.J. McMichael *et al.* Oxford University Press.

Larché, M., Lamb, J.R., O'Hehir, R.E., Imami-Shita, N., Zanders, E.D., Quint, D.E., Moqbel, R. and Ritter, M.A. (1988a) Functional evidence for a monoclonal antibody that binds to the human interleukin 4 receptor. *Immunology* (in press).

Larché, M., Lamb, J.R., Ritter, M.A. (1988b) A novel T- lymphocyte molecule that may function in the induction of self-tolerance and MHC-restriction within the human thymic microenvironment. *Immunology* **64**, 101-105.

Lo, D. and Sprent, J. (1986) Identity of cells that imprint H-2-restricted T cell specificity in the thymus. Nature **319**, 672-675.

Lowenthal, J.W., Ransom, J., Howard, M. and Zlotnik, A. (1988) Up-Regulation of Interleukin 4 Receptor Expression on Immature (Lyt-2$^-$/L3T4$^-$) Thymocytes. *The Journal of Immunology* **140**, 474-478.

McMichael, A.J., Beverley, P.C.L., Cobbold, S., Crumpton, M.J., Gilks, W., Gotch, F.M., Hogg, N., Horton, M., Ling, N., MacLennan, I.C.M., Mason, D.Y., Milstein, C., Spiegelhalter,

D. and Waldmann, H. (Editors) (1987) *Leucocyte Typing III - White Cell Differentiation Antigens*. Oxford University Press.

Malkovsky, M. and Medawar, P.B. (1984) Is immunological tolerance (non-responsiveness) a consequence of interleukin 2 deficit during the recognition of antigen? *Immunology Today* **5**, 340-343.

Merlino, G.T., Xu, Y.H., Ishii, S., Clark, A.J.L., Sembra, K., Toyoshima, K., Yamamoto,T and Pastan, I. (1984) Amplification and enhanced expression of the epidermal growth factor receptor gene in A431 human carcinoma cells. *Science* **224**, 417-419.

Mizutani, S., Watt, S.M., Robertson, D., Hussein, S., Healy, L.E., Furley, A.J.W. and Greaves, M.F. (1987) Cloning of human thymic subcapsular cortex epithelial cells with T-lymphocyte binding sites and hemopoietic growth factor activity. *Proceedings of the National Academy of Sciences USA* **84**, 4999-5003.

Mosmann, T. (1988) Directional release of lymphokines from T cells. *Immunology Today* **9**, (10), 306-307.

O'Hehir, R.E., Bal, V., Quint, D., Moqbel, R., Kay, A.B., Zanders, E.D. and Lamb, J.R. (1988a) IgE induction by human cloned T lymphocytes specific for house dust mite is IL-4 dependent. *FASEB J* **2**, A1442 (Abs 6653).

Paliard, X., de Waal M.R., de Vries, J.E. and Spits, H. (1988) Interleukin-4 mediates CD8 induction on human $CD4^+$ T-cell clones. *Nature* **335**, 642-644.

Park, L.S., Friend, D., Sassenfeld, H.M., Urdal, D.L. (1987) Characterisation of the human B cell stimulatory factor-1 receptor. *Journal of Experimental Medicine* **166**, 476-487.

Ready, A.R., Jenkinson, E.J., Kingston, R. and Owen, J.J.T. (1984) Successful transplantation across major histocompatibility barrier of deoxyguanosine-treated embryonic thymus expressing class II antigens. *Nature* **310**, 231-233.

Ritter, M.A. and Haynes, B.F. (1987) Summary of Thymic Epithelium Workshop In: *Leucocyte Typing III*. Edited by A.J. McMichael *et al.* Oxford University Press p.247.

Roehm, N.W., Leibson, H.J., Zlotnik, A., Kappler, J., Marrack, P. and Cambier, J.C. (1984) Interleukin-induced increase in Ia expression by normal mouse B cells. Journal of Experimental Medicine **160**, 679-694.

Sideras, P., Funa, K., Quintana-Zalcberg, I., Xanthopoulos, K.G., Kisielow, P and Palacios R. (1988) Analysis by *in situ* hybridization of cells expressing mRNA for interluekin 4 in the developing thymus and in peripheral lymphocytes from mice. *Proceedings of the National Academy of Sciences USA* **85**, 218-221.

Snapper, C.M. and Paul, W.E. (1987) Interferon-gamma and B cell stimulatory factor-1 reciprocally regulate Ig isotype production. *Science* **236**, 944-947.

Spits, H., Yssel, H., Takebe, Y., Arai, N., Yokota, T., Lee, F., Arai, K., Banchereau, J. and De Vries, J.E. (1987) Recombinant interleukin 4 promotes the growth of human T cells. *Journal of Immunology* **130**, 1142-1147.

Stutman, O., Yunis, E.J. and Good, R.A. (1969) Thymus: an essential factor in lymphoid repopulation. *Transplantation Proceedings* **1**, 614.

Willcox, H.N.A., Schluep, M., Ritter, M.A., Schuurman, H-J., Newsom-Davis, J., Christensson, B. (1987) Myasthenic and nonmyasthenic thymoma: an expansion of a minor cortical epithelial subset? *American Journal of Pathology* **127**, 447-460.

CHAPTER 5

INNERVATION OF THE THYMUS

DAVID L. FELTEN AND SUZANNE Y. FELTEN

Department of Neurobiology and Anatomy, University of Rochester, School of Medicine, Rochester, New York, U.S.A.

I. INTRODUCTION TO NEURAL-IMMUNE INTERACTIONS

A. Evidence for a Link Between the Brain and the Immune System

Evidence from many fields, including immunology, neurobiology, endocrinology, and behavioral medicine, supports the existence of bidirectional channels of communication between the nervous system and the immune system. A wide range of environmental and psychosocial factors can exert an influence over immune functions through the central nervous system (Ader, 1981; Ader, Cohen and Felten, 1987). Specific immune responses can be conditioned in the direction of either enhancement or suppression by classical techniques (Ader and Cohen, 1985). Immune parameters are altered by numerous stressors, and the outcome of disease also is affected. Stereotaxic lesions in specific regions of the brain, particularly the hypothalamus and limbic structures, result in transient or long-lasting alterations in immune responses (reviewed by Felten and Felten, 1987). Recent evidence has demonstrated that neurotransmitter signals and neuroendocrine signals can exert significant influences over immune responses, and that cytokines can influence the activities of cells of the nervous system (see *J. Immunol.* **135**: Suppl., 1985; *J. Neurosci. Res.* **18**, 1987; *Immunol. Rev.* **100**, 1987). Normal development of the hypothalamo-pituitary axis depends upon the functional integrity of the thymus (Besedovsky and Sorkin, 1977; Pierpaoli and Besedovsky, 1975), and the development of a competent immune system depends upon the normal development and differentiation of the hypothalamo-pituitary axis (Pierpaoli and Besedovsky, 1975).

The brain can exert influence over the immune system through two major outflow channels: (1) the neuroendocrine system; and (2) the autonomic nervous system. Neuroendocrine effects on immune functions have been described for virtually all of the pituitary hormones and hormones of target organs (Berczi, 1986; Berczi, 1987). The presence of autonomic nerve fibers in organs of the immune system provides an anatomical link between the nervous system and immune system that can translate CNS messages into neurotransmitter signals that interact with receptors on cells of the immune system and influence immune functions (see D. Felten *et al.* 1987b for review).

B. General Patterns of Innervation of Primary and Secondary Immune Organs

Although early histological studies and histochemical investigations reported nerve fibers in primary lymphoid organs (Calvo, 1968; Pines and Majman, 1929; Sergeeva, 1974) and secondary lymphoid organs (Reilly *et al.* 1979; Reilly, McCuskey and Meineke, 1976; Galindo and Imaeda, 1962; Anderson and Anderson, 1975), the emphasis on the functional role of such innervation was directed towards contraction of the smooth muscle of the capsular and trabecular systems, and regulation of blood flow (Reilly, 1985; Reilly and McCuskey, 1977). However, numerous reports have described the presence of nerve fibers in the parenchyma of lymphoid organs, distributed within specific compartments (Dahlström and Zetterstrom, 1965; Williams and Felten, 1981; Utterback, 1944; Zetterstrom *et al.* 1973; Reilly *et al.* 1979; Blue and Weiss, 1981; Moore, Mumaw and Schoenberg, 1964; Galindo and Imaeda, 1962; D. Felten *et al.* 1987a, 1987b; S. Felten *et al.* 1987. S. Felten and Olschowka (1987) found direct contacts between tyrosine-hydroxylase-positive (noradrenergic) sympathetic nerve terminals and lymphocytes in the periarteriolar lymphatic sheath of the adult rat spleen; these neuro-effector junctions exhibited 6 nm appositions (2 nm = gap junction; 20 nm = classical CNS synapse). These findings suggested the possibility of autonomic innervation of lymphoid organs with cells of the immune system as targets. The neurotransmitters utilized by these nerves have been investigated only recently. Although there is some evidence for peptidergic innervation of immune organs (eg. Felten *et al.* 1985; Fink and Weihe, 1988), the most extensive chemical system innervating all lymphoid organs is the noradrenergic sympathetic postganglionic system.

In all lymphoid organs, the noradrenergic (NA) nerve fibers distribute into specific cellular compartments, and end in close apposition to lymphocytes and macrophages (D. Felten *et al.* 1987a, 1987b). In secondary lymphoid organs, NA fibers particularly distribute into zones of lymphocyte entry, antigen capture, antigen presentation and lymphocyte activation, and lymphocyte egress (Ackerman *et al.* 1987). Innervation of both primary and secondary lymphoid organs derives from sympathetic chain ganglion cells or from prevertebral (collateral) ganglion cells, travels to the organ with the vasculature, and further distributes with smooth muscle systems or into the parenchyma in the target organ. Preganglionic neurons from the intermediolateral column of the thoracic and upper lumbar spinal cord innervate these ganglion cells, establishing a classical two neuron link between the CNS and organs of the immune system.

NA fibers innervating the bone marrow travel with the vasculature *via* the nutrient foramena, and accompany arterioles into the marrow, where they continue with the vasculature and send branches directly into the parenchyma.

NA fibers innervating the thymus enter with the vasculature, form plexuses with the vasculature beneath the capsule, distribute throughout the cortex and paracortex, and distribute branches into the parenchyma, where they end among thymocytes and other cell types.

NA fibers innervating the spleen enter the hilar region with the vasculature, distribute with the capsular/trabecular system, follow vasculature into the spleen, and travel in close association with the white pulp. In the white pulp, the NA fibers distribute with the central arteriolar system, and branch abundantly into the periarteriolar lymphatic sheath (PALS), along the marginal sinus, and in parafollicular sites. Scattered fibers are found in the marginal zone and within the follicles. In the PALS and along the marginal sinus, NA terminals form direct appositions with lymphocytes. The red pulp is very sparsely innervated in most species.

NA fibers innervating lymph nodes enter at the hilar region, course through the medullary cords among lymphocytes and macrophages, and distribute into the cortical and paracortical zones where T lymphocytes reside. The follicles are not supplied by NA fibers.

NA fibers innervating gut-associated lymphoid tissue (GALT) travel with the autonomic plexuses in the muscular layers, course between the nodules towards Peyer's patches or lamina propria of lymphoid zones, travel directly through the T-dependent zones, and arborize extensively in the lamina propria among lymphocytes, enterochromaffin cells, and plasma cells.

C. A Functional Role for Noradrenergic Innervation of Immune Organs

In order for norepinephrine to act as a neurotransmitter with lymphoid cells as targets, it must fulfill the traditional criteria for neurotransmission: (1) presence in nerve fibers; (2) release and availability at the site of target cells; (3) existence of receptors on the target cells, with intracellular second messenger links; and (4) pharmacologically predictable functional role. In the spleen, norepinephrine (NE) has been examined in detail for fulfillment of these criteria (see D. Felten *et al.* 1987b for detailed review): (1) NA fibers are present among specific immune cells in lymphoid compartments of the spleen, and form direct contacts with them; (2) NA is released from the nerves in spleen following stimulation, and has been measured directly in the spleen with neurochemical and *in vivo* dialysis methods; (3) Lymphocytes (both T and B), macrophages and monocytes, and neutrophils possess beta-adrenoceptors on their surfaces, linked through second messengers (often cAMP) with intracellular metabolic processes; (4) Surgical and pharmacological manipulation of the NA nerves results in predictable alterations of immune responses. Thus, it appears that NA has fulfilled the criteria for neurotransmission with immunocytes as targets. The criteria for NA neurotransmission in the thymus will be examined in detail below.

Extensive denervation studies from our laboratories (D. Felten *et al* 1987b; D. Felten *et al* 1985; D. Felten *et al* 1984; Livnat *et al* 1985, 1987, 1988) have shown that ablation of NA innervation of lymphoid organs results in diminished primary immune responses (in C3H mice, 80% decrease with systemic challenge of spleen, and 97% decrease with foot pad challenge in popliteal lymph nodes), diminished secondary immune responses, diminished delayed hypersensitivity response to a contact sensitizing agent, diminished cytotoxic T cell response with concommitant decrease in IL2 production, enhanced baseline and poly I:C stimulated natural killer (NK) activity, enhanced antigen-stimulated B cell proliferation in inguinal lymph nodes of mice, and diminished T lymphocyte proliferation from murine lymph nodes. We suggest that NE interacts with receptors on the surface of many cell types in both primary and secondary immune organs, affects proliferation, differentiation, and effector functions, and also exerts organ-specific and state-dependent influences on the interactions of subsets of immune cells (see D. Felten *et al.* 1987b for discussion). In addition, we have preliminary evidence that the state of NA innervation of lymph nodes, and the expression of beta-adrenoceptors on lymphocytes, can influence lymphocyte traffic and the distribution of lymphocytes to immune organs.

II. NORADRENERGIC SYMPATHETIC INNERVATION OF THE THYMUS

A. Anatomical Distribution

The pattern of NA sympathetic distribution has been studied in most detail in rats and mice (Williams and Felten, 1981; Williams *et al* 1981; Livnat *et al* 1985; S. Felten *et al* 1988; D. Felten *et al* 1985; D. Felten *et al.* 1987b; Bulloch and Pomerantz, 1984; Fujiwara, Muryobeyaski and Shenamoto, 1966; Sergeeva, 1974; Bellinger, Felten and Felten, 1989) although the pattern of distribution appears to be consistent among all mammals examined thus far; the following description is a summary of these findings from our laboratory. NA fibers coursed into the thymus in nerve bundles that travelled with surface arteries, or entered through the thymic capsule. Large plexuses of varicosities travelled with the vasculature in the subcapsular and associated septal regions. Numerous varicose fibers branched from these plexuses into the cortical and paracortical parenchyma and were found adjacent to fields of thymocytes. Small clusters of NA terminals also were found adjacent to yellow cortical autofluorescent cells, mast cells, and other cell types within that region. Most of the NA nerve fibers were confined to the cortical and paracortical regions, and were present throughout these zones, with a slightly higher density near the cortical-medullary junction.

At the cortical-medullary boundary, NA fibers were found along the medullary venous sinuses; the NA plexuses that travelled with the septal system were continuous from the septa to the large venous sinuses. These venous sinuses were abundant at the cortical-medullary boundary. NA fibers extended from these venous plexuses into the paracortical parenchyma. Dense vascular plexuses also were found along the thymic arteries and arterioles in the medulla. Scattered individual varicose fibers branched from both the arteriolar and venous plexuses into the medulla.

B. A Functional Role for Noradrenergic Innervation of the Thymus

The anatomical studies demonstrated the presence of NA nerve fibers in the thymus, associated with the arteriolar and venous vasculature and with the parenchyma; NA varicosities also were found adjacent to thymocytes. Neurochemical measurement with high performance liquid chromatography with electrochemical detection (LCEC) demonstrated chemical presence and availability in the thymus. Thus, the first two criteria for neurotransmission have been fulfilled.

The third criterion for neurotransmission is the presence of appropriate receptors on the putative target cells. The work of Singh and colleagues (Singh, 1979a, 1979b, 1984, 1985; Singh *et al* 1979; Singh and Owen, 1976) has provided evidence for the presence of beta-adrenoceptors on developing thymocytes. These receptors appear to be mainly of the beta-2 subclass, and are linked through adenylate cyclase to cyclic AMP (cAMP) as a second messenger.

Evidence also points to a predictable functional role for NA interactions with thymocytes. Singh and colleagues demonstrated sensitivity of developing thymocytes to catecholamines. Stimulation of the beta-adrenoceptors on these cells with catecholamines resulted in diminished proliferation, and enhanced the expression of antigens associated with differentiation on thymocytes. The demonstration of sympathetic NA nerve fibers in the developing neonatal thymus by Singh (1984) and by our laboratories (Ackerman *et al.*, unpublished) is consistent with the proposal that the NA sympathetic nervous system can influence thymic activities associated with thymocyte proliferation and differentiation. Additional support for this hypothesis has come from denervation studies. Singh (1985) has demonstrated that non-lymphoid rudiments of fetal thymus from nude mice support the development of thymocytes when transplanted into the anterior chamber of a surgically sympathectomized eye. He also has shown lymphopoiesis in the nude fetal mouse thymus following sympathectomy (Singh, 1985). Preliminary evidence from our laboratories (D. Felten *et al.* 1987b) showed that thymocytes from 10 day old rats, sympathectomized at birth, demonstrate enhanced proliferative responses, consistent with a functional role for NE in the inhibition of proliferation and the enhancement of differentiation of thymocytes. It should be noted clearly that such a role cannot be generalized to include mature T lymphocytes and B lymphocytes (D. Felten *et al.* 1987b); indeed, many T-dependent immune responses are decreased markedly following denervation, with opposite directionality compared with proliferative responses of thymocytes from denervated thymus. Such findings can only apply to a specific cell type in a specific immune organ and microenvironment, with unique conditions of age, sex, species, and type of challenge. It seems no more reasonable to propose a single, unidirectional response for NE in the immune system than it does to propose a similar response for IL1 or IL2.

A possible additional role for NA innervation of the thymus is the control of cell migration. Stimulation of the sympathetic trunk enhanced the egress of reticulocytes from the bone marrow into the blood (Webber *et al.* 1970). Catecholamine also stimulated the release of activated lymphocytes from the spleen (Ernström and Sandberg, 1973, 1974; Ernström and Soder, 1975) and lymph nodes (Moore, 1984) into peripheral blood. We suggest the possibility of a similar regulatory role for NA nerve fibers and their neurotransmitter in controlling lymphocyte traffic.

C. Innervation of the Thymus in Aging and Involution

In the rat, the thymus showed a progressive diminution in weight and an involution from 3 months to 27 months of age. However, the NA nerve fibers remained intact in the thymus despite the marked involution (Bellinger, Felten and Felten, 1989). The total quantity of NA per thymus remained relatively constant, at approximately 400 pmoles, while the concentration per mg protein increased considerably. Fluorescence histochemical observations showed a dramatic increase in the density of NA varicosities, particularly in the cortical and paracortical zones, at 17, 21, and 27 months of age. In the involuted thymus, there were scattered lobules with diminished cortex and diminished thymocyte numbers, with infiltration of fatty and connective tissue. Thus, the NA innervation retained its compartmentation, even in the presence of marked loss of thymocytes and a changing geometry of the involuting thymus, suggesting a remarkable plasticity. The remaining thymocytes in the involuted thymus were present in a microenvironment where the density of NA terminals, the concentration of NE, and the ratio of MHPG: NE (a relative measure of turnover) is considerably greater than in the young thymus, suggesting an enhanced potential for NA interactions with remaining cells of the involuted thymus. Of course, such interactions depend upon the continued presence and efficacy of beta-adrenoceptors on such target cells. This issue has not been investigated yet.

The maintenance of NA innervation of the involuted thymus is in marked contrast to the aged rodent spleen and lymph nodes, where NA innervation is diminished extensively with age (Bellinger *et al.* 1987; S. Felten *et al.* 1987). We have suggested that the loss of NA innervation may contribute to diminished T-dependent functions in these secondary immune organs, particularly in light of such diminished responses in young adults denervated with chemical sympathectomy (D. Felten *et al.* 1984, 1985, 1987; Livnat *et al.* 1985). Thus, the thymus maintains its NA innervation with age while secondary lymphoid organs do not. We suggest that this may be due to the effect of cytokines, released during ongoing immune responses, that can influence the turnover of NA (Besedovsky *et al.* 1979, 1981, 1987). For a more detailed discussion of the implications of this hypothesis, see Bellinger *et al.* (1988).

III. OTHER CHEMICALLY SPECIFIC NEURAL SYSTEMS IN THE THYMUS

A. Evidence for Lack of Cholinergic Innervation of the Thymus

Because the earlier discovery of adrenoceptors on thymocytes and lymphocytes was followed by evidence of abundant NA innervation of lymphoid organs where these cells reside, it seemed reasonable that the finding of muscarinic receptors on thymocytes and lymphocytes (Hadden, 1977; Strom, Lundin and Carpenter, 1977; Strom, Lane and George, 1981; Gordon, Cohen and Wilson, 1978; Maslinski *et al.* 1988) should be accompanied by the existence of cholinergic innervation of these same organs. Our laboratory has been unable to confirm such innervation.The main evidence for putative cholinergic innervation of the thymus is based upon two lines of investigation (Bulloch and Moore, 1981; Bulloch and Pomerantz, 1984): (1) the report of direct connections from the brain stem (nucleus ambiguus-retrofacial area) *via* the vagus nerve and from cervical spinal cord anterior horn cells to the thymus in rodents; and (2) the presence of a network of acetylcholinesterase (AChE)-positive fiber-like profiles in the thymus, consistent with silver-stained profiles reported as neural fibers in early studies (such as Pines and Majman, 1929). Bulloch and Moore (1981) used retrograde tracing with horseradish peroxidase (HRP) injected into the thymus; brain stem and spinal cord neurons were described with HRP label, as were some cells in the superior cervical ganglion, indicative of the previously described NA innervation. In a retrograde tracing study by Nance and colleagues (Nance, Hopkins and Biegler, 1987), utilizing lectin-bound HRP, labeled cells were found in the superior cervical ganglion and upper chain ganglia, but were not found in the brain stem and spinal cord. They interpreted the findings of Bulloch and Moore (1981) as an artefact of diffusion, and attributed the labeled cells in brain stem and anterior horn to transport of diffused HRP in nerves innervating musculature in the mediastinum. As Fox and Powley (1986) demonstrated in a carefully controlled and detailed investigation of retrograde transport from the pancreas, visceral organs show tremendous diffusion of such tracers following injection, with subsequent leeching of the tracer into surrounding regions, producing even worse artefact than direct injection of HRP into the peritoneum. Since Bulloch and Moore (1981) used very large injections of concentrated HRP, a relatively diffusible tracer, the possibility cannot be dismissed that the labeling seen in brain stem and spinal cord is due to diffusion from the thymus into adjacent innervated structures of the mediastinum, such as the surrounding musculature, consistent with the localization of those neurons; indeed, this impression is reinforced by the observation in that study that injection of HRP directly into the mediastinum resulted in a similar pattern of light labeling in these same neurons of the brain stem and spinal cord. These authors interpreted this finding as evidence that HRP diffused into the thymus from the mediastinal injection, rather than the other way around. Fox and Powley (1986) consider the use of a diffusion barrier agent in all such tracing studies as essential, in order to avoid false positives from the use of highly diffusable retrograde tracing agents. The use of lectin-bound HRP in small amounts and concentrations, sufficient to permit excellent retrograde labeling of the sympathetic superior cervical ganglion, but without evidence of brain stem or spinal cord labeling, throws considerable doubt on the existence of such connections directly from CNS to the thymus (Nance, Hopkins and Biegler, 1987). Thus, although the report of direct innervation from the brain stem and spinal cord to the thymus has been cited widely, the actual existence of this putative connection is in need of careful confirmation using diffusion barrier agents and using anterograde as well as retrograde labeling. We have injected wheat-germ agglutanin HRP in large quantities into the retrofacial/rostral nucleus ambiguus area of the rat brain stem, and have not been able to find anterograde labeling of fibers in the thymus, although the esophagus demonstrates some labeled fibers.

The second line of evidence for putative cholinergic innervation of the thymus is the presence of AChE-positive profiles that have been interpreted to be nerves. Since lymphoid organs contain many cell types, such as macrophages, dendritic cells, reticular cells, and others types, whose processes closely resemble neural profiles, the appearance of light microscopic AChE staining is insufficient evidence for describing any stained profile as neural. As an additional step, observations can be made at the electron microscopic level; however, this also can be difficult, since processes of non-neural cells in immune organs can demonstrate vesicles and intracellular inclusions at the EM level that closely resemble neural inclusions. A second approach is to carry out denervation and ablation studies that demonstrate experimentally that such profiles actually are neural. Our preliminary EM observations suggest that much of the AChE staining in thymus is non-neural, although some AChE-positive neural profiles were found. However, 6-hydroxydopamine sympathectomy resulted in diminution of some AChE staining in neural-like profiles in the thymus, suggesting that AChE may be co-localized with norepinephrine in sympathetic nerves, as it is in gut-associated lymphoid tissue (D. Felten *et al.* 1981) and elsewhere. AChE is a notoriously unreliable marker for peripheral cholinergic fibers unless experimental confirmation of its existence in acetylcholine-containing nerves in that specific system can be obtained. In a study of AChE in the rat spleen, we have found both neural and non-neural AChE profiles, despite the total absence of the synthetic enzyme for acetylcholine, choline acetyltransferase (ChAT), again suggesting the abundant presence of AChE in non-cholinergic structures.

In order to resolve whether or not cholinergic innervation of lymphoid organs exists, Robert Hamill has undertaken a detailed collaborative study with our laboratories to measure ChAT activity and total acetyltransferase activity in the thymus and spleen. In the rat spleen, no such activity was found. In the rat thymus, a very low level of ChAT activity was found that was less than twice the blank, usually considered as evidence for its absence in an organ; such diminishingly low ChAT activity is inconsistent with the much higher ChAT activity found in other organs with proven cholinergic innervation. If there are very sparse cholinergic fibers present in the thymus, the ChAT activity is sufficiently low to raise serious doubts that it could account for significant cholinergic activity; such a low level of activity is not consistent with the extent of putative brain stem and spinal cord anterior horn cell innervation suggested by the study of Bulloch and Moore (1981). In the absence of ChAT activity, it is difficult to devise an alternate hypothesis for the existence of cholinergic neurotransmission in the thymus. We therefore conclude that cholinergic innervation in the thymus is miniscule, if it exists at all. We further conclude that evidence for brain stem and spinal cord direct connections to the thymus is open to serious question, and must be confirmed by more carefully controlled and detailed retrograde, anterograde, and ablation studies; if such innervation does exist, it is highly unlikely to be cholinergic.

B. Possible Peptidergic Innervation of the Thymus

The possibility of peptidergic nerves in the thymus has not been addressed in detail. We found preliminary evidence for VIP-like immunoreactive profiles that resembled nerve fibers in the thymus (D. Felten *et al.* 1985); however, such findings are in need of further confirmation and experimental demonstration to ascertain whether the labeled structures were nerve fibers. Our laboratories and Olschowka's laboratory have found the presence of neuropeptide Y (NPY) and substance P in nerves in the rodent spleen, with the NPY nerve profiles closely overlapping the distribution and compartmentation of tyrosine hydroxylase-positive NA nerves, suggesting possible co-localization. Whether NPY is co-localized in NA nerves in the thymus remains to be demonstrated. In addition, numerous neuropeptides have been found in nerve-like profiles in lymph nodes (including neuropeptide Y, vasoactive intestinal polypeptide, peptide histidine isoleucine, substance P, and calcitonin gene-related peptide (Fink and Weihe,

1988) and in gut-associated lymphoid tissue associated with mast cells (including substance P and calcitonin gene-related peptide) (Stead *et al.* 1987). However, as in the case for all immune organs, light microscopic immunocytochemistry alone, without experimental confirmation that those profiles are actually nerves, is insufficient to confirm peptidergic innervation with confidence. The neuropeptides oxytocin and vasopressin have been found in the thymus (Geenen *et al.* 1986; Geenen *et al.* 1987; Marwick, Lolait and Funder, 1986; Clements and Funder, 1986), but appear to be in epithelial cells, nurse cells, or other non-neural elements. It certainly is possible that such cells could release these peptides, perhaps even in response to neural stimulation or interactions. The issue of peptidergic innervation of the thymus barely has been addressed, although it shows great promise for further elucidation of neural signal molecules that can convey information from the CNS to the thymus.

IV. POSSIBLE MODES OF INTERACTION OF NEUROTRANSMITTERS WITH THYMOCYTES

The finding that the basic criteria for neurotransmission for norepinephrine have been fulfilled with lymphocytes and thymocytes as targets, raises the question of modes of communication utilized by the neurotransmitter to act upon target cells which are assumed to be mobile. The most obvious possibility is that the neurotransmitter is present in nerves, released in response to a neural signals from the brain *via* the two neuron autonomic connection, diffuses to the surface of the target cell, interacts through a classical ligand-receptor binding, initiates a second messenger response in the target cell, which then proceeds to change the function, metabolism, gene expression, secretion, or mobility of that cell. This is a classical process of peripheral neurotransmission for smooth muscle, cardiac muscle, and exocrine glands, the traditional targets for autonomic innervation; now, the same criteria for neurotransmission have been fulfilled with cells of the immune system as targets. However, other possibilities also exist for communication. The neurotransmitter might act by direct uptake into the target immunocyte, still exerting an effect but not through a classical ligand-receptor interaction. This might be the case for serotonin interactions with macrophages (Jean Jackson, personal communication). The neurotransmitter also might act on an accessory cell or a local secretory cell to influence the release of other signal molecules (eg. serotonin or histamine from mast cells) or to regulate other activities (eg. antigen processing or presentation by macrophages or other antigen-presenting cells). A further possibility is that the neurotransmitter may modulate (eg. potentiate, exert synergistic effects, or counteract other such effects) the activity of other molecules of immune or neural origin. Direct and indirect evidence supports the notion that NE can modulate the effects of IL1, IL2, and interferon-gamma (Frohman *et al.* 1988; Boyle *et al.* 1988; Beckner and Farrar, 1988), interacting with what traditionally was viewed as exclusively an immune network of communication.

Evidence for these multiple modes of biological signaling raises the dismaying spectre that neurotransmitter communication with immunocytes, and cytokine communication with neurons, may involve all of the complexities of both of these systems, including primary effects, secondary effects, and modulatory effects, both receptor mediated and non-receptor mediated, with the added complication of the local neuroendocrine milieu that probably influences both neurotransmitter and cytokine synthesis, secretion, and activities. In addition, the neurotransmitters may regulate blood flow, lymphocyte trafficing, and ingress and egress of cells from a specific organ, systemic immunoregulatory activities that are not explored easily with *in vitro* systems. Although the task seems daunting, the challenge is exceptionally rich for the opportunity to expand our understanding of neural-immune communication in both directions, and to give mechanistic substance to the understanding of neural-immune interactions, and their role in behavioral medicine.

Acknowledgements

Supported by N00014-84-K-0488 from the Office of Naval Research, RO1 NS25223 from NIH; RO1 MH42073 from NIMH, and a MacArthur Foundation Prize Fellowship to D.L.F. We thank John Housel for excellent technical assistance.

REFERENCES

Ackerman, K.D., Felten, S.Y., Bellinger, D.L., Livnat, S. and Felten, D.L. (1987) Noradrenergic sympathetic innervation of spleen and lymph nodes in relation to specific cellular compartments. *Prog. Immunol.* **6**, 588-600.

Ader, R. (1981) *Psychoneuroimmunology*, New York: Academic Press.

Ader, R. and Cohen, N. (1985) CNS-Immune system interactions: Conditioning phenomena. *Behav. Brain Sci.* **8**, 379-395.

Ader, R., Cohen, N. and Felten, D.L. (1987) Brain, behavior, and immunity. *Brain Behav. Immun.* **1**, 1-6.

Anderson, A.O. and Anderson, N.D. (1975) Studies on the structure and permeability of the microvasculature in normal rat lymph nodes. *Am. J. Pathol.* **80**, 387-418.

Beckner, S.K. and Farrar, W.L. (1988) Potentiation of lymphokine-activated killer cell differentiation and lymphocyte proliferation by stimulation of protein kinase C or inhibition of adenylate cyclase. *J. Immunol.* **140**, 208-214.

Bellinger, D.L., Felten, S.Y., Collier, T.J. and Felten, D.L. (1987) Noradrenergic sympathetic innervation of the spleen: IV. Morphometric analysis in adult and aged F344 rats. *J Neurosci Res.* **18**, 55-63.

Bellinger, D.L., Felten, S.Y. and Felten, D.L. (1989) Maintenance of noradrenergic sympathetic innervation in the involuted thymus of the aged Fischer 344 rat. *Brain Behav. Immun.* (in press) .

Berczi, I. (1986) *Pituitary Function and Immunity,* Boca Raton: CRC Press.

Berczi, I. (1987) *Hormones and Immunity: MTR Press.*

Besedovsky, H.O., Da Prada, M., del Rey, A.E. and Sorkin, E. (1981) Immunoregulation by sympathetic nervous system. *Trends Pharmacol. Sci.* **2**, 236-238.

Besedovsky, H.O., del Rey, A.E., Sorkin, E., Burri, R., Honegger,C.G., Schlumpf, M., *et al.* (1987) T lymphocytes affect the development of sympathetic innervation of mouse spleen. *Brain Behav. Immun.* **1**, 185-193.

Besedovsky, H.O., del Rey, A.E., Sorkin, E., Da Prada, M. and Keller, H.H. (1979) Immunoregulation mediated by the sympathetic nervous system. *Cell. Immunol.* **48**, 346-355.

Besedovsky, H.O. and Sorkin, E. (1977) Network of immune neuroendocrine interactions. *J. Clin. Exp. Immunol.* **27**, 1-12.

Blue, J. and Weiss, L. (1981) Electron microscopy of the red pulp of the dog spleen including vascular arrangements, periarterial macrophage sheaths (ellipsoids), and the contractile, innervated reticular meshwork. *Am. J. Anat.* **161**, 189-218.

Boyle, M., Yamamoto, G., Chen, M., Rivier, J. and Vale, W. (1988) Interleukin 1 prevents loss of corticotropic responsiveness to β- adrenergic stimulation *in vitro. Proc. Natl. Acad. Sci. (USA)* **85**, 5556-5560.

Bulloch, K. and Moore, R.Y. (1981) Innervation of the thymus gland by brain stem and spinal cord in mouse and rat. *Am. J. Anat.* **162**, 157-166.

Bulloch, K. and Pomerantz, W. (1984) Autonomic nervous system innervation of thymic related lymphoid tissue in wild-type and nude mice. *J. Comp. Neurol.* **228**, 57-68.

Calvo, W. (1968) The innervation of the bone marrow in laboratory animals. *Am. J. Anat.* **123**, 315-328.

Calvo, W. and Forteza-Vila, J. (1969) On the development of bone marrow innervation in newborn rats as studied with silver impregnation and electron microscopy. *Am. J. Anat.* **126**, 355-359.

Clements, J.A. and Funder, J.W. (1986) AVP and AVP-like immunoreactivity in peripheral tissues. *Endocrin. Rev.* **7**, 449-460.

Dahlström, A.B. and Zetterstrom, B.E.M. (1965) Noradrenaline stores in nerve terminals of the spleen: Changes during hemorrhagic shock. *Science* **147**, 1583-1585.

Ernström, U. and Sandberg, G. (1973) Effects of alpha- and beta-receptor stimulation on the release of lymphocytes and granulocytes from the spleen. *Scand. J. Haematol.* **11**, 275-286.

Ernström, U. and Sandberg, G. (1974) Stimulation of lymphocyte release from the spleen by theophylline and isoproterenol. *Acta Physiol. Scand.* **90**, 202-209.

Ernström, U. and Soder, O. (1975) Influence of adrenaline on the dissemination of antibody-producing cells from the spleen. *Clin. Exp. Immunol.* **21**, 131-140.

Felten, D.L., Ackerman, K.D., Wiegand, S.J. and Felten, S.Y. (1987a) Noradrenergic sympathetic innervation of the spleen: I. Nerve fibers associate with lymphocytes and macrophages in specific compartments of the splenic white pulp. *J. Neurosci. Res.* **18**, 28-36.

Felten, D.L. and Felten, S.Y. (1988) Immune interactions with specific neural structures. *Brain Behav. Immun. (in press)*

Felten, D.L., Felten, S.Y., Bellinger, D.L., Carlson, S.L., Ackerman, K.D., Madden, K.S., *et al.* (1987b) Noradrenergic sympathetic neural interactions with the immune system: structure and function. *Imm. Rev.* **100**, 225-260.

Felten, D.L., Felten, S.Y., Carlson, S.L., Olschowka, J.A. and Livnat, S. (1985) Noradrenergic and peptidergic innervation of lymphoid tissue. *J. Immunol.* **135**, 755s-765s.

Felten, D.L., Livnat, S., Felten, S.Y., Carlson, S.L., Bellinger, D.L. and Yeh, P. (1984) Sympathetic innervation of lymph nodes in mice. *Brain Res. Bull.* **13**, 693-699.

Felten, D.L., Overhage, J.M., Felten, S.Y. and Schmedtje, J.F. (1981) Noradrenergic sympathetic innervation of lymphoid tissue in the rabbit appendix: further evidence for a link between the nervous and immune systems. *Brain Res. Bull.* **7**, 595-612.

Felten, S.Y., Bellinger, D.L., Collier, T.J., Coleman, P.D. and Felten, D.L. (1987) Decreased sympathetic innervation of spleen in aged Fischer 344 rats. *Neurobiol. Aging* **8**, 159-165.

Felten, S.Y., Felten, D.L., Bellinger, D.L., Carlson, S.L., Ackerman, K.D., Madden, K.S., *et al.* (1988) Noradrenergic sympathetic innervation of lymphoid organs. *Prog. Allergy* **43**, 14-36.

Felten, S.Y. and Olschowka, J.A. (1987) Noradrenergic sympathetic innervation of the splen: II. Tyrosine hydroxylase (TH) -positive nerve terminals form synaptic-like contacts on lymphocytes in the splenic white pulp. *J. Neurosci. Res.* **18**, 37-48.

Fink, T. and Weihe, E. (1988) Multiple neuropeptides in nerves supplying mammalian lymph nodes: messenger candidates for sensory and autonomic neuroimmunomodulation?. *Neurosci. Lett.* **90**, 39-44.

Fox, E.A. and Powley, T.L. (1986) Tracer diffusion has exaggerated CNS maps of direct preganglionic innervation of pancreas. *J. Autonom. Nerv. Syst.* **15**, 55-69.

Frohman, E.M., Vayuvegula, B., van den Noort, S. and Gupta, S. (1988) Norepinephrine inhibits gamma-interferon-induced MHC class II (Ia) antigen expression on cultured brain astrocytes. *J. Neuroimmunol.* **17**, 89-101.

Fujiwara, M., Muryobeyaski, T. and Shenamoto, K. (1966) Histochemical demonstration of monoamines in the thymus of rats. *Jpn. J. Pharmacol.* **16**, 493-494.

Galindo, B. and Imaeda, T. (1962) Electron microscopic study of the white pulp of the mouse spleen. *Anat. Rec.* **143**, 399-415.

Geenen, V., Legros, J.J., Franchimont, P., Baudrihaye, M., Defresne, M.P. and Boniver, J. (1986) The neuroendocrine thymus: Coexistence of oxytocin and neurophysin in the human thymus. *Science* **232**, 508-510.

Geenen, V., Legros, J.J., Franchimont, P., Defresne, M.P., Boniver, J., Ivell, R., *et al.* (1987) The thymus as a neuroendocrine organ: Synthesis of vasopressin and oxytocin in human thymic epithelium. *Ann. N.Y. Acad. Sci.* **496**, 56-66.

Giron, L.T., Crutcher, K.A. and Davis, J.N. (1980) Lymph nodes - a possible site for sympathetic neuronal regulation of immune responses. *Ann. Neurol.* **8**, 520-522.

Gordon, M.A., Cohen, J.J. and Wilson, I.G. (1978) Muscarinic cholinergic receptors in murine lymphocytes - Demonstration by direct binding. *Proc. Natl. Acad. Sci. (USA)* **79**, 5097-5100.

Hadden, J.W. (1977) Cyclic nucleotides in lymphocyte proliferationand differentiation. In: *Immunopharmacology*, edited by J.W. Hadden, R.G. Coffey and F. Spreafico. pp.1-28. New York: Plenum.

Livnat, S., Eisen, J., Felten, D.L., Felten, S.Y., Irwin, J., Madden, K.S., *et al.* (1988) Behavioral and sympathetic neural modulation of immune function. In: *Progress in Catecholamine Research, Part A: Basic Aspects and Peripheral Mechanisms*, edited by A Dahlström, R.H. Belmaker and M. Sandler, pp. 539–546. New York: Alan R. Liss..

Livnat, S., Felten, S.Y., Carlson, S.L., Bellinger, D.L. and Felten, D.L. (1985) Involvement of peripheral and central catecholamine systems in neural-immune interactions. *J. Neuroimmunol.* **10**, 5-30.

Livnat, S., Madden, K.S., Felten, D.L. and Felten, S.Y. (1987) Regulation of the immune system by sympathetic neural mechanisms. *Prog. Neuro-psychopharmacol. & Biol. Psychiat.* **11**, 145-152.

Marwick, A.J., Lolait, S.J. and Funder, J.W. (1986) Immunoreactive arginine vasopressin in the rat thymus. *Endocrinology* **119**, 1690-1696.

Maslinski, W., Kullberg, M., Nordstrom, O. and Bartfai, T. (1988) Muscarinic receptors and receptor-mediated actions on rat thymocytes. *J. Neuroimmunol.* **17**, 265-274.

Moore, R.D., Mumaw, V.R. and Schoenberg, M.C. (1964) The structure of the spleen and its functional implications. *Exp. Molec. Path.* **3**, 31-50.

Moore, T.C. (1984) Modification of lymphocyte traffic by vasoactive neurotransmitter substances. *Immunol.* **52**, 511-518.

Nance, D.M., Hopkins, D.A. and Biegler, D. (1987) Re-investigation of the innervation of the thymus gland in mice and rats. *Brain Behav. Immun.* **1**, 134-147.

Pierpaoli, W. and Besedovsky, H.O. (1975) Role of the thymus in programming neuroendocrine functions. *Clin. Exp. Immunol.* **20**, 323- 338.

Pines, L. and Majman, R. (1929) The innervation of the thymus. *J.Nerv. Ment. Dis.* **69**, 361-384.

Reilly, F.D. (1985) Innervation and vascular pharmacodynamics of the mammalian spleen. *Experientia* **41**, 187-192.

Reilly, F.D., McCuskey, P.A., Miller, M.L., McCuskey, R.S. and Meineke, H.A. (1979) Innervation of the periarteriolar lymphatic sheath of the spleen. *Tissue and Cell* **11**, 121-126.

Reilly, F.D. and McCuskey, R.S. (1977) Studies of the hemopoietic microenvironment VII. Neural mechanisms in splenic microvascular regulation in mice. *Microvascular Res.* **14**, 293-302.

Reilly, F.D., McCuskey, R.S. and Meineke, H.A. (1976) Studies of the hemopoietic microenvironment VIII. Adrenergic and cholinergic innervation of the murine spleen. *Anat. Rec.* **185**, 109-118.

Sergeeva, V.E. (1974) Histotopography of catecholamines in the mammalian thymus. *Bull. Exp. Bio. Med.* **77**, 456-458.

Singh, U. (1979a) Effect of catecholamines on lymphopoiesis in fetal mouse thymic explants. *Eur. J. Immunol.* **14**, 757-759.

Singh, U. (1979b) Effect of catecholamines on lymphopoiesis in fetal mouse thymic explants. *J. Anat.* **129**, 279-292.

Singh, U. (1984) Sympathetic innervation of fetal mouse thymus. *Eur. J. Immunol.* **14**, 757-759.

Singh, U. (1985) Effect of sympathectomy on the maturation of fetal thymocytes grown within the anterior eye chambers in mice. *Adv. Exp. Biol. Med.* **186**, 349-356.

Singh, U. (1985) Lymphopoiesis in the nude fetal mouse thymus following sympathectomy. *Eur. J. Immunol.* **93**, 222-228.

Singh, U., Millson, S., Smith, P.A. and Owen, J.J.T. (1979) Identification of beta-adrenergic adrenoceptors during thymocyte ontogeny in mice. *Eur. J. Immunol.* **9**, 31-35.

Singh, U. and Owen, J.J.T. (1976) Studies on the maturation of thymus stem cells - The effects of catecholamines, histamine, and peptide hormones on the expression of T alloantigens. *Eur. J.Immunol.* **6**, 59-62.

Stead, R.H., Tomioka, M., Quinonez, G., Simon, G., Felten, S.Y. and Bienenstock, J. (1987) Intestinal mucosal mast cells in normal and nematode-infected rat intestines are in intimate contact with peptidergic nerves. *Proc. Natl. Acad. Sci. (USA)* **84**, 2975-2979.

Strom, T.B., Lane, M.A. and George, K. (1981) The parallel, time- dependent, bimodal change in lymphocyte cholinergic binding activity and cholinergic influence upon lymphocyte mediated cytotoxicity after lymphocyte activation. *J. Immunol.* **127**, 705-710.

Strom, T.D., Lundin, A.P. and Carpenter, C.B. (1977) The role of cyclic nucleotides in lymphocyte activation and function. *Prog. Clin. Immunol.* **3**, 115-153.

Utterback, R.A. (1944) The innervation of the spleen. *J. Comp. Neurol.* **81**, 55-68.

Walcott, B. and McLean, J.R. (1985) Catecholamine-containing neurons and lymphoid cells in a lacrimal gland of the pigeon. *Brain Res. 328*, 129-137.

Webber, R.H., DeFelice, K.R., Ferguson, R.J. and Powell, J.P. (1970) Bone marrow response to stimulation of the sympathetic trunk in rats. *Acta Anat.* **77**, 92-97.

Williams, J.M. and Felten, D.L. (1981) Sympathetic innervation of murine thymus and spleen: A comparitive histofluorescence study. Anat. Rec. **199**, 531-542.

Williams, J.M., Peterson, R.G., Shea, P.A., Schmedtje, J.F., Bauer, D.C. and Felten, D.L. (1981) Sympathetic innervation of murine thymus and spleen: Evidence for a functional link between the nervous and immune systems. *Brain Res. Bull.* **6**, 83-94.

Zetterstrom, B.E.M., Hökfelt, T., Norberg, K.A. and Olsson, P. (1973) Possibilities of a direct adrenergic influence on blood elements in the dog spleen. *Acta Chir. Scand.* **139** 117-121.

CHAPTER 6

THE THYMUS IN MHC DEFICIENCY SYNDROME. IMPLICATIONS FOR IMMUNE FUNCTION

HENK-JAN SCHUURMAN, Ph.D.,* and BEN J.M. ZEGERS, Ph.D.+

**Division of Immunopathology, Departments of Internal Medicine and Pathology, University Hospital, Catharijnesingel 101, 3511 GV Utrecht, The Netherlands, phone (0)30-379111, ext. 2604/688; +Department of Immunology, University Children's Hospital "Het Wilhelmina Kinderziekenhuis", Nieuwe Gracht 137, 3512 LK Utrecht, The Netherlands, phone (0)30-320330*

ABSTRACT

In 1978 and 1979 children presenting with a combined immunodeficiency were described, in whom the deficiency was ascribed to a defective expression of MHC class I and/or class II antigens on blood mononuclear cells. This type of immunodeficiency was designated as "Bare lymphocyte" syndrome, and is nowadays called "MHC deficiency" syndrome. Two categories exist, MHC class I and MHC class II deficiency. MHC class II deficiency has been extensively studied by the authors' groups. The defective MHC class II deficiency results in the inability of cellular interaction *in vitro* between T cells and antigen-presenting cells in initiating immune responses. The defective expression is not only present in blood mononuclear cells, but also concerns antigen-presenting cells in skin (Langerhans cells) and cells in the placenta (obtained at birth of a MHC class II deficient baby). The defective class II expression appears to be linked to defects in genes encoding *trans*-acting factors involved in regulation of class II expression. In thymic biopsies taken during life a defective expression of class I and a complete lack of class II antigen was observed. But, in other (immuno)histologic and functional assays the gland did not differ from the normal childhood thymus. We conclude that the MHC expression has no apparent impact in T cell differentiation processes in the thymus, visualized in (immuno)histology.

This does not exclude the possibility that processes involving class II antigens do not occur, which have a privileged occurrence inside the gland (e.g., education of self-MHC class II restriction). In this discussion, we draw parallels between the child with MHC class II deficiency and laboratory animal constructs prepared by neonatal injection of mice with anti-MHC class II antibody.

INTRODUCTION

The so-called "Bare lymphocyte" syndrome was first described in The Netherlands and France in a few children who showed a combined immunodeficiency associated with defective expression of class I and/or class II major histocompatibility complex (MHC) products on blood mononuclear cells (Touraine *et al.*, 1978; Schuurman *et al.*, 1979). The name of the syndrome has recently been revised as MHC class I and/or class II deficiency syndrome (Rosen *et al.*, 1986). This is a better designation of the syndrome because non-MHC antigens are expressed normally on the cells. MHC deficiency patients actually represent "experiments of nature" which illustrate the important role of MHC antigen expression in various phases of the immune response. All patients have T lymphocytes, B lymphocytes and antigen-presenting cells (monocytes) in blood and lymphoid organs, but antigen-specific immune responses such as delayed-type hypersensitivity and antibody formation after immunization with T-dependent antigens are absent or severely deficient. In this aspect the various patients studied are strikingly similar. Apparently the lack of MHC membrane recognition structures precludes cellular interactions between antigen-presenting cells, T cells and B cells which are required to induce effective immunity.

The study of these "experiments of nature" is also relevant to understanding lymphoid development in man. Studies on animal models have shown the important role of MHC in intrathymic T-cell processing, amongst others in terms of generation of restriction specificity and allotolerance (reviewed by Schuurman, 1988; Von Boehmer, 1988). In this respect it is worth emphasizing the abundant expression of MHC class I and class II antigens in the normal thymus. It is obvious that the study of MHC deficiency syndrome enables the validation of concepts developed in animal models. In this report we review the clinical and immunological features of the syndrome. We were able to study thymic biopsies from three patients taken during life. The results of these investigations will be presented in detail, and interpreted in the context of the present knowledge on the relevance of MHC antigen expression in the thymus.

FEATURES OF THE SYNDROME

Expression of MHC Antigens on Blood Mononuclear Cells

The first patients described revealed a defective expression of HLA A,B,C, and β2 microglobulin on blood mononuclear cells (Touraine *et al.*, 1978; Schuurman *et al.*, 1979). Later on, patients were found in whom the expression of MHC class II antigens was more severely affected than that of class I antigens (Kuis *et al.*, 1981; Lisowska-Grospierre *et al.*,1983; Hadam *et al.*, 1984). These two patient categories will be referred to as MHC class I deficiency and MHC class II deficiency, respectively. Analyses for MHC antigen expression on cell membranes were done using conventional methods such as cytotoxicity and immunofluorescence microscopy. More recently, FACS analysis enabling the quantitative assessment of antigen expression was performed (Hadam *et al.*, 1984; Rijkers *et al.*, 1987). In two Dutch MHC class II deficient patients, members of the same family, a decreased expression of class I (HLA A,B, C, and β2 microglobulin) and a complete lack of class II (HLA-DR, DP, DQ) was consistently found. But, early in life the patients revealed HLA-DR expression of low intensity on blood B lymphocytes and mitogen-activated T cells. Both parents showed decreased expression of MHC class I compared with normal healthy adults. The expression of class II antigens was normal (Rijkers *et al.*, 1987).

The defective expression of membrane proteins in the patients is restricted to MHC antigens. Non-MHC antigens including T-cell markers CD2, CD3, CD4 and CD8, and the B-cell marker CD20 are expressed normally. This also concerns the presence of the α/β chain of the T cell receptor, labelled by the antibody WT31, and the γ/δ chain of the receptor. Also α and β chains of the LFA1;CR3;gp150,95 cell-adhesion glycoprotein family show normal expression on blood lymphocytes, monocytes and polymorphonuclear granulocytes. In absolute amounts, however, the Dutch MHC class II deficient patients showed lowered numbers of T cells in blood, and a reversed CD4/CD8 ratio. For B cells in blood, a high sIgM/sIgD ratio is found encompassing the neonatal membrane phenotype (Hadam *et al.*, 1984; Rijkers *et al.*, 1987; Clement *et al.*, 1988).

Expression of MHC Antigens on Placenta and Skin

For one of the Dutch MHC class II deficient patients placental tissue was sampled at birth, after an uneventful pregnancy. Immunohistochemistry on frozen tissue sections showed that the villi were negative for MHC class II, and only faintly expressed MHC class I and β2 microglobulin. In contrast, the normal full-term placenta comprises (endothelial) cells labelled with anti-MHC class I and II antibodies (Schuurman *et al.*, 1985a). Skin biopsies were obtained from both Dutch MHC deficient patients. The specimens showed Langerhans cells in the epidermis at a normal or enhanced density, e.g. identified by CD1 staining. These cells, as well as keratinocytes and endothelial cells, expressed MHC class I antigens and β2 microglobulin similar to the normal skin, but at a reduced labelling intensity. A variety of antibodies to MHC class II (HLA-DR, HLA-DP, HLA-DQ) gave no labelling of components in the biopsy, whereas in the normal skin these reagents stained Langerhans cells and endothelial cells with a high intensity and keratinocytes with a low intensity (Zegers *et al.*, 1986).

The observations in tissue fit completely with the data on peripheral blood. They show in addition that the defect is not restricted only to blood cells, but occurs on other cell types at other sites of the body as well. The class II defect also concerns antigen-presenting cells *in situ*, i.e. Langerhans cells in skin.

Clinical and Immunological Findings

Patients with MHC deficiency show the clinical characteristics of combined immune deficiency disease. Most have an early onset of infections, i.e. in the first months of life. Oral candidiasis, *Pneumocystis carinii* infections, persistent viral infections of the gastrointestinal tract resulting in severe mucosal damage and malabsorbtion have been described (Touraine *et al.*, 1978; Schuurman *et al.*, 1979; Kuis *et al.*, 1981; Touraine, 1981; Lisowska-Grospierre *et al.*, 1983; Hadam *et al.*, 1984; Rijkers *et al.*, 1987). Failure to thrive is a prominent feature. Some patients with a MHC class I deficiency of blood mononuclear cells have been described without clinical symptoms of combined immune deficiency disease (Payne *et al.*, 1983; Maeda *et al.*, 1985; Sullivan *et al.*, 1985).

The immunologic defects can be quite heterogeneous between individual patients, but the patients all share defects in antigen-specific immune responses. Regarding humoral immunity, MHC class II deficient cases almost invariably show a complete lack of serum IgG and IgA, and trace amounts of IgM. In bone marrow, pre-B and B-cells carrying the various immunoglobulin isotypes IgM, IgD, IgA and IGg are present in normal numbers. But plasma cells are almost completely lacking: the very few present only contain IgM. Following vaccination or infection, antibody formation *in vivo* is absent, although a few exceptions have been noted in some patients producing anti-viral antibodies after infection. Immunization with Pneumovax may generate anti-pneumococcal polysaccharide antibodies. After *in vitro* stimulation of blood mononuclear cells with polyclonal B-cell mitogens such as formalinized *Staphylococcus aureus Cowan I*, plasmablasts may develop but these cells secrete only IgM (Rijkers *et al.*, 1987).

Regarding cellular immunity, there is no delayed-type skin reactivity to anamnestic antigens like *Candida albicans*. The *in vitro* T-cell proliferative responses to mitogens Phytohaemagglutinin and Concanavalin A can be quite variable depending on the clinical condition, i.e. the presence of severe infections. But, the *in vitro* response after antigenic stimulation invariably is negative. Elaborate analysis of the interactions between antigen-presenting monocytes and T-helper cells revealed the inability of monocytes from MHC class II deficient patients to present antigen to HLA-DR identical T cells from normal healthy adults (Zegers *et al.*, 1986). Conversely, T-lymphocytes from these patients could not be activated by antigen-presenting monocytes from haplo-identical parents. Apparently recognition structures on T cells are lacking which preclude cell activation. Also the ability to recognize allogeneic MHC antigens in mixed leukocyte cultures is absent, or severely impaired in class II deficient patients (Zegers *et al.*, 1983; 1986). Thus, for effective interaction between T-helper cells and antigen-presenting cells the concomittant presence of (identical) MHC class II antigens at the cell surface of both cell types is required. This may not apply to patients with a class I deficiency: a normal response to allogeneic stimulation has been noted for these patients (Schuurman *et al.*, 1979). Cell interaction studies with purified leucocyte subsets have not been performed for class I deficient patients.

Thymus (Immuno)histology

We were able to study thymus biopsies obtained during life from three patients, one of the Dutch patients with a class II deficiency (at 7 months of age) and two other patients with a class I deficiency (2 and 3 years of age) (Schuurman *et al.*, 1985b;1985c). Conventional (immuno)histology and electron microscopy revealed no gross abnormalities. The specimens exhibited a normal lobulated architecture, with small-sized interlobular septa in between. A cortex-medulla distinction was clearly present, and the vasculature was well-developed. The subcapsular cortex revealed large lymphoblasts and some mitotic figures. The cortex was densily packed by lymphocytes expressing the cortex-thymocyte phenotype (i.e., CD1-positive, peanut-agglutinin binding). The medulla contained the normal density of medium-sized thymocytes and normal phenotype expression, i.e. the majority were CD4-positive and a minor part were CD8-positive. Membrane CD3 expression was observed on scattered cells through the cortex and on all medullary lymphocytes. These data do not differ from the normal picture in the uninvoluted childhood thymus (Janossy *et al.*, 1986; Haynes and Lobach, 1987). For the class II deficient case the immunohistologic phenotyping was confirmed by phenotype analysis of thymocytes in suspension (fluorescence microscopy) (Kuis *et al.*, 1981; Schuurman *et al.*, 1985b). Again, the stromal component revealed no abnormalities in morphology. Light microscopy revealed well-differentiated epithelial cells and Hassall's corpuscles, including newly-formed ones, in the medulla. Immunotyping of epithelial cells at distinct locations was done with reagents of the epithelium panel (De Maagd *et al.*, 1985; Ritter and Haynes, 1987); a normal expression of antigens on cortical epithelium (i.e. antibodies MR3 and MR6), subcapsular/medullary epithelium (reagents MR10, MR19 or RFD4), and Hassall's corpuscles (reagents MR13, 4A6) was found. Electron microscopy showed the presence of all epithelial subtypes defined on the basis of morphologic appearance (Van de Wijngaert *et al.*, 1984; Kendall, 1988). The major epithelial cell population in the cortex comprised "pale" cells with a euchromatic nucleus with marginal heterochromatin and a prominent nucleolus, and well-developed cytoplasmic components including rough endoplasmic reticulum and mitochondria. Cells with a heterochromatic nucleus, dilated cisternae of the nucleus and rough endoplasmic reticulum, swollen mitochondria and vacuoles in the cytoplasm, and "dark" cells with a more pronounced degenerative aspect were more frequently observed in the biopsy of one of the class I deficient patients. Lympho-epithelial cell interactions, visualized in ultrathin sections by aggregates called "Nurse Cells" (Van de Wijngaert *et al.*, 1983) were present in all three biopsies. In the medullary area, large-sized epithelial cells were prominent, especially around Hassall's corpuscles. Also cells with an undifferentiated morphology were observed, and many interdigitating cells were present. The functional activity of the epithelial component was evident by the presence of large numbers of cells throughout the medulla containing thymosin fraction 5, using immunohistology with an anti-thymosin antibody. In addition, an epithelial culture was established from the biopsy of the class II deficient patient: the supernatant of this culture had an enhancing effect on the mitogen-induced proliferative response of mouse thymocytes, similar to pooled normal thymus epithelial culture supernatant (Schuurman *et al.*, 1985b). We concluded that the histology and immunohistology of the thymus, both in class I and class II MHC deficiency, was identical to that of the normal thymus in childhood.

A different picture emerged from the analysis of MHC antigens. In the normal thymus, MHC class I and β2 microglobulin are present on stromal cells including epithelium and blood vessels, in the cortex and medulla: thymocytes in the cortex showed a low intensity of staining and those in the medulla, a high expression. MHC class II expression is restricted to epithelium in the cortex, blood vessels, macrophages, and dendritic cells in the medulla. Epithelium in the subcapsular and medullary compartment is weakly positive or completely negative for HLA DR, DP or DQ antigens. The presence of class II antigen on medulla dendritic cells, but not on epithelium in the normal thymus medulla was documented by two-colour immunohistology, using combinations with anti-keratin labelling (Schuurman *et al.*, 1985c; Janossy *et al.*, 1986). The patient's biopsies deviated markedly from these normal observations: interestingly, defects in either class I or class II expression on the epithelium were quite similar for both types of MHC defect (Table I). The cortex lacked the dendritic staining of epithelium for class I and class II. Anti-MHC class I and anti-β2-microglobulin antibodies gave a faint labelling of blood vessels, and anti-MHC class II antibodies did not give any staining at all at this location. In the medulla, lymphocytes and stromal cells were weakly labelled by anti-MHC class I reagents. Medullary dendritic cells in the MHC class I deficient patients were positive for MHC class II antigens, but these cells in the MHC class II deficient case did not show class II antigens. The latter observation is in accord with the absence of class II on Langerhans cells in skin and on blood monocytes, all of which belong to the same bone marrow-derived cell lineage.

Thus, in MHC deficiency, the defective expression of class I on blood cells is accompanied by a low expression on bone marrow-derived cell types in the thymus (lymphocytes, dendritic cells), and with an undetectable class I

TABLE I. Expression of MHC Class I (HLA A,B,C and β_2 Microglobulin) and Class II (HLA-DR and HLA-DC) Antigens in Normal Thymus and Thymus in MHC Deficiency

Component	Normal thymus		MHC I deficiency		MHC II deficiency	
	class I	class II	class I	class II	class I	class II
Cortex						
epithelium	+	+	-	-	-	-
blood vessels	+	+	±	-	±	-
lymphocytes	±	-	-	-	-	-
macrophages	+	+	-	-	-	-
Medulla						
epithelium	+	-	±	-	±	-
blood vessels	+	+	±	-	±	-
dendritic cells	+	+	±	+[a]	±	-
lymphocytes	+	-	±	-	±	-

[a] In one of the specimens investigated dendritic cells labeled only with anti-HLA-DR antibodies.

expression on epithelium, especially in the cortex. The absence of class II antigens on cortical epithelium in patients with class I deficiency is remarkable, because for example monocytes and B lymphocytes of these patients do express class II antigens. This variation in expression on different cell types may be related to the embryonic origin of the cell types involved. On the other hand it may be related to exogeneous factors involved in the induction of MHC expression. For instance, studies on fetal thymus cultures in mice have disclosed the inductive effect of macrophage/dendritic cells on MHC class II antigen expression on epithelial cells (Lo and Sprent, 1986a).

Genetic Aspects

Family studies have disclosed that MHC class I and II deficiency syndrome follows an autosomally recessive inheritance pattern. DNA and RNA studies on freshly isolated cells from patients with defective MHC class II antigen expression have revealed the presence of class II genes by Southern blotting, but mRNA encoding for α/β chains was lacking (Hadam *et al.*, 1985; Lisowska-Grospierre *et al.*, 1985; Marcadet *et al.*, 1985). It has been recognized that γ-interferon upregulates the expression of class I antigen in these patients (Lisowska-Grospierre *et al.*, 1983; Hadam *et al.*, 1986; Zegers *et al.*, 1986). However, class II antigen expression cannot be upregulated by γ-interferon (Hadam *et al.*, 1986; Zegers *et al.*, 1986: De Préval *et al.*, 1988). Family studies have indicated that the defect in this syndrome is not linked to genes encoding the class II antigens, but rather to a *trans*-acting regulatory gene which regulates class II antigen expression (De Préval *et al.*, 1985). Evidence for the existence of such genes has been presented (Accolla *et al.*, 1985; Salter *et al.*, 1985). Recently, cell fusion experiments have been performed with an Epstein-Barr virus transformed B cell line of the MHC class II deficient patient being one partner, and a mutant class II negative human B cell line being the other partner (Yang *et al.*, 1988). Heterokaryons were formed which expressed class II antigen. Presumably a mutation may exist in the patient's cells genome encoding a transacting factor regulating class II gene expression (De Préval *et al.*, 1988; Yang *et al.*, 1988).

COMMENT

The findings in MHC deficiency syndrome have implications for the present understanding of the role of MHC expression in intrathymic T cell processing. The normal thymic architecture, including phenotypic expression of markers associated with maturation stage of the cell, indicates that there is no block in this process. This has only been judged using changes in membrane markers, including the expression of common determinants on the α/β and γ/δ chain of the T cell receptor (assessed on T cells in the peripheral blood), and the membrane expression of the CD3 molecule associated with the receptor. It is not known whether the antigen-recognition repertoire exerted by α/β receptor chains is fully developed in the patients. The cellular immune system is unable to respond to antigenic stimuli *in vitro* and *in vivo*. For such a response T cells not only need an antigen-recognizing structure, but also the capacity to recognize antigen in association with self-MHC molecules. Functional studies on blood mononuclear cells in MHC class II deficiency showed that on one hand, this inability is due to the absence of MHC expression on antigen-presenting cells, but on the other hand, to T cells which appear to be defective in recognizing antigen in association with MHC products. It is therefore tempting to speculate that the immature T cell population is not educated for MHC restriction-specificity.

The primary role of the thymus in the education of restriction-specificity in immature T cells is well-established. This only concerns that of self-MHC class II antigen recognition, e.g. by helper/inducer T cells in the initiation of immune reactions (reviewed by Schuurman, 1988; Von Boehmer, 1988). The learning of self-MHC class I antigen recognition in association with antigen recognition, e.g. exerted by cytotoxic T cells, is not a privileged function of the thymus, but can also be learned outside the gland. Thus, in extrapolating these data obtained in animal models to the human situation, MHC class II deficiency is the most interesting as patients with this syndrome live without any class II molecule in their body.

Current hypotheses state that cell-cell communication is required in learning self class II restriction-specificity. This communication involves the recognition of the polymorphic class II determinants. Both epithelial cells (Ron *et al.* 1986) and (medullary) dendritic cells (Rock *et al.*, 1984; Longo *et al.*, 1985) have been claimed as candidates in the education of self-MHC restriction specificity. This process follows the rearrangement of genes encoding for the α/β chains of the T cell receptor. The latter intrathymic T cell maturation processes also require lymphocyte-stromal interactions, but recognition of MHC molecules is not involved. In studies on the interaction between thymocytes and cultured thymic epithelial cells, CD2 and LFA-3 molecules are involved (Singer *et al.*, 1986; Vollger *et al.*, 1986): this interaction is not blocked by antibodies to MHC molecules. *In vivo*, thymic reconstitution of cultured thymic fragments after implantation occurs equally well for allogeneic and syngeneic combinations (Hong *et al., 1979; Schuurman et al.*, 1985d, 1986; Jenski and Miller, 1988; Jenski *et al.*, 1988) which indicates a significant absence of polymorphic MHC determinants in early intrathymic events. In repopulation studies of the thymus with bone marrow cells, *in situ* lympho-stromal interactions such as rosetting with macrophages, and lympho- epithelial cell complexes ("Thymic Nurse Cells") develop equally well in allogeneic or in syngeneic combinations (Kyewski *et al.* 1984). Thus, lymphocyte-stroma interactions early in intrathymic T-cell processing apparently do not require recognition of MHC molecules. This differs from steps later on whereby T cells "learn" self-MHC class II recognition by cell communication with class II-bearing stromal cell types.

The relevance of class II expression by the thymic microenvironment has also been approached by blocking studies using anti-class II antibodies. In mouse fetal thymus organ cultures, anti-class II antibodies inhibited the expansion of precursor T cells as assessed by the cellularity of the cultured organ (DeLuca, 1986; DeLuca and Mizel, 1986). This result has not been confirmed by others (P. Brekelmans, personal communication, 1988). The discrepancy may be ascribed to differences in experimental design and analysis. In the *in vivo* approach, neonatal mice were treated with anti-class II antibody. Such animals develop peripheral T cells which are unable to respond in allogeneic stimulation assays and which are unable to exert T-helper cell function in MHC class II- restricted antigen stimulation. But, the peripheral T cells are able to respond in mitogen stimulation assays (Kruisbeek *et al.*, 1983, 1985). It is tempting to suggest that in these animals, early intrathymic T-cell maturation proceeds normally, i.e. resulting in mitogen-responsive cells, but lympho-stromal interactions resulting in self-MHC restriction do not occur. The same may apply for allo-antigen recognition or, in the reverse definition, self-MHC tolerance, as this is also learned inside the thymus (Jenkinson *et al.*, 1985; Lo and Sprent, 1986b; Schuurman *et al.*, 1986).

How do the findings in the MHC class II deficient patients fit with the data derived from animal models? We propose that these patients resemble the neonatal mice treated with anti-class II antibody. Lympho-stromal interactions in the thymus cortex occur normally (e.g., visualized by the presence of "Thymic Nurse Cells" in the tissue section). T-cell generation associated with the sojourn in the cortex, e.g. rearrangement of genes encoding for the α and β chain of the T cell receptor and changes in immmunophenotype, are not disturbed. But, processes involving lympho-stromal interactions with recognition of class II molecules, i.e. the education of self-MHC restriction specificity, do not occur. This implies that thymic emigrants (mature T cells) in MHC class II deficiency differ from those in the normal situation. The absence of learning self-MHC restriction specificity, i.e. the "selection" and expansion of relevant clones associated with this process, apparently is not visible by (immuno)histology. Thus, T cell clones are expanding presumably in close contact with the thymus stroma, but not in an MHC- restricted manner.

The situation is quite different for patients with MHC class I deficiency. Here we found defective class II deficiency on thymus cortex epithelium, but the samples investigated showed MHC class II antigens on medullary dendritic cells (Table I). Thus, self-restriction specificity can be learned in this situation, assuming that this is mediated by medullary dendritic cells. As stated above, there is a discrepancy in the literature with respect to the function of epithelial or dendritic cells in this process. A definite function of MHC class I expression on thymus components has not yet been established. Treatment of neonatal mice with anti-MHC class I antibodies resulted in a deficiency of CD4-CD8+ cells and precursors of T-cytotoxic cells: but, is was not documented whether the thymus was involved in the generation of this deficiency (Marusic-Galesic *et al.*, 1988).

In principle, the absence of MHC class-II restricted expansion of T cells may have additional consequences. Induction of tolerance to self antigens is thought to occur inside the thymus and in association with class II recognition (Singer *et al.*, 1987; Kappler *et al.*, 1988). Also the choice of the variable gene repertoire encoding for T cell receptor chains may be altered by the absent intrathymic MHC-restricted interaction process (Kappler *et al.*, 1987; McDuffie *et al.*, 1987). The absence of MHC-supervised surveillance may yield the potential for auto-immune reactions. But, none of the patients manifest overt auto-immune disease. This does not exclude the above hypothesis as the induction of autoimmunity requires similar MHC-restricted cell-cell communication processes as the normal immune response, and such cellular co-operation evidently is lacking in these patients.

To conclude, patients with MHC class II deficiency show a defective immune response due to defects in MHC class II-restricted cell-cell communication. This defect may be based on a random expansion of immature T cells in the thymus which is not governed by MHC class II antigens. When occurring, such a random expansion apparently is not visualized by alterations in thymus (immuno)histology. At present there are no tools available to test the random expansion hypothesis. We have to wait for the availability of molecular biology reagents to analyse the V-gene use of the α and β chain of the T cell receptor repertoire. Alternatively, two animal models are worth studying to elaborate the MHC defect: either MHC transgenic mice or mice treated in the neonatal period by anti-class II antibodies. In this way, the "experiments of nature" presented in this report, may become of utmost relevance for understanding processes in normal thymus histophysiology.

ACKNOWLEDGEMENTS

The authors gratefully acknowledge the collaboration with Dr. C.J. Heijnen, Dr. W. Kuis, Dr. G.T. Rijkers and Dr. J.J. Roord, University Children's Hospital "Het Wilhelmina Kinderziekenhuis"; Professor J. Huber and Dr. L.H.P.M. Rademakers, Institute for Pathology, Utrecht; Dr. Th.M. Vroom, Slotervaart Hospital, Amsterdam; Dr. R.K.B. Schuurman, Dr. F. Koning, and Dr. H. van de Berg, University Hospital, Leiden. They thank Dr. P. Brekelmans, Rotterdam, for fruitful discussions.

REFERENCES

Accolla, R.S., Scarpellino, L., Carra, G. and Guardiola, J. (1985) *Trans*-acting element(s) operating across species barriers positively regulate expression of major histocompatibility complex class II genes. *The Journal of Experimental Medicine* **162**, 1117-1133.

Clement, L.T., Plaeger-Marshall, S., Haas, A., Saxon, A. and Martin, A.M. (1988) Bare lymphocyte syndrome. Consequences of absent class II major histocompatibility antigen expression for B lymphocyte differentiation and function. *The Journal of Clinical Investigation* **81**, 669- 675.

DeLuca, D. (1986) Ia-positive nonlymphoid cells and T cell development in murine fetal thymus organ cultures: monoclonal anti-Ia antibodies inhibit the development of T cells. *The Journal of Immunology* **136**, 430-439.

DeLuca, D. and Mizel, S.B. (1986) I-A-positive nonlymphoid cells and T cell development in murine fetal thymus organ cultures: interleukin 1 circumvents the block in T cell differentiaton induced by monoclonal anti-I-A antibodies. *The Journal of Immunology* **137**, 1435-1441.

De Maagd, R.A., MacKenzie, W.A., Schuurman, H.-J., Ritter, M.A., Price, K.M., Broekhuizen, R. *et al*. (1985) The human thymus microenvironment: heterogeneity detected by monoclonal anti- epithelial cell antibodies. *Immunology* **54**, 745-754.

De Préval, C., Lisowska-Grospierre, B., Loche, M., Griscelli, C. and Mach, B. (1985) A *trans*-acting class II regulatory gene unlinked to the MHC controls expression of HLA class II genes. *Nature* **318**, 291-293.

De Préval, C., Hadam, M.R. and Mach, B. (1988) Regulation of genes for HLA class II antigens in cell lines from patients with severe combined immunodeficiency. *The New England Journal of Medicine* **318**, 1295-1300.

Hadam, M.R., Dopfer, R., Peter, H.-H. and Niethammer, D. (1984) Congenital agammaglobulinaemia associated with lack of expression of HLA-D-region antigens. In *Progress in Immunodeficiency Research and Therapy I,* edited by J. Vossen and C. Griscelli, pp. 43-50. Amsterdam: Elsevier Science Publishers.

Hadam, M.R., Dopfer, R., Dammer, G., De Préval, C., Peter, H.H., Niethammer, D. *et al.* (1985) Establishment of MHC class II negative lymphoblastoid B cell lines from patients with congenital agammaglobulinaemia and malabsorbtion. *Journal of Leucocyte Biology* **38**, 118.

Hadam, M.R., Dopfer, R., Dammer, G., Peter, H.-H., De Préval, C., Mach, B. *et al*. (1986) Immunodeficiency and aberrant expression of MHC antigens: a reappraisal. In *Primary Immunodeficiency Diseases,* edited by M.M. Eibl and F.S. Rosen, pp. 35-39. Amsterdam: Elsevier Science Publishers.

Haynes, B.F. and Lobach, D.F. (1987) Ontogeny of the human thymus during fetal development. *The Journal of Clinical Immunology* **7**, 81-97.

Hong, R., Schulte-Wissermann, H., Jarrett-Toth, E., Horowitz, S.D. and Manning, D.D. (1979) Transplantation of cultured thymic fragments. II. results in nude mice. *The Journal of Experimental Medicine* **149**, 398-415.

Janossy, G., Bofill, M., Trejdosiewicz, L.K., Willcox, H.N.A. and Chilosi, M. (1986) Cellular differentiation of lymphoid subpopulations and their microenvironments in the human thymus. *Current Topics in Pathology* **75**, 89-125.

Jenkinson, E.J., Jhittay, P., Kingston, R. and Owen, J.J.T. (1985) Studies of the role of the thymic environment in the induction of tolerance to MHC antigens. *Transplantation* **39**, 331-333.

Jenski, L.J., Belloni, M.L. and Miller, B.A. (1988) Thymic influence on the T-lymphocyte self MHC repertoire. I. Helper T-lymphocyte precursors. *Thymus* **11**, 135-150.

Jenski, L.J. and Miller, B.A. (1988) Thymic influence on the T-lymphocyte self MHC repertoire. II. Cytotoxic T-lymphocyte precursors. *Thymus* **11**, 151-171.

Kappler, J.W., Roehm, N. and Marrack, P. (1987) T cell tolerance by clonal elimination in the thymus. *Cell* **49**, 273-280.

Kappler, J.W., Staerz, U., White, J. and Marrack, P.C. (1988) Self-tolerance eliminates T cells specific for Mls-modified products of the major histocompatibility complex. *Nature* **332**, 35-45.

Kendall, M.D. (1988) Anatomical and physiological factors influencing the thymic microenvironment. *Thymus Update* **1**, 27-65.

Kruisbeek, A.M., Fultz, M.J., Sharrow, S.O., Singer, A. and Mond, J.J. (1983) Early development of the T cell repertoire. *In vivo* treatment of neonatal mice with anti-Ia antibodies interferes with differentiation of I-restricted T cells but not K/D-restricted T cells. *The Journal of Experimental Medicine* **157**, 1932-1946.

Kruisbeek, A.M., Bridges, S., Carmen, J., Longo, D.L. and Mond, J.J. (1985) *In vivo* treatment of neonatal mice with anti-I-A antibodies interferes with the development of the class I, class II a Mls-reactive proliferating cell subset. *The Journal of Immunology,* **134**, 3597-3604.

Kuis, W., Roord, J., Zegers, B.J.M., Schuurman, R.K.B., Heijnen, C.J., Baldwin, W.M., *et al.* (1981) Clinical and immunological studies in a patient with the "bare lymphocyte" syndrome. In *Bone marrow transplantation in Europe. Vol. II,* edited by J.-L. Touraine, E. Gluckman and C. Griscelli, pp. 201-208. Amsterdam: Excerpta Medica Elsevier.

Kyewski, B.A., Travis, M. and Kaplan, H.S. (1984) Intrathymic lymphopoiesis: stromal cell-associated proliferation of T cells is independent of lymphocyte genotype. *The Journal of Immunology* **133**, 1111-1116.

Lisowska-Grospierre, B., Durandy, A., Virelizier, J.-L., Fischer, A. and Griscellli, B. (1983) Combined immunodeficiency with defective expression of HLA: modulation of an abnormal HLA synthesis and functional studies. In *Primary Immunodeficiency Diseases. Birth defects: original article Series Vol. 19,* edited by R.J. Wedgwood, F.S. Rosen and N.W. Paul, pp. 87-93. New York: Alan Liss.

Lisowska-Grospierre, B., Charron, D.J., De Préval, C., Durandy, A., Griscelli, C. and Mach, B. (1985) A defect in the regulation of major histocompatibility complex class II gene expression in human HLA-DR negative lymphocytes from patients with combined immunodeficiency syndrome. *The Journal of Clinical Investigation* **76**, 381-385.

Lo, D. and Sprent, J. (1986a) Exogenous control of I-A expression in fetal thymus explants. *The Journal of Immunology* **137**, 1772-1775.

Lo, D. and Sprent, J. (1986b) Identity of cells that imprint H-2-restricted T-cell specificity in the thymus. *Nature* **319**, 672-675.

Longo, D.L., Kruisbeek, A.M., Davis, M.L. and Matis, L.A. (1985) Bone marrow-derived thymic antigen- presenting cells determine self-recognition of Ia-restricted T lymphocytes. *Proceedings of the National Academy of Sciences USA* **82**, 5900-5904.

Maeda, H., Hirata, R., Chen, R.F., Suzaki, H., Kudoh, S. and Tohyama, H. (1985) Defective expression of HLA class I antigens. A case of the bare lymphocyte without immunodeficiency. *Immunogenetics* **21**, 549-558.

Marcadet, A., Cohen, D., Dausset, J., Fischer, A., Durandy, A. and Griscelli, C. (1985) Genotyping with DNA probes in combined immunodeficiency syndrome with defective expression of HLA. *The New England Journal of Medicine* **312**, 1287-1292.

Marusic-Galesic, S., Stephany, D.A., Longo, D.L. and Kruisbeek, A.M. (1988) Development of $CD4^-CD8^+$ cytotoxic T cells requires interactions with class I MHC determinants. *Nature* **333**, 180- 183.

McDuffie, M., Roehm, N., Born, W., Marrack, P. and Kappler, J.W. (1987) T cell receptor/MHC interactions in the thymus and the shaping of the T cell repertoire. *Transplantation Proceedings 19* **Supplement 7,** 111-116.

Payne, R., Brodsky, F.M., Peterlin, B.M. and Young, L.M. (1983) "Bare Lymphocytes" without immunodeficiency. *Human Immunology* **6**, 219-227.

Ready, A.R., Jenkinson, E.J. Kingston, R. and Owen, J.J.T. (1984) Successful transplantation across major histocompatibility barrier of deoxyguanosine-treated embryonic thymus expressing class II antigens. *Nature* **310,** 231-233.

Rijkers, G.T., Roord, J.J., Koning, F., Kuis, W. and Zegers, B.J.M. (1987) Phenotypical and functional analysis of B lymphocytes of two siblings with combined immunodeficiency and defective expression of major histocompatibility complex (MHC) class II antigens on mononuclear cells. *The Journal of Clinical Immunology* **7**, 1-9.

Ritter, M.A. and Haynes, B.F. (1987) Summary of Thymic Epithelium Workshop. In *Leucocyte Typing III,* edited by A. McMichael and 13 others, pp. 247-248. Oxford: Oxford University Press.

Rock, K.L., Yeh, E.T.H. and Benacerraf, B. (1984) Selection of thymocyte MHC restriction specificity *in vitro. The Journal of Molecular and Cellular Immunology* **1**, 311-320.

Ron, Y., Lo, D. and Sprent, J. (1986) T cell specificity in twice-irradiated F1-parent bone marrow chimeras: failure to detect a role for immigrant marrow-derived cells in imprinting intrathymic H-2 restriction. *The Journal of Immunology* **137,** 1764-1771.

Rosen, F.S., Wedgwood, R.J. and Eibl, M. (1986) Primary immunodeficiency diseases: report of a World Health Organisation scientific group. *Clinical Immunology and Immunopathology* **40**, 166-196.

Salter, R.D., Alexander, J., Levine, F., Pious, D. and Cresswell, P. (1985) Evidence for two trans- acting genes regulating HLA class II antigen expression. *The Journal of Immunology* **135,** 4235-4239.

Schuurman, H.-J. (1988) Biological functions of the thymic microenvironment. *Thymus Update* **1,** 67-99.

Schuurman, H.-J., Huber, J., Zegers, B.J.M. and Roord, J.J. (1985a) Placental diagnosis of the "bare lymphocyte" syndrome. *The New England Journal of Medicine* **313**, 757.

Schuurman, H.-J., Van De Wijngaert, F.P., Huber, J., Schuurman, R.K.B., Zegers, B.J.M., Roord, J.J., *et al.* (1985b) The thymus in "Bare Lymphocyte" Syndrome: Significance of expression of major histocompatibility complex antigens on thymic epithelial cells in intrathymic T-cell maturation. *Human Immunology* **13**, 69-82.

Schuurman, H.-J., Van De Wijngaert, F.P., Huber, J., Zegers, B.J.M., Schuurman, R.K.B., Roord, J.J., *et al.* (1985c) The thymus in "Bare Lymphocyte" syndrome. *Advances in Experimental Medicine and Biology* **186**, 921-928.

Schuurman, H.J., Vos, J.G., Broekhuizen, R., Brandt, C.J.W.M. and Kater, L. (1985d) In vivo biological effect of allogeneic cultured thymic epithelium on thymus-dependent immunity in athymic nude rats. *Scandinavian Journal of Immunology* **21**, 21-30.

Schuurman, H-J., Vaessen, L.M.B., Vos, J.G., Hertogh, A., Geertzema, J.G.N., Brandt, C.J.W.M., *et al.* (1986) Implantation of cultured thymic fragments in congenitally athymic nude rats. Ignorance of thymic epithelial haplotype in generation of alloreactivity. *The Journal of Immunology* **137**, 2440-2447.

Schuurman, R.K.B., Van Rood, J.J., Vossen, J.M., Schellekens, P.TH.A., Feltkamp-Vroom, TH.M., Doyer, E., *et al* (1979) Failure of lymphocyte-membrane HLA-A and -B expression in two siblings with combined immunodeficiency. *Clinical Immunology and Immunopathology* **14**, 418-434.

Singer, A., Munitz, T.I. and Gress, R.E. (1987) Specificity of thymic selection and the role of self antigens. *Transplantation Proceedings* **19, Supplement 7**, 107-110.

Singer, K.H., Wolf, L.S., Lobach, D.F., Denning, S.M., Tuck, D.T., Robertson, A.L., *et al.* (1986) Human thymocytes bind to autologous and allogeneic cells *in vitro*. *Proceedings of the National Academy of Sciences USA* **83**, 6588-6592.

Sullivan, K.E., Stobo, J.D. and Peterlin, B.M. (1985) Molecular analysis of the bare lymphocyte syndrome. *The Journal of Clinical Investigation* **76**, 75-79.

Touraine, J.-L. (1981) The bare-lymphocyte syndrome. Report on the registry. *Lancet* **1**, 319-321.

Touraine, J.-L., Bétuel, H., Souillet, G. and Jeune, M. (1978) Combined immunodeficiency disease associated with absence of cell-surface HLA-A and -B antigens. *The Journal of Pediatrics* **93**, 47-51.

Van De Wijngaert, F.P., Rademakers, L.H.P.M., Schuurman, H.-J., De Weger, R.A. and Kater, L. (1983) Identification and *in situ* localization of the "thymic nurse cell" in man. *The Journal of Immunology* **130**, 2348-2351.

Van De Wijngaert, F.P., Kendall, M.D., Schuurman, H.-J., Rademakers, L.H.P.M. and Kater, L. (1984) Heterogeneity of epithelial cells in the human thymus. An ultrastructural study. *Cell and Tissue Research* **237**, 227-237.

Vollger, L.W., Tuck, D.T., Springer, T.A., Haynes, B.F. and Singer, K.H. (1987) Thymocyte binding to human thymic epithelial cells is inhibited by monoclonal antibodies to CD-2 and LFA-3 antigens. *The Journal of Immunology* **138**, 358-363.

Von Boehmer, H. (1988) The developmental biology of T lymphocytes. *Annual Review of Immunology* **6**, 309-326.

Yang, Z., Accolla, R.S., Pious, D., Zegers, B.J.M. and Strominger, J.L. (1988) Two distinct genetic loci regulating class II gene expression are defective in human mutant and patient cell lines. *EMBO Journal*, in press.

Zegers, B.J.M., Heijnen, C.J., Roord, J.J., Kuis, W., Schuurman, R.K.B., Stoop, J.W., *et al.* (1983) Defective expression of mononuclear cell membrane HLA antigens associated with combined immunodeficiency: impaired cellular interactions. In *Primary Immunodeficiency Diseases. Birth defects: Original Article Series Vol. 19,* edited by R.J. Wedgwood, F.S. Rosen and N.W. Paul, pp. 93- 96. New York: Alan Liss.

Zegers, B.J.M., Heijnen, C.J., Roord, J.J., Kuis, W., Stoop, J.W. and Ballieux, R.E. (1984) Combined immunodeficiency with defective expression of HLA-antigens: analysis of the nature of the defective monocyte-T cell interaction. In *Progress in Immunodeficiency Research and Therapy I,* edited by C. Griscelli and J.M. Vossen, pp. 35-42. Amsterdam, Excerpta Medica Elsevier.

Zegers, B.J.M., Rijkers, G.T., Roord, J.J., Koning, F., Schuurman, H.J., Heijnen, C.H., *et al.* (1986) Defective expression of MHC antigens on mononuclear cells in two siblings with combined immunodeficiency: delineation of the class II antigen defect. In *Progress in Immunodeficiency Research and Therapy II,* edited by J. Vossen and C. Griscelli, pp. 97-105. Amsterdam: Elsevier Science Publishers.

CHAPTER 7

THE THYMUS IN MYASTHENIA GRAVIS PATIENTS, AND THE *IN VIVO* EFFECTS OF CORTICOSTEROIDS ON ITS CELLULARITY, HISTOLOGY AND FUNCTIONS

NICHOLAS WILLCOX

Neuroscience Group, Institute for Molecular Medicine, John Radcliffe Hospital, Headington, Oxford OX3 9DU.

ABSTRACT

The thymus of young onset myasthenia gravis (MG) patients usually includes invading lymph node-type T cell areas and germinal centres (GC) in addition to typical cortex and medulla. We have compared thymuses from patients pre-treated with corticosteroids (CS) with those of untreated cases to study their *in vivo* actions on cellularity, histology and culture responses. While cortical thymocytes appeared the most sensitive to depletion (as also in MG thymoma), the medulla was substantially depopulated too. Most strikingly, however, these effects were only seen in half of the treated patients: they were undetectable in about 30% and borderline in 20%. Being less depleted than the cortex, the medulla was correspondingly enriched and, together with its T cell zones and GC, appeared relatively healthy both structurally and functionally. Per 10^6 surviving cells, T cell proliferative responses to autoantigen, and production of specific autoantibody and total IgG were all in the normal range; effects of therapy were seen only on the total productivity of the entire thymus.

Thus:-

1) the major therapeutic effects of corticosteroids are on total cell numbers in the periphery rather than on function per surviving cell, and

2) the thymic effects are very variable between patients; in some the cortex is apparently resistant, as others have found for mature responding T cells in the periphery.

INTRODUCTION

There have been numerous studies in laboratory rodents on the effects of corticosteroids - both *in vitro* and *in vivo* - on immunological function as well as histology (reviewed by Claman, 1972; Bach and Strom, 1985). However, there are major differences in sensitivity to steroids between species (Claman, Moorhead and Benner, 1971; Claman, 1972). They are used to treat an extraordinarily wide range of human autoimmune, immune complex, vasculitic, and other disorders as well as to prevent graft rejection, and we clearly need to to investigate their in vivo actions in these diseases in man, however logistically difficult that may be. For obvious reasons, *in vitro* effects have been much more extensively studied, though the results are surprisingly confusing (reviewed by Cupps and Fauci, 1982; Cupps *et al*, 1984).

The main effects of corticosteroids can be grouped as immunoregulatory, anti-inflammatory and endocrine. While these are all relevant to therapy, this review will concentrate on their *in vivo* immunoregulatory effects which we have been able to study in the thymus of patients with the classic autoimmune condition myasthenia gravis (MG). We will first briefly review the background of this disease.

Myasthenia gravis is mediated by autoantibodies to the acetylcholine receptor (AChR) of striated muscle which cause complement-mediated damage to the motor endplate and ultimately lead to muscle weakness. Their production is T cell-dependent in laboratory animals and probably in man too (reviewed by Willcox and Vincent, 1988). In most MG patients, serum anti-AChR levels and muscle weakness both decline with steroid therapy. Patients can be classified into four main groups, two of them with thymic abnormalities. About 10% of Caucasian MG cases have a thymoma. Another 50% have medullary follicular hyperplasia (Castleman and Norris, 1949; see below); these are predominantly young females with onset before 40 years of age and a strong HLA-B8-DR3 bias (Compston *et al*, 1980). Their MG tends to improve after thymectomy which is widely practised in the hope of avoiding the use of corticosteroids; however, others favour the reverse, or else to pre-treat with steroids before thymectomy. In the third main MG subgroup, onset is after age 40, there is a modest excess of males and of HLA-B7-DR2, and the thymus is often atrophic. In the fourth, weakness is restricted to eye and eyelid movements, and there is no obvious HLA or sex bias. The weakness in the old-onset and thymoma subgroups responds particularly well to steroids (Newsom-Davis, 1986).

THYMOMA IN MG AND THE EFFECTS OF CORTICOSTEROIDS

There is no obvious age, sex or HLA bias in this subgroup. The patients normally have antibodies to striational antigens in muscle that apparently cross-react with the thymoma tissue (Gilhus *et al*, 1984); there may also be epitopes on the AChR that are shared with the thymomatous epithelial cells (Kirchner *et al*, 1988). That seems more than coincidence, and strongly suggests that the thymoma tissue itself autosensitises against both antigens. Alternatively, the generation of T cells in such a disorganised environment (see below) might lead to defects in self-tolerance induction. The thymoma is normally removed because of its potential malignancy, but that rarely ameliorates the MG, which responds much better to steroids (Newsom-Davis, 1986).

The tumour is an epithelial neoplasm (Rosai and Levine, 1976) that, in untreated cases, typically generates enormous numbers of thymocytes (Fig. 1), and has the appearance of disorganised cortex. In our survey of about 35 patients, there were usually about 10 times more lymphocytes than in the hyperplastic MG thymus at the same donor age (Fig. 1, and Fig. 2; Willcox *et al*, 1987), most of them with the phenotype of cortical thymocytes (Lauriola *et al*, 1983, Willcox *et al*, 1987). Among them are clusters of epithelial cells that vary greatly in morphology both within and between cases; they may form a relatively complete network, or more often are seen in fragmented islands with no organised architecture discernible (Kirchner and Müller-Hermelink, 1988). By nuclear morphology, they are predominantly of cortical epithelial type, and, furthermore, a high proportion proved to be positive for all the markers of the normally distinct cortical and subcapsular/medullary epithelia (Willcox *et al*, 1987): minority subsets expressed some of these markers in unusual combinations. In some cases, HLA class II (a cortical marker) and HNK-1 (characteristic of subcapsular epithelium) tended to be weak or absent. Obvious medullary differentiation was unusual and Hassall's corpuscles were especially rare. These findings led us to propose a common ancestry of cortical and subcapsular epithelia (Willcox *et al*, 1987), which may also apply to medullary epithelium (Lampert and Ritter, 1988).

(a)

(b)

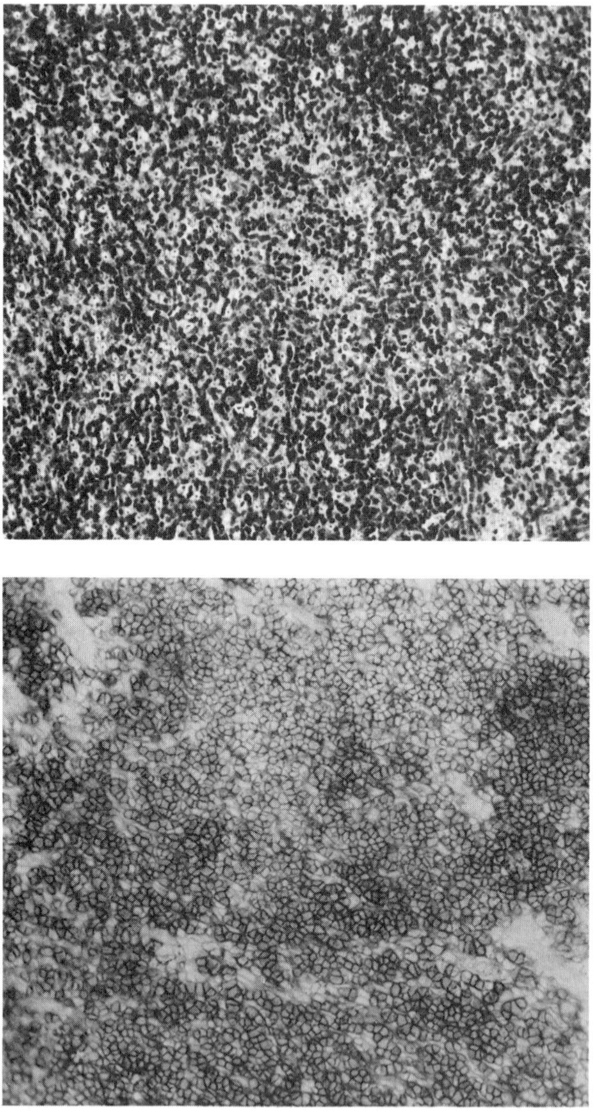

Figure 1. Sections of thymoma samples from an untreated MG patient (a and b) and a corticosteroid-treated myasthenic (c and d). (a) and (c) were stained with Haematoxylin and Eosin, and (b) and (d) with a monoclonal anti-CD1 antibody by immunoperoxidase. Note the

(c)

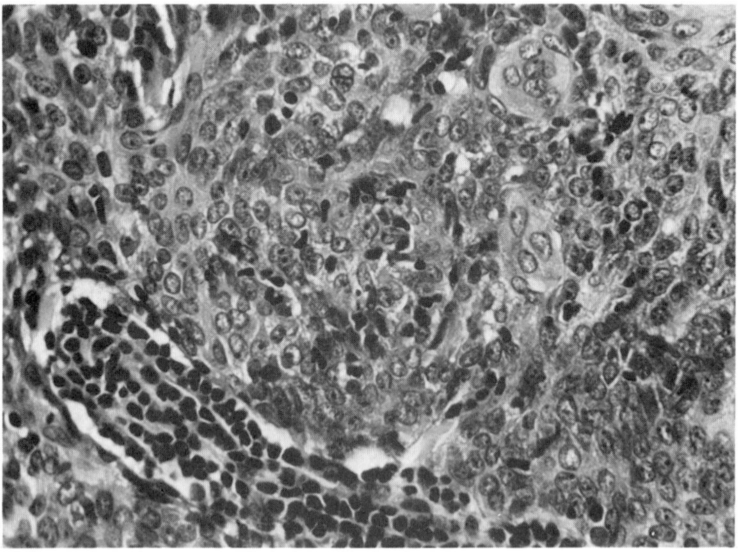

(d)

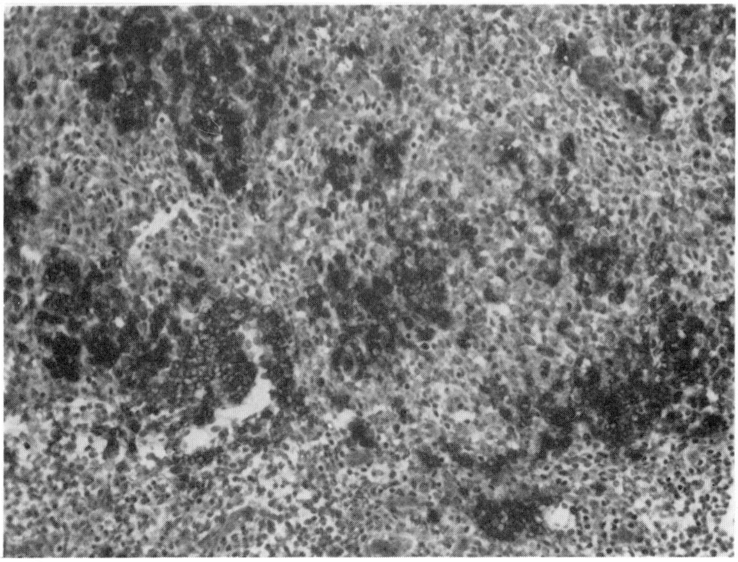

depletion of $CD1^+$ cells after treatment (d), compared with their dense packing in (b), even in a field selected to show some residual labelling.
(Reproduced from Willcox *et al* 1987 with permission).

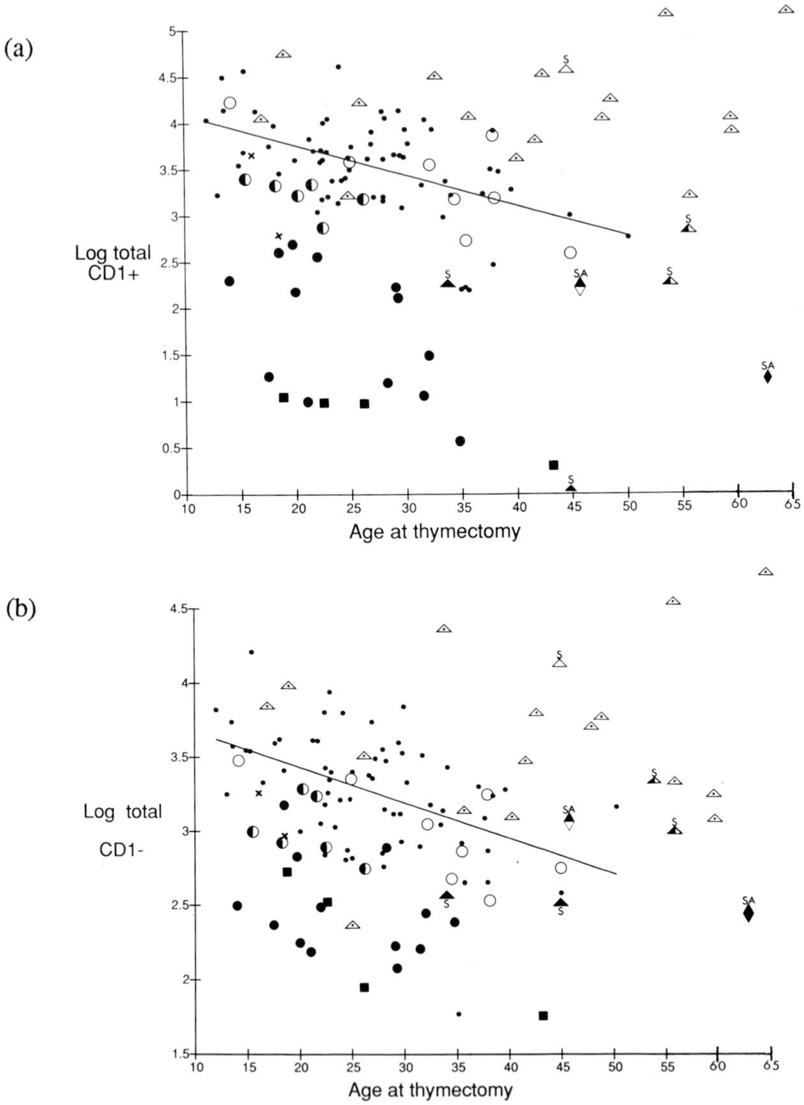

Figure 2. The total yield (log_{10}) of cortical or medullary cell suspension from the entire MG thymus (or thymoma) is plotted against age at thymectomy. (a) yield of $CD1^+$ cells and (b) of $CD1^-$ cells. The regression lines (of yield on age) for untreated non-thymoma MG cases are drawn as solid lines.

non-thymoma		thymoma
•	untreated MG	△ (dotted)
x	pretreated with Aza alone	
○	pretreated with CS alone (- non-depleted	△
◐	(- intermediate	◭
●	(- depleted	▲
■	pretreated with CS + Aza	◆

We have now studied ten further thymoma samples from MG cases pre-treated with corticosteroids (with or without azathioprine (Aza)). Only one of these resembled the untreated, with abundant lymphocytes and epithelial cells that were almost purely cortical in type (Willcox *et al*, 1987). In the other 9, total cellularity was drastically reduced and $CD1^{+}$ frequencies were low (0-40% compared with the 70-85% typical of untreated thymoma or normal infant thymus, Fig. 2). The remaining lymphocytes were mainly $CD3^{+}$ mature T cells in perivascular spaces, (in one case predominantly $CD8^{+}$), and were often richer in AChR-reactive T cells than were the same patients' blood T cells - especially after corticosteroid pre-treatment (N. Sommer *et al*, in preparation).

In most of these cases, the thymoma consisted of almost pure epithelium arranged in lobules separated by connective tissue (Fig. 1). The latter formed the bulk of the tumour in some cases and typically included many HLA-class II^{+} macrophages and small blood vessels. The epithelium usually appeared relatively healthy, but sometimes gave unsatisfactory immunolabelling where there was extensive lymphocyte necrosis. If one allowed for these differences, the epithelial cells appeared morphologically and phenotypically similar to those in untreated samples, except that HNK-1 labelling tended to be weak or absent (Willcox *et al*, 1987).

Moreover, although thymic epithelial cells may express steroid receptors (Dardenne *et al*, 1986), these tumours apparently continue to grow in spite of therapy and require removal. As a striking example, one patient was treated with steroids plus azathioprine for the 5 years after resection of her primary tumour. Over this period, a pleural metastasis grew to a bulk of $55cms^{3}$ and was then removed. She died after 2.5 years' further therapy with at least 3 more secondary deposits. Interestingly, the epithelium in the first metastasis expressed all the cortical and subcapsular markers (except HNK-1), and contained moderate numbers of $CD1^{+}$ lymphocytes (Willcox *et al*, 1987). It has also been observed radiologically that a thymoma may transiently become much smaller after steroid therapy, only to recrudesce later (presumably because of continued epithelial cell growth after initial thymocyte depletion, Shellito *et al*, 1978).

THYMIC HYPERPLASIA AND THE EFFECTS OF CORTICOSTEROIDS

Untreated cases

In the hyperplastic MG thymus, the medulla is expanded (at the expense of a relatively normal though somewhat disrupted cortex) by the presence of germinal centres (GC) and T cell zones (Fig. 3A) (Bofill *et al*, 1985; Schluep *et al*, 1988). These apparently originate in perivascular areas (Levine and Rosai, 1978; Bofill *et al*, 1985), and closely resemble those in peripheral lymphoid tissue both in structure and in function. The centres of the GC contain IgM^+ immunoblasts, some T cells, and a meshwork of follicular dendritic cell processes plus the immune complex material that they bind and present. The surrounding corona consists mainly of smaller IgM^+IgD^+B lymphocytes. The adjacent T cell areas contain numerous interdigitating reticular (ID) cells and high endothelial venules as well as abundant mature T cells (Bofill *et al*, 1985). These invading lymph node-type elements compress the residual true thymic medulla into characteristic bands and arches where the epithelium shows some selective hyperplasia (Schluep *et al*, 1988).

(a)

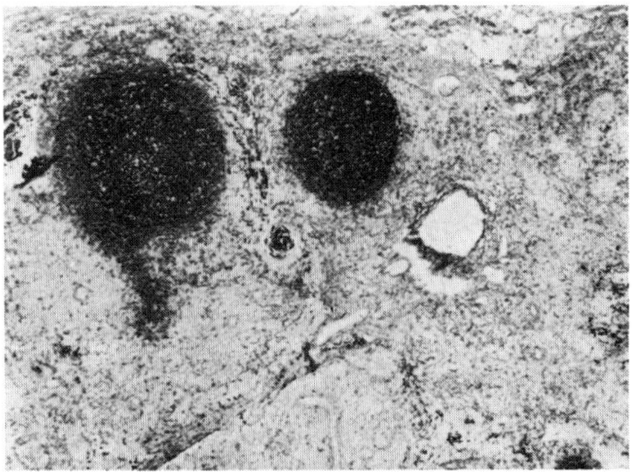

(b)

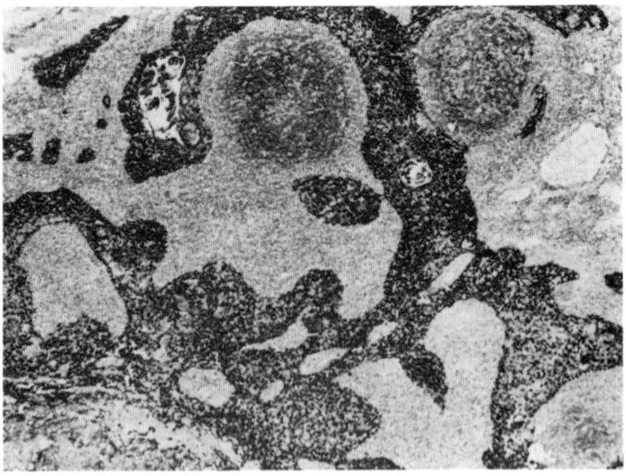

Figure 3A. Typical MG thymic hyperplasia in an 18 year old patient (given azathioprine alone), evident also in most of the corticosteroid- and corticosteroid+azathioprine-treated cases studied. The same area has been stained with various monoclonal antibodies in adjacent

(c)

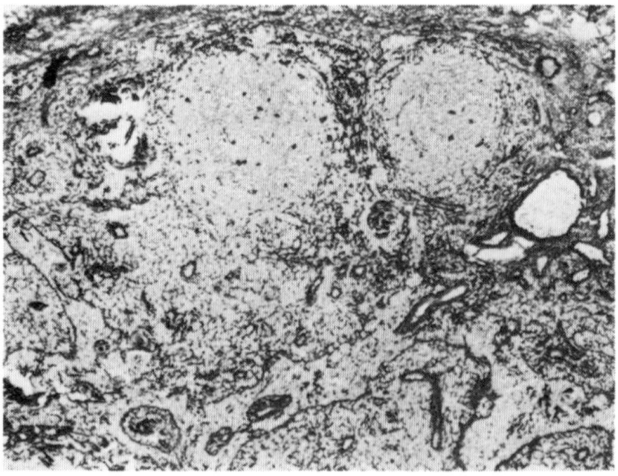

(d)

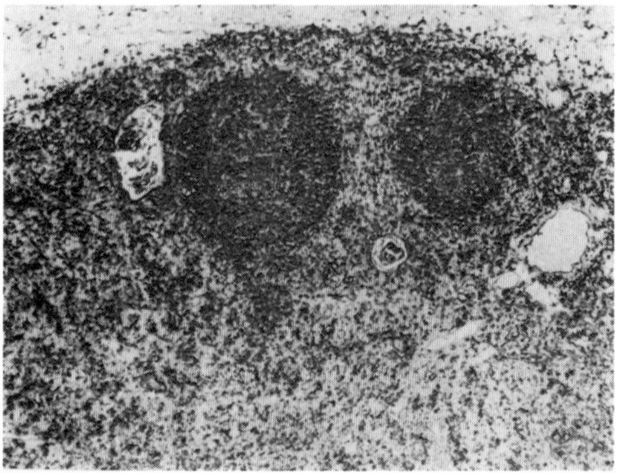

(e)

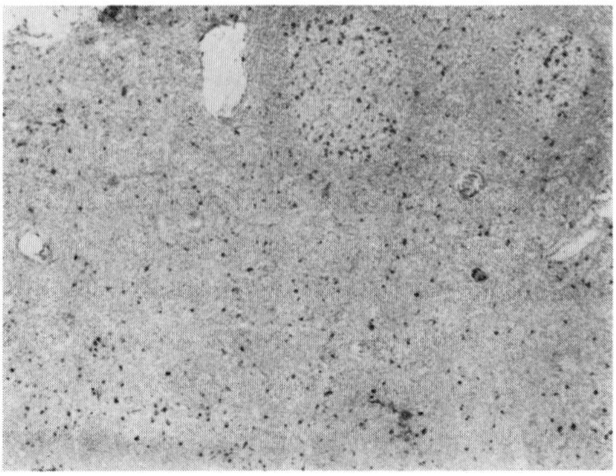

sections; RFB4$^+$ germinal centres (a) are surrounded by bands of hyperplastic medullary epithelial cells (MR19$^+$,RFD4$^+$;(b)) and by a fine fibronectin+ network (c) rich in blood vessels. The germinal centres contain many B cells, and the adjacent T cell areas many interdigitating cells, which are all strongly HLA-DR$^+$ (d). There a few HNK-1$^+$ cells in the germinal centres (e).

the residual true thymic medulla into characteristic bands and arches where the epithelium shows some selective hyperplasia (Schluep *et al*, 1988).

Cultured cell suspensions from hyperplastic (but not neoplastic) MG thymus typically produce anti-AChR spontaneously. This is evidently preferentially activated *in vivo*, as the IgG produced is 10-100 times richer in anti-AChR than in the serum (Scadding *et al*, 1981) or culture supernatants from lymph node or bone marrow cells from the same patient (Fujii *et al*, 1985). There are usually also many other B cells present, some of them specific for extraneous antigens such as influenza, but these only make detectable antibody after mitogen stimulation (Thomas *et al*, 1982). Our indirect evidence is that the anti-AChR is mainly produced by plasma cells in the T cell areas and it is a very consistent and MG-specific feature (Schluep *et al*, 1988). By contrast the remaining B cells and the IgG they produce correlate better with the indices of GC; these vary unpredictably from patient to patient - possibly because they depend on circulating immune complexes.

A special relationship between MG and the thymus has long been discussed (e.g. Wekerle and Ketelsen, 1977). The evidence in its favour, in young-onset cases, includes the hyperplastic changes and the high specific activity of the thymic cell anti-AChR antibody just described, as well as the frequent amelioration of the myasthenia after thymectomy. Moreover, there are rare muscle-like myoid cells in the medulla that express AChR in culture (Wekerle and Ketelsen, 1977) and *in situ* (Schluep *et al*, 1987). An intriguing hypothesis is that this AChR is responsible for the autosensitisation in MG (Wekerle and Ketelsen, 1977). However, a serious difficulty is that virtually all of the histological and functional observations reviewed above have been replicated in the uninvolved thymus adjacent to MG thymomata (Willcox *et al*, 1987), and therefore appear to be effects of an autoimmune response established elsewhere, rather than its initial cause (for further discussion see Schluep *et al*, 1988).

Pre-treated cases

The MG thymus is a self-contained organ, and in untreated patients has a somewhat predictable total cell yield. Furthermore its culture activities correlate well with serum anti-AChR titre or with GC or B cell frequencies. In the course of these studies, we have also collected thymus from 37 patients who had been pre-treated with corticosteroids and/or azathioprine, and have thus been able to examine the effects of this therapy on peripheral as well as central lymphoid tissue, and on function as well as cellularity and histology using these generally comparable untreated cases as positive controls (Willcox *et al*, 1989). The former were all referred by other neurologists for thymectomy and their treatment regime was far from standard, but consisted in most of alternate day prednisolone for very variable durations. Systematic follow-up has not been possible. About 10-15% of the thymus was studied immunohistologicaly in 16 cases, and, in nearly all, cell suspension was prepared from the remaining 80-90% for cell marker and culture studies.

Total Cellularities

In the untreated cases the yields of cortical ($CD1^+$) medullary ($CD1^-$) and total lymphocytes show the expected decline with age, but vary over a roughly tenfold range (Fig. 2). After steroid pre-treatment, they were strikingly variable. They were at least tenfold below this range in about half of the patients ("depleted" cases), well within it in about 30% ("non-depleted"), and borderline in a further 20% ("intermediate cases"). Frequencies of cortical thymocytes were usually selectively reduced (Berrih *et al*, 1984), and sometimes almost all the remaining cells were mature $CD3^+$ T cells or B cells. However, although these cells were enriched, absolute yields of cortical cells, of medullary cells, and of all thymus cells showed a remarkably similar pattern of depletion (Fig. 2); only in three to four cases were medullary cells relatively spared. Within the cortex, it was often the TdT^+ $CD1^+$ cells that were depleted most, and the TdT^+ $CD1^-$ cells relatively less so (Willcox *et al*, 1989).

Immunohistology

The most striking abnormality in depleted cases was the reduction and necrosis of cortical lymphocytes (as in Craven *et al*, 1981), and consequent loss of cortico-medullary architecture. The cortical epithelium was also reduced in bulk, but less so than the lymphocytes (Fig. 3B) and it often stained rather weakly for keratin. In addition, septal and adipose tissue was greatly expanded, and epithelial cysts of varying size were common. The medulla was substantially enriched, though somewhat depleted in lymphocytes, and the epithelial cords and bands were prominent. At the same time, GC and T cell areas appeared relatively healthy, and that is supported by functional studies (see below). Non-depleted samples were very similar to those from untreated patients: some intermediate samples showed a slightly abnormal disruption/necrosis in the cortex, but otherwise resembled advanced untreated cases (Schluep *et al*, 1988).

It was very striking that the effects on histology and cellularity both correlated very poorly with the duration of therapy and total dosage given; a depletion index based on total cell numbers gave correlation coefficients of -0.30 with both. Some patients were quite severly depleted after 21, 14 or even 5 days' treatment, whereas others treated for months or years showed no detectable ef-

(a)

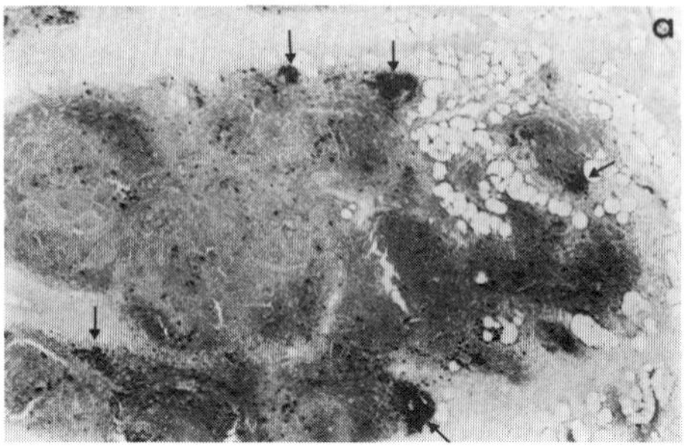

(b)

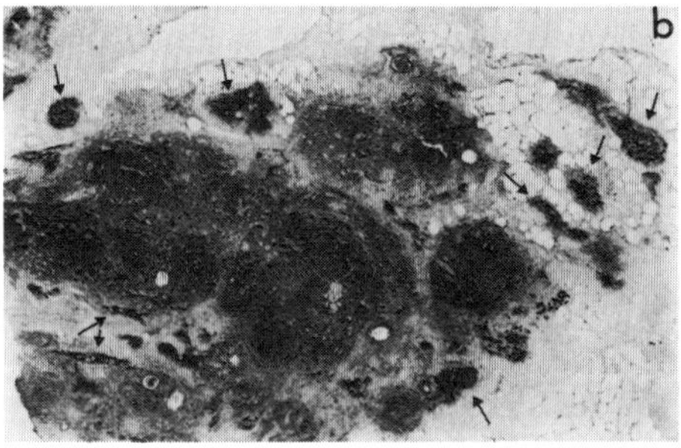

(c)

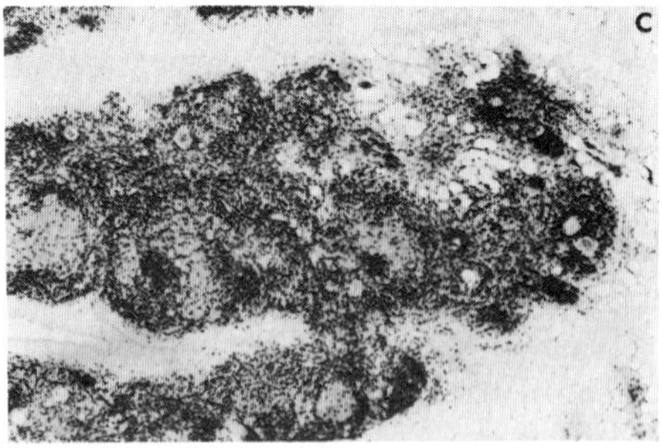

Figure 3B. Adjacent immunoperoxidase stained sections of the thymus from a 19 year old MG patient pretreated with corticosteroids and azathioprine for 6 weeks. Labelling with anti-CD1 (a) reveals a few small patches of cortical thymocytes (arrows). Staining for cortical epithelium is more extensive in these and other areas, both with MR6 (b) and anti-MHC class II (d)

(d)

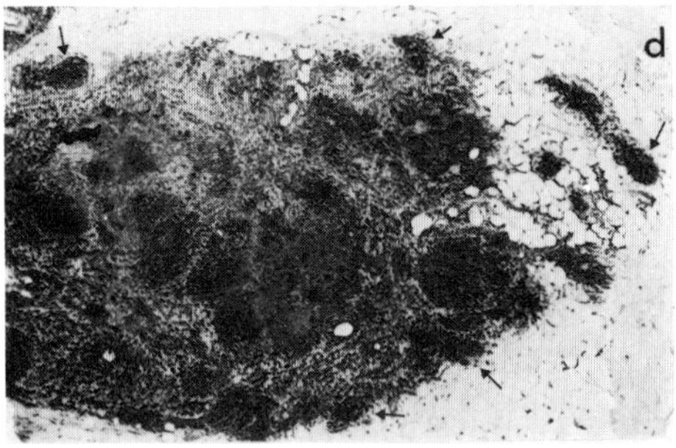

(e)

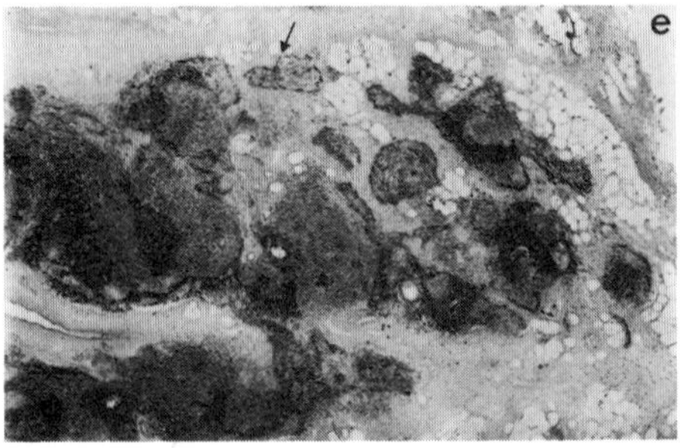

(f)

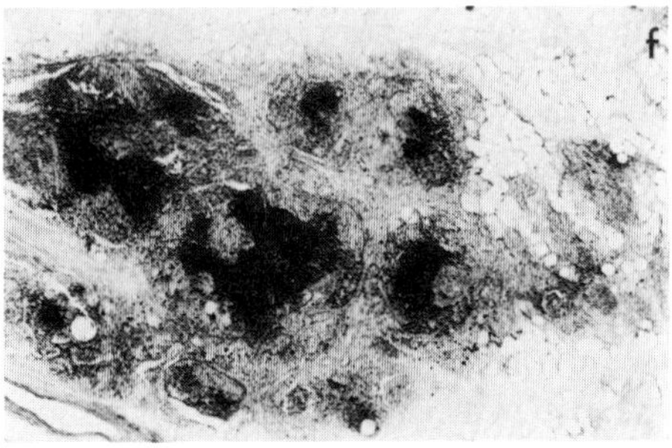

monoclonal antibodies. Anti-CD8 also labels the T cell areas (c). MR19 (e) stains the residual subcapsular epithelium (arrows) and the medullary epithelial bands between germinal centres; follicular dendritic cells and B cells are strongly $RFB6^{+}$ in the latter (f). The septal/adipose areas surrounding and penetrating the lobules are abnormally large for a patient of this age.

fects. The parameters that discriminated best between the subgroups were the durations of MG prior to therapy or to thymectomy (r = -0.40 and -0.47 respectively, P 0.05) - i.e. depletions were greater after shorter durations. That might imply that a persistently hyperplastic thymus is more steroid resistant or, as we suspect, that patients are more likely to be referred for thymectomy if steroid treatment has failed: in other words, our long-term treated cases may be a selected sample. We could not, of course, exclude non-compliance as another explanation, though perhaps unlikely in as many as 8 (or even 13) out of 27.

Functional Studies

Culture responses of thymic cells from untreated patients are generally enchanced after *in vitro* removal of cortical T cells (e.g. by complement killing; Willcox *et al*, 1984), because the responder cells are more concentrated and culture conditions are improved. The same applies after *in vivo* depletion by steroids (see below). So to give a fairer assessment of lymphocyte function, we expressed responses either per 10^6 medullary cells, or summed for the entire yield of thymus cells (Fig. 4).

In untreated cases, spontaneous anti-AChR production by thymus cells correlates very strongly with serum anti-AChR titre, spanning about a 30 fold range at any one titre. Assessed per 10^6 cells, productivity fell below this range surprisingly rarely - only in 3 of the 16 depleted cases; in most of the non- depleted and intermediate cases it was in its upper half (Willcox *et al*, 1989). Because of the loss of total cells, the calculated productivity of the entire thymus was below the typical range in 7 of the 16 depleted cases and in its lower half in the others (Willcox *et al*, 1989). In the non-depleted and intermediate cases it was close to the normal average.

Thus one major *in vivo* effect of corticosteroids is to reduce total cell numbers rather than the function of the surviving cells (cf Fauci and Dale, 1975; Gillis, Crabtree and Smith, 1979). It may be no surprise that the anti-AChR producing plasma cells are steroid-resistant. However, the same conclusion applies to three other largely independent cell functions. Total IgG is apparently produced spontaneously by GC blasts and to some extent by plasma cells elsewhere, and, after pokeweed mitogen (PWM) stimulation, by the B cells in the GC corona (Schluep et al, 1988). Expressed per 10^6 cells, both of these correlate loosely with B cell frequencies in untreated thymus cultures (Fig. 4A; Schluep *et al* 1988); after corticosteroid treatment they were often above the typical range and were rarely depressed. Because of the low cell yields, depleted samples often gave a somewhat decreased total productivity per entire thymus (Fig 4B). Finally, the proliferative responses to Torpedo AChR and to PPD presumably depend on T cells and antigen presenting cells in the T cell zones. Stimulation by AChR is detected in about 60% of untreated MG cases and by PPD in about 30%, both after enrichment of low density thymic cells (N. Sommer *et al*, in preparation). The degree of response correlates well with the frequency of $CD3^+$ cells and, interestingly, T cells from thymoma tended to respond even better

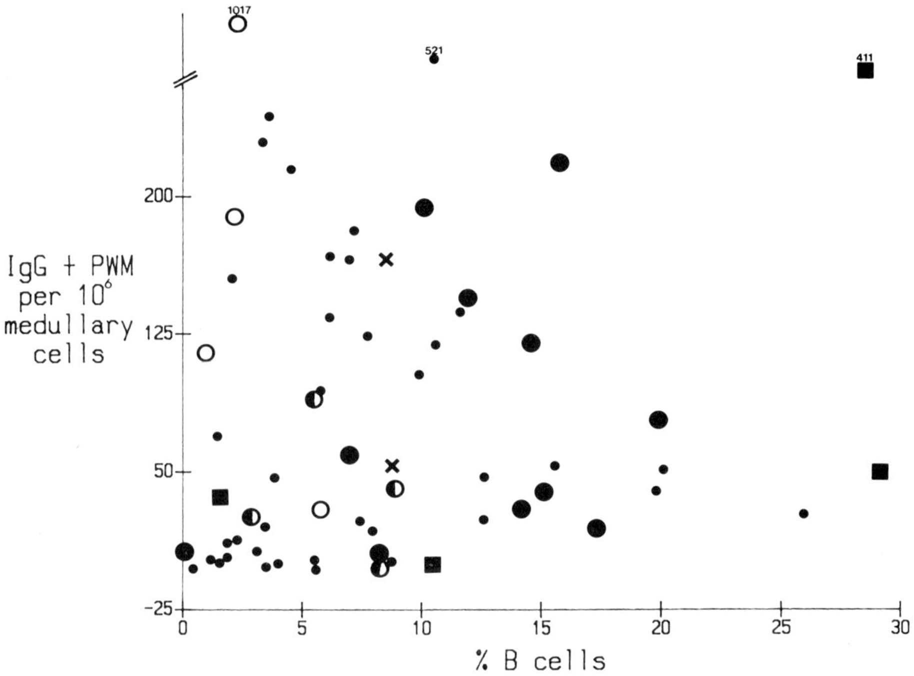

Figure 4A. Scatter plot of PWM-stimulated total IgG production per 10^6 $CD1^-$ cells, against thymic B cell frequency (%). Symbols are as in Figure 2.

than those from hyperplastic thymus. After steroid pre-treatment, responses per 10^5 cells were often similarly high, in keeping with the higher $CD3^+$ frequencies (Willcox *et al*, 1989).

CONCLUSIONS

Our main conclusions are that:

1) The human thymic cortex is especially sensitive to corticosteroids *in vivo* (as also reported by Craven *et al*, 1981; and Berrih *et al*, 1984). That is almost certainly a direct cytolytic effect on the cortical thymocytes in view of their known high sensitivity both in rodents and man to *in vitro* corticosteroids (Homo-Delarche, 1984) and of the rapidity of their *in vivo* destruction (Ishidate

and Metcalf, 1963; De Sousa and Fachet, 1972). This subset has a particularly high natural death rate *in vivo*, possibly because of the cell selection that occurs there, and corticosteroids have been invoked as one mediator for that (Sprent *et al*, 1987). However, these cells are also a major target of stress atrophy (Dourov, 1986), and of other toxins (Snoeij *et al*, 1985). Their death after steroid treatment depends on a specialised suicidal metabolic pathway (Munck and Crabtree, 1981). Even in intermediate cases, we could find no sign of a cytostatic effect of corticosteroids; there was a normal high proportion of cells in cycle (assessed by the Ki67 marker; Gerdes *et al*, 1984) among the remaining cortical T cells

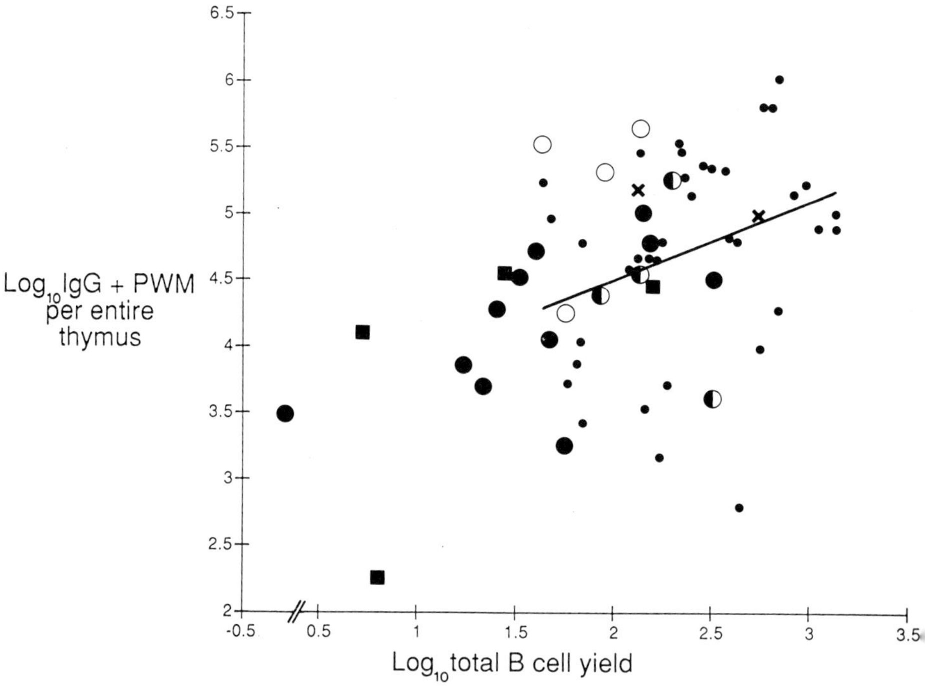

Figure 4B. Plot of PWM-induced IgG production summed for the entire thymus ($\log_{10}$) versus the total thymic B cell yield ($\log_{10}$). Thymoma cultures have not been assayed. Symbol and regression line are as in Figure 2.

(Willcox *et al*, 1989).

2) The cells in the medulla may be reduced because of a diminished output from the cortex, where they are probably generated (Fowlkes, Schwartz and Pardoll, 1988; Janossy *et al*, 1986). In addition they too may be directly killed by the steroids, and there could also be a reduced traffic of circulating cells, which probably explains the lower total numbers of thymic B cells (Fig. 4B).

Whatever the mechanism, our second main conclusion is that, at the functional level, the major effect of corticosteroids is clearly on total cell numbers rather than activity per cell. Conceivably the cells might only respond when liberated from these drugs *in vitro*, but that seems unlikely in view of recent findings on the effects of alternate day corticosteroids (Fauci and Dale, 1975). Our evidence is that AChR-reactive T cells, resting B cells and GC blasts as well as AChR-specific plasma cells all survive this therapy, and their overall activity declines only as a result of loss of total cell numbers. Cupps *et al* (1984) have reported a complex variety of abnormalities in blood lymphocyte functions at various stages after brief *in vivo* steroid treatment of healthy subjects, including activation of spontaneous IgG secretion and depression of either the T cells or the B cells involved in the IgG response to PWM. The latter have not been obvious in our study.

If their main relevant action is indeed on total cell numbers, how do steroids selectively reduce specific anti-AChR rather than total IgG levels in the serum? We suggest that the autoimmune response is continually being boosted by recurrent release of AChR-antibody complexes from damaged endplates, and that the corticosteroids interfere with this reactivation by inhibiting IL-2 production or the response to it (Gillis, Crabtree and Smith, 1979; Cupps and Fauci 1981; Walker, Potter and House, 1987), thus preferentially depressing this particular response.

3) The effects varied greatly from patient to patient: they were undetectable in one of ten treated cases with thymoma, and 8 of 27 without, and in a further 5 of the 27 were only partial. That is very reminiscent of a most important series of recent papers, where 30-40% of healthy subjects or patients with renal failure were classified as steroid-insensitive (Dumble *et al*, 1981; Langhoff and Ladefoged, 1987). Their phytohaemagglutinin- stimulated T cells secreted IL-2 at a particularly high rate that was relatively refractory to inhibition *in vitro* by corticosteroids, a property that was stable over a period of months or years (Walker, Potter and House, 1987). Moreover it had very significant prognostic value; nearly all the patients with non-functioning kidneys 3-6 months post-transplantation had previously typed as resistant in this or similar assays. It has also been noted by others in MG that around 30% of cases - especially of young onset - give a poor clinical response to steroids, requiring higher doses or longer courses of therapy for meagre benefits (e.g. Tindall, 1981; Sghirlanzoni *et al*, 1984). It is tempting to speculate that our non-depleted cases fall into the same resistant category, though we were not able to test that. If true, that implies that the same variations in sensitivity of mature functional T cells also apply to very immature precursor T cells in the thymic cortex.

4) As previously thymectomised patients may show very pronounced therapeutic benefits from corticosteroids, their thymic effects are clearly not

essential. Their actions on peripheral lymphoid tissues and cells, and perhaps on the complement components involved in endplate damage, are presumably more relevant.

ABBREVIATIONS

AChR, acetylcholine receptor: Aza, azathioprine: CS, corticosteroids: GC, germinal centre(s): MG, myasthenia gravis: PPD, purified protein derivative, PWM, pokeweed mitogen.

ACKNOWLEDGEMENTS

I am extremely grateful to my co-workers for their original work on which this review is based, and to many colleagues for referring patients. This work was supported by grants from the Medical Research Council and the Sir Jules Thorn Charitable Trust.

REFERENCES

Bach J-F, Strom TB (1985) *The Mode of Action of Immunosuppressive Agents*, 2nd edition. Elsevier, Amsterdam, NY, Oxford.

Berrih S, Safar D, Levasseur P, Gaud C, Bach J-F (1984) The in vivo effects of corticosteroids on thymocyte subsets in myasthenia gravis. *Journal of Clinical Immunology* **4**: 92-97.

Bofill M, Janossy G, Willcox N, Chilosi M, Trejdosiewicz LK, Newsom-Davis J (1985) Microenvironments in the normal thymus and the thymus in myasthenia gravis. *American Journal of Pathology* **119**: 462-473.

Castleman B, Norris EH. (1949) The pathology of the thymus gland in myasthenia gravis. A study of 35 cases. *Medicine (Baltimore)*, **28:** 27-58.

Claman H. (1972) Corticosteroids and lymphoid cells. *New England Journal of Medicine* **287**: 388-397.

Claman HN, Moorhead JW, Benner WH. (1971) Corticosteroids and lymphoid cells *in vitro*. I. Hydrocortisone lysis of human, guinea pig and mouse thymus cells. *Journal of Laboratory and Clinical Medicine*, **78**: 499-507.

Compston DAS, Vincent A, Newsom-Davis J, Batchelor JR (1980) Clinical, pathological, HLA antigen and immunological evidence for disease heterogeneity in myasthenia gravis. *Brain* **103**: 579-601.

Craven C, Reddy PK, Ringel SP, Rutherford RB (1981) Effect of corticosteroids on the thymus in myasthenia gravis. *Muscle and Nerve*, **4**: 425-428.

Cupps TR, Fauci AS (1982) Corticosteroid-mediated immunoregulation in man. *Immunological Reviews* **65**: 133-155.

Cupps TR, Edgar LC, Thomas CA, Fauci AS (1984) Multiple mechanisms of B cell immunoregulation in man after administration of *in vivo* corticosteroids. *Journal of Immunology* **132**: 170-175.

Dardenne M, Itoh T, Homo-Delarche F (1986) Presence of glucocorticoid receptors in cultured thymic epithelial cells. *Cellular Immunology* **100**: 112-118.

De Sousa M, Fachet J (1972) The cellular basis of the mechanism of action of cortisone acetate on contact sensitivity to oxazolone in the mouse. *Clinical and Experimental Immunology* **10**: 673-684.

Dourov N (1986) Thymic atrophy and immune deficiency in malnutrition. *Current Topics in Pathology* **75**: 127-150.

Dumble LJ, Macdonald IM, Kincaid-Smith P, Clunie GJA. (1981) Correlation between ADCC resistance to *in vitro* steroid and renal allograft failure. *Transplantation Proceedings,* **13**: 1569-1571.

Fauci AS, Dale DC. (1975) Alternate day prednisone therapy and human lymphocyte subpopulations. *Journal of Clinical Investigation* **55**: 22-32.

Fowlkes BJ, Schwartz RH, Pardoll DM (1988) Deletion of self- reactive thymocytes occurs at a $CD4^{+}8^{+}$ precursor stage. *Nature* **334**: 620-623.

Fujii Y, Hashimoto J, Monden Y, Ito T, Nakahara K, Kawashima Y. (1986) Specific activation of lymphocytes against acetylcholine receptor in the thymus in myasthenia gravis. *Journal of Immunology* **136**: 887-891.

Gerdes J, Lemke H, Baisch H, Wacker H-H, Schwab U, Stein H. (1984) Cell cycle analysis of a cell proliferation-associated human nuclear antigen defined by the monoclonal antibody Ki67. *Journal of Immunology* **133**: 1710-1715.

Gilhus NE, Aarli JA, Christensson B, Matre R. (1984) Rabbit antiserum to a citric acid extract of human skeletal muscle staining thymomas from myasthenia gravis patients. *Journal of Neuroimmunology* **7**: 55-64.

Gillis S, Crabtree R, Smith KA. (1979) Glucocorticoid-induced inhibition of T cell growth factor production I. The effect on mitogen-induced lymphocyte proliferation. *Journal of Immunology* **123**: 1624-1631.

Harcourt GC, Sommer N, Rothbard J, Willcox HNA, Newsom-Davis J. (1988) A juxta-membrane epitope on the human acetylcholine receptor recognised by T cells in myasthenia gravis. *Journal of Clinical Investigation,* **82**: 1295-1300.

Homo-Delarche F. (1984) Glucocorticoid receptors and steroid sensitivity in normal and neoplastic human lymphoid tissues; a review. *Cancer Research* **44**: 431-437.

Ishidate M, Metcalf D. (1963) The pattern of lymphopoiesis in the mouse thymus after cortisone administration or adrenalectomy. *Australian Journal of Experimental Biology and Medical Science* **41**: 637-649.

Janossy G, Bofill M, Trejdosiewicz LK, Willcox HNA, Chilosi M. (1986) Cellular differentition of lymphoid subpopulations and their microenvironments in the human thymus. *Current Topics in Pathology* **75**, 89-125.

Kirchner T, Tzartos S, Hoppe F, Schalke B, Wekerle H, Müller- Hermelink H-K. (1988) Pathogenesis of myasthenia gravis: acetylcholine receptor-related antigenic determinants in tumour- free thymuses and thymic epithelial tumours. *American Journal of Pathology* **130**: 268-279.

Kirchner T, Müller-Hermelink H-K. (1988) New approaches to the diagnosis of thymic epithelial tumours. In: *Progress in Surgical Pathology X* edited by Fenoglio-Preizer CM, Wolff M, Rilke F. Field and Wood, Inc, Philadelphia. *In press.*

Lampert I, Ritter MA. (1988) The origin of the diverse epithelial cells of the thymus: is there a common stem cell? *Thymus Update* 1: 2-25.

Langhoff E, Ladefoged J. (1987) The impact of high lymphocyte sensitivity to glucocorticoids on kidney graft survival in patients treated with azathioprine and cyclosporine *Transplantation,* **43**: 380-384.

Lauriola L, Musiani P, Ranelletti FP, Maggiano N, Piantelli M. (1983) Human thymoma lymphocyte mitogenesis: glucocorticoid inhibitory capacity as a function of the size of the more mature T cell subset. *Clinical and Experimental Immunology* **52**: 477-484.

Munck A, Crabtree GR. (1981) *Glucocorticoid-induced lymphocyte death In: Cell Death in Biology and Pathology*, edited by ID Bowen, RA Lockskin. London, New York, Chapman and Hall.

Pla M, Zakany J, Fachet J. (1976) H_2 influence on corticosteroid effects on thymus cells. *Folia Biologica (Praha)* **22**: 49-50.

Rosai J, Levine GD. (1976) *Tumours of the thymus. Atlas of Tumour Pathology*, 2nd series, fascicle 13. U.S. Armed Forces Institute of Pathology.

Scadding GK, Vincent A, Newsom-Davis J, Henry K (1981) Acetylcholine receptor antibody synthesis by thymic lymphcoytes: correlation with thymic histology. *Neurology (NY)* **31**: 935-943.

Schluep M, Willcox N, Vincent A, Dhoot GK, Newsom-Davis J. (1987) Acetylcholine receptors in human thymic myoid cells *in situ*: an immunohistological study. *Annals of Neurology* **22**: 212-222.

Schluep M, Willcox N, Ritter MA, Newsom-Davis J, Larché M, Brown AN (1988) Myasthenia gravis thymus: clinical, histological and culture correlations. *Journal of Autoimmunity* **1**: in press.

Sghirlanzoni A, Peluchetti D, Mantegazza R, Fiacchino F, Cornelio F. (1984) Myasthenia gravis: prolonged treatment with steroids. *Neurology* **34**: 170-174.

Shellito J, Khandekar JK, McKeever WP, Vick NA. (1978) Invasive thymoma responsive to oral corticosteroids. *Cancer Treatment Reports* **62**: 1397-1400.

Snoeij NJ, Van Iersel AAJ, Penninks AH, Seinen W. (1985) Toxicity of tri-organotin compounds: comparative *in vivo* studies with a series of trialkyl tin compounds and triphenyltin chloride in male rats. *Toxicology and Applied Pharmacology*, **81**: 274-286.

Sprent J, Lo D, Gao E-K, Ron Y. (1988) T cell selection in the thymus. *Immunological Reviews* **101**: 173-190.

Thomas JA, Willcox HNA, Newsom-Davis J. (1982) Immunohistological studies of the thymus in myasthenia gravis. Correlation with clinical state and thymocyte culture responses. *Journal of Neuroimmunology* **3**: 319-335.

Tindall RSA. (1980) Humoral immunity in myasthenia gravis: effect of steroids and thymectomy. *Neurology,* **30**: 554-557.

Walker KB, Potter JM, House AK. (1987) Interleukin 2 synthesis in the presence of steroids: a model of steroid resistance. *Clinical and Experimental Immunology*, **68**: 162-167.

Wekerle H, Ketelsen U-P. (1977) Intrathymic pathogenesis and dual genetic control of myasthenia gravis. *Lancet*, **i**: 678-680.

Willcox N, Newsom-Davis J, Calder LR. (1984) Cell types required for anti-acetylcholine receptor antibody synthesis by cultured thymocytes and blood lymphocytes in myasthenia gravis. *Clinical and experimental Immunology* **58**: 97-106.

Willcox N, Schluep M, Ritter MA, Schuurman H-J, Newsom-Davis J, Christensson B. (1987) Myasthenic and nonmyasthenic thymoma: an expansion of a minor cortical epithelial cell subset? *American Journal of Pathology*, **127**: 447-460.

Willcox N, Vincent A. (1988) Myasthenia gravis as an example of organ-specific autoimmune disease. In: *B Lymphocytes in Human Disease*, edited by AG Bird and JE Calvert. Oxford University Press, Oxford. 469-506.

Willcox N, Schluep M, Sommer N, Campana D, Janossy G, Brown AN, Newsom-Davis J. Variable corticosteroid sensitivity of thymic cortex and medullary peripheral-type lymphoid tissue in myasthenia gravis patients: structural and functional effects. Submitted for publication.

PART 2: CURRENT CONCEPTS AND REPORTS FROM RECENT MEETINGS

CYTOKINES

ANDREW J.H. GEARING

Immunobiology, National Institute for Biological Standards and Control, South Mimms, Herts EN6 3QG.

Introduction

It is the aim of this review to cover current concepts of cytokines as members of overlapping and interdependent families of molecules with critical roles in physiology and pathology (Sporn and Roberts, 1988; Oppenheim *et al.* 1987). The consequences of this interdependence and overlap in measuring cytokine levels, and in ascribing functions to individual cytokines will also be discussed.

Definitions

Immunologists are quite clear as to which molecules they would regard as cytokines, a simple definition of what cytokines are is not so easy to obtain. Cytokines are proteins or glycoproteins, they can be produced by lymphocytes or monocytes, they interact in pg-ng/ml concentrations with target cells via receptor proteins, and they are involved in homeostatic control of the immune system and in interactions between the immune system and other organ systems. Many endocrinologists would however claim, with some justification, that cytokines are merely hormones which immunologists discovered first!

Most work on cytokine function prior to the 1980's was subjected to uncertainty due to the use of impure reagents, and bioassays of unknown specificity. Cytokines were originally defined as biological activities found in the supernatants of particular cell cultures. Over 200 different activities had been described by the 1970's. Gradually it became apparent that this number could be reduced as purification procedures, reproducible bioassays, antibodies and finally recombinant DNA technology were applied.

Nomenclature

Cytokine nomenclature is currently a mixture of confusion and order. Twenty-two different molecules which have been cloned and purified are generally included in the cytokine family, these are listed with their biological properties in Table I.

TABLE I Cloned and purified Cytokines and their properties

Factor	Actions
Interleukin 1 α and β	Multiple: inflammatory, fever, cytotoxicity, interleukin 2 production, haemopoietic activity.
Interleukin 2	T and B cell proliferation and differentiation, macrophage activation, NK cell activation.
Interleukin 3	Progenitor growth and differentiation, mast cell growth.
Interleukin 4	T and B cell proliferation and differentiation, IgE and IgE receptor induction.
Interleukin 5	Eosinophil differentiation (B cell growth).
Interleukin 6	B cell differentiation, acute phase response, T cell growth.
Interleukin 7	Pre B cell growth, T and B cell stimulation.
Interferon α and β	Anti-viral, growth inhibition, differentiation.
Interferon γ	Anti-viral, macrophage activation, MHC antigen expression, growth inhibition.
Granulocyte CSF	Growth and differentiation of granulocytes.
Macrophage CSF	Growth and differentiation of monocyte/macrophages.
Granulocyte macrophage CSF	Growth and differentiation of granulocytes/monocytes.
Tumour Necrosis Factor α and β	Multiple: cytotoxicity, cachectia, haemolytic necrosis, bone resorption, fever.
Low MW B cell growth factor	B cell proliferation.
Erythropoietin	Erythropoiesis.
Leukaemia inhibitory factor	Macrophage differentiation.
Monocyte derived chemokinetic factor.	Chemokinesis of neutrophils.
Macrophage inflammatory protein	Neutrophil activation and attraction.
Transforming growth factor β	Growth inhibition, tissue repair.

There are five major subdivisions: Interleukins (Dinarello, 1984; Robb, 1984; Yang *et al.*, 1986; Yokota *et al.*, 1986; Campbell *et al.*, 1987; Hirano *et al.*, 1986; Namen *et al.*, 1988); Colony Stimulating Factors (Clark and Kamen, 1987; Metcalf, 1984); Interferons (Burke, 1985; Pestka, 1986); Tumour Necrosis Factors (Old, 1985); and others (Gearing *et al.*, 1987; Davatelis *et al.*, 1988; Sharma *et al.*, 1987; Sporn *et al.*, 1986; Lin *et al.*, 1985; Yoshimure *et al.*, 1987). These divisions reflect the historical isolation of researchers working on different types of cytokines. It seems unlikely that the present system will be altered in the near future. The five groups and their individual members are listed in Tables II-VI along with their major properties including defining bioassay, physical properties, and chromosomal location of the gene. As each of these molecules has either been cloned or is available in highly purified form this excludes identity with other known molecules. It is certain that other cytokines are yet to be discovered, just as it is certain that the full list of activities which the known cytokines possess is not yet complete.

TABLE II Interleukins

Name	Type Assay	Amino Acids	Mw[a]	Conformation[b]	Glycosylation[c]	Disulphide Bonds	Chromosome
1α	T cell activation	159	17.5	M	-	0	2
1β	Fever	153	17.5	M	-	1	2
2	T cell growth	133	15	M	+(0)	1	4
3	Colonies-multiple	133	15-28	M	+(N)	1	5
4	B cell growth	129	25	M	+(N)	3	?
5	Eosinophil growth	112	20-30	D	+(N)	1	5
6	Hepatocyte-acute phase; B cell-antibody/growth	184	21-22	M	+(N)	2	7
7	pre-B cell growth	129	15	M	+(N)	3?	?

[a]MW = Molecular Weight x1000
[b]M = Monomers; D = Dimers
[c]O = O-linked; N= N-linked

TABLE III Interferons

Name	Type Assay	Amino Acids	Mw[a]	Conformation[b]	Glycosylation	Disulphide Bonds	Chromosome
α subclass I	anti-viral	165/166	18-26	M	-	2	9
α subclass II	anti-viral	172	18.5	M	+	2	9
β	anti-viral	166	22	M/D?	+	2	9
γ	anti-viral, macrophage activation	143	20/25	D	+	0	12

[a]MW = Molecular Weight x1000
[b]M = Monomers, D = Dimers

TABLE IV Colony Stimulating Factors

Name	Type Assay	Amino Acids	Mw[a]	Conformation[b]	Glycosylation[c]	Disulphide Bonds	Chromosome
G	Granulocyte growth and differentiation	174	18-22	M	+(0)	2	17
M	Macrophage growth and differentiation	192	18-26	D	+(N)	3	5
GM	Granulocyte/ macrophage growth and differentiation	127	14-35	M	+(N)	4	5

[a]MW = Molecular Weight x1000
[b]M = Monomers; D = Dimers
[c]O = O-linked; N = N-linked

TABLE V Tumour Necrosis Factors

Name	Type Assay	Amino Acids	Mw[a]	Conformation[b]	Glycosylation[c]	Disulphide Bonds	Chromosome
α	Cytotoxicity Cachectia	157	17	T	-	1	6
β	Cytotoxicity Cachectia	171	25	D/T	+(N)	0	6

[a]MW = Molecular Weight x1000
[b]T = Trimers; D/T = Dimers and Trimers
[c]N = N-linked

TABLE VI Other Cytokines

Name	Type Assay	Amino Acids	Mw[a]	Conformation[b]	Glycosylation[c]	Disulphide Bonds	Chromosome
low mw BCGF	B Cell Growth	124	12	M	+(N)	1?	?
Leukaemia Inhibitory Factor	Macrophage differentiation	179	20-60	M	+(N/0)	?	22
Monocyte Derived Chemokinetic Factor	Chemotaxis/ chemokinesis	77	7.5-11	M	-	1	?
Transforming Growth Factorβ	Growth Inhibition	112	12.5	D	-	4	?
Erythropoietin	Erythroid growth differentiation	166	34	M	+(N/0)	2	7

[a]MW = Molecular Weight x1000
[b]M = Monomers; D = Dimers
[c]O = O-linked; N= N-linked

TABLE VII Cytokine Receptors

	MW[a]	Glycosylated	AA[b]	Ig Superfamily	IC Domains[c]	Chromosome
IL-1α/β	80	Yes	536	Yes	217	-
IL-2	H 75	-	-	-	-	-
	L 55	Yes	272	No	13	10
IL-3	70	-	-	-	-	-
IL-4	60-75(mouse) 138-145(man)	-	-	-	-	-
IL-5	70	-	-	-	-	-
IL-6	55-95	Yes	468	Yes	82	-
IL-7	-	-	-	-	-	-
MCSF	165	Yes	-	Yes	-	5
GCSF	180	-	-	-	-	-
GMCSF	70,132	-	-	-	-	-
TNF	70	-	-	-	-	-
IFNα/β	120	Yes	-	-	-	21
IFN	90	Yes	-	-	-	6
EPO	85-105	-	-	-	-	-
TGFβ	500-600	-	-	-	-	-

[a]MW = Molecular Weight x1000
[b]AA = Amino Acids
[c]IC Domain = Number of amino acids in intracytoplasmic domain

What do cytokines do?

A number of approaches have been used to find out what cytokines do. For any physiological or pathological system cytokines can be applied and any changes measured, cytokines can be inhibited and changes measured, or different systems can be observed and cytokine levels can be measured and correlated with other changes. Until recently these approaches have been difficult or impossible to achieve, due to lack of appropriate reagents or assays.

Purification of cytokines from cells or cell line supernatants had proven difficult. Many cytokines are similar in size, glycosylation, charge and other characteristics, and therefore multiple separation steps are needed for purification. As cytokines are only normally present in pg-ng/ml concentrations in culture fluids this necessitates the use of very large volumes of starting material to overcome losses on separation. Yields from such systems are insufficient for full investigation of cytokine function. Since recombinat DNA methods were introduced, cytokines have become available in virtually unlimited quantities of pure reagents (see Fig. 1). Recombinant cytokines have been used to produce polyclonal and monoclonal antibodies which can be used to neutralise biological activity or for detection of cytokines (Fig. 2).

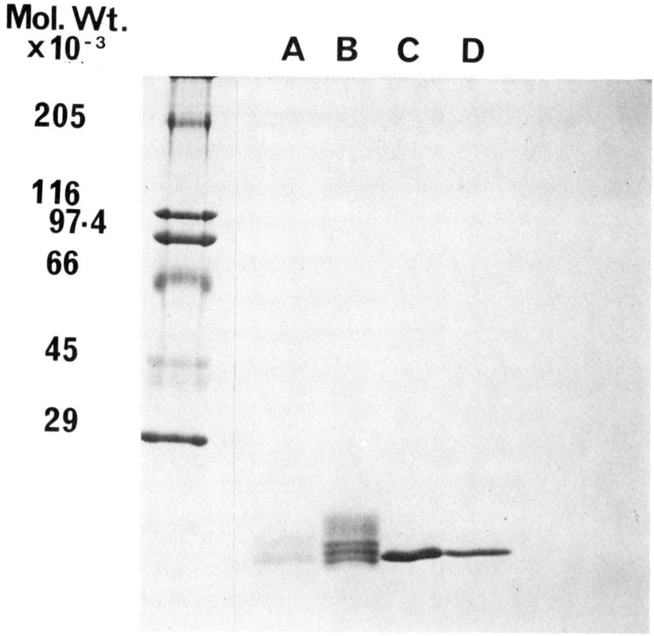

Figure 1. SDS polyacrilamide gel, run under reducing conditions and stained with Coomasie Blue. The first track contains molecular weight markers, track A GMCSF, track B Il-3, tracks C and D GCSF.

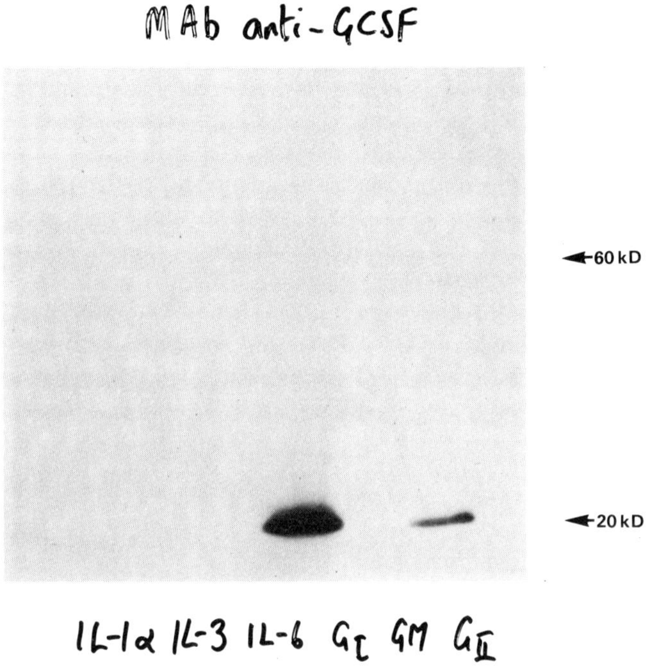

Figure 2. Immunoblot of IL-1α, IL-3, Il-6, GCSF I, GCSF and a different GCSF II stained with a monoclonal antibody raised against GCSF I.

Once cytokines and neutralising antibodies became widely available it quickly became apparent that most of the bioassays in common use were at best responsive to at least two and up to five or six different cytokines (Clemens *et al.*, 1987). Many cytokines were also shown to be active in a wide range of assays based on different cell types. It also became clear that some cytokines exhibited considerable overlap in activities, and with improvements in assay sensitivity and specificity it also became obvious that many cytokines could be produced by many cell types, often in response to stimulation with cytokines. The case of Interleukin 1 (IL-1) best illustrates these points.

Interleukin 1

Originally IL-1 was defined as a monocyte product which stimulated the proliferation of murine thymocytes which had been activated with a sub-mitogenic dose of lectin (the lymphocyte activating factor assay). IL-1 was also observed to cause a febrile response *in vivo* (pyrogen assay)

(Oppenheim *et al.*, 1986; Dinarello, 1987). When IL-1 activity was cloned, two different gene products were obtained, which only shared 26% amino acid homology (March *et al*, 1985). The two molecules IL-1α and IL-1β have almost identical biological activities and bind to the same cell surface receptor. In the last year it has been reported that the LAF assay is also responsive to IL-2, IL-4, IL-6 and TNF. In addition the pyrogen assay responds to IL-6, TNF and IFNα. Both assays also respond to endotoxin, a frequent contaminant of many samples. Endotoxin appears to function by causing the release of cytokines such as IL-1, TNF or IL-6.

The search for specific bioassays for IL-1, and indeed for other cytokines, has so far proved fruitless. The best assays are T cell based, either relying on IL-1 to stimulate IL-2 or Il- 2 receptor production e.g. by LBRM33 or EL4 clones, (Gearing et al, 1987) or to stimulate proliferation of lines such as D10 (Symons *et al.*, 1987). The use of such murine cell lines has the advantage that certain cytokines particularly Il-4, IFN, IL-3 and GMCSF are species restricted, the human factors not working on mouse cells. In addition murine cells are approximately ten-fold less sensitive to human than murine TNF. Only the D10 line is unresponsive to murine or human TNF.

Ironically the only specific assays for IL-1 are too specific. Immunoassays will only detect either IL-1α or IL-1β, and not total IL-1 biological activity, and they are usually restricted to detecting IL-1 from closely related species (see Fig. 3). In addition, in common with other immunoassays they can be capable of detecting immunologically intact, but biologically inactive forms of the molecules (Hopkins *et al.*, 1988). Despite these drawbacks immunoassays configured as ELISA (Enzyme Linked Immunosorbant Assay) or IRMA (Immunoradiometric Assay) seem to hold most promise for investigation of cytokine levels. This is particularly valid for biological fluids, particularly serum where inhibitory or toxic molecules normally obscure bioassay results.

Using T cell based IL-1 assays for detection, and antibodies for confirming specificity, it became obvious that IL-1 could be produced by many different cell types, including keratinocytes, T cells, B cells, polymorphonuclear leucocytes, fibroblasts and glial cells (Oppenheim, 1986). Receptors for IL-1 could also be identified on most nucleated cell types. Receptor distribution correlated with the wide tissue range of IL-1 action on lymphocytes, neutrophils, monocytes, fibroblasts, epithelial and endothelial cells, muscle, hepatocytes, osteoclasts and chondrocytes. In general most cells have less than two hundred receptor sites, although some of the more sensitive T cell lines such as EL-4 and NOB 1 could have a few thousand (Bird *et al.*, 1988).

Once the full spectrum of IL-1 activities was delineated, other cytokines were added to the same assay systems. By 1987 it became apparent that most of the properties of IL-1 could be shown by TNF (Dinarello, 1987b; Le and Vilcek,

1987). The two molecules are remarkably similar in many aspects of their function although they share no amino acid or gene sequence homologies, and operate through different cell surface receptors. TNF also exists as two forms α and β which share 51% amino acid homology and bind to the same receptor. There are only a few systems where TNF and IL-1 are not both active, e.g. neutrophil superoxide production by TNF. In many systems TNF and IL-1 act synergistically (Waage and Espevik, 1988).

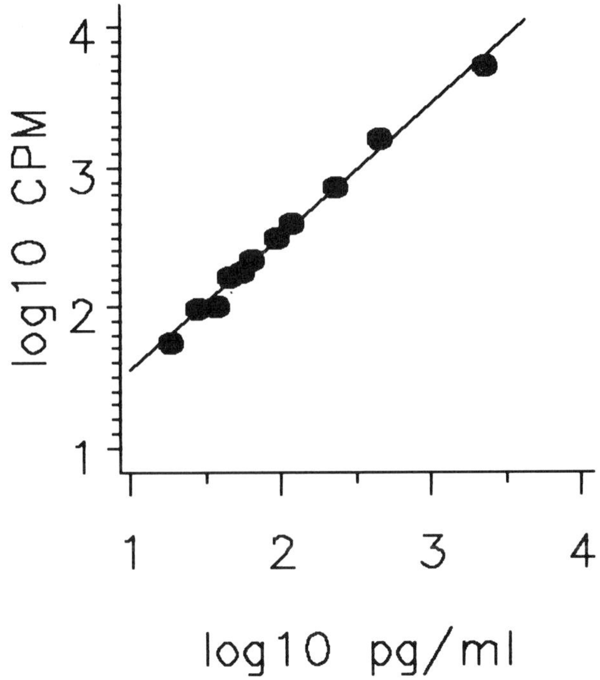

Figure 3. Standard curve for a two site immunoradiometric assay to detect human IL-1α. The asssay uses a monoclonal antibody to 'catch' the IL-1α and a radio labelled sheep polyclonal antiserum to detect bound IL-1α. This assay shows no response to human IL- 1β, murine IL-1α, TNF, IL-6, nor to any other recombinant cytokine.

A further complication is that IL-1 and TNF were found not only to induce secretion of cytokines such as IL-2, IL-4, IL-6 and GMCSF but also IL-1 and TNF themselves (Fibbe *et al.*, 1988; Ghezzi and Dinarello, 1988). This obviously makes it more difficult to investigate the direct effects of individual cytokines, particularly on systems involving more than one cell type.

Despite these difficulties it is now obvious that cytokines have properties consistent with central roles in the immune system, particularly in haemopoiesis, and in local and systemic inflammation.

Haemopoiesis

The haemopoietic system generates all of the mature blood cells including erythrocytes, platelets, granulocytes, monocytes and lymphocytes from multipotent stem cells (Metcalf, 1984). The process by which these mature cells form *in vivo* can be mimicked *in vitro* by the application of appropriate combinations of different cytokines. Some cytokines are active in all stages of this process from stem cells onwards, for example IL-1 or IL-6, some such as Il-3 and GMCSF are slightly more restricted in action, whilst others are restricted to only one or two lineages, e.g. IL-2, IL-5, IL-7, GCSF, MCSF and some of the human B cell growth factors. Some cytokines, for example TGFβ and TNF, are inhibitory for haemopoiesis (Trinchieri *et al.*, 1987; Keller *et al.*, 1988) (see Fig. 4).

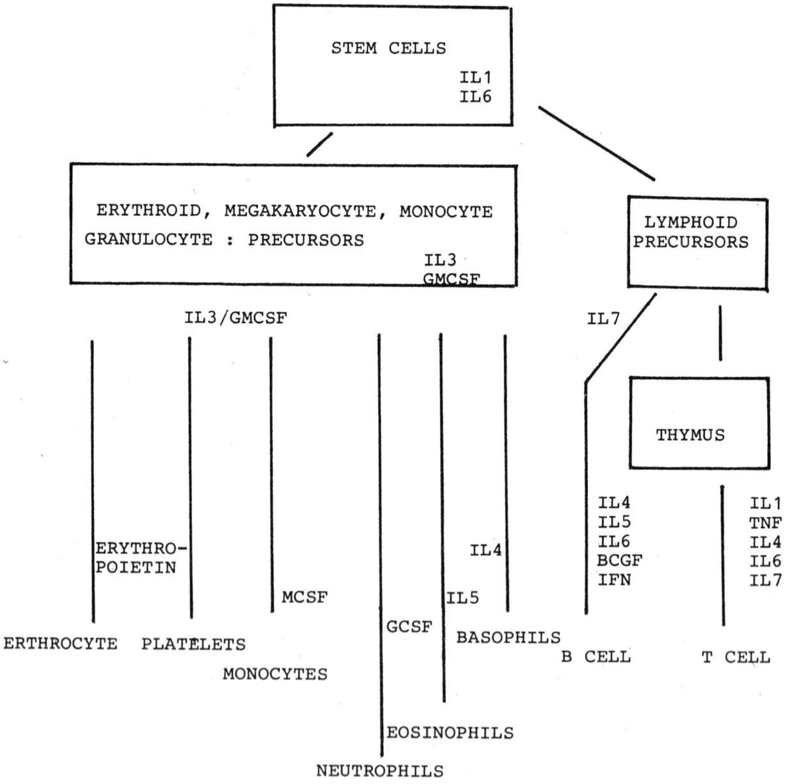

Figure 4. Simplified diagram of cytokine involvement in haemopoiesis.

A great deal of current research is being devoted to the isolation of other haemopoietic growth factors, particularly those involved in generating erythroid cells and megakaryocytes, or in regulating the survival or differentiation of stem cells. Administration of haemopoietic growth factors such as GCSF, GMCSF or IL-3 to animals and also to humans has resulted in increases in circulating levels of granulocytes or monocytes. These increases are associated with decreased incidence of infections in immunocompromised patients, such as after chemotherapy (Vadhan- Raj *et al.*, 1987; Morstyn *et al.*, 1988). High levels of CSFs can be maintained in transgenic mice carrying an unregulated GMCSF gene or in mice infected with retroviruses containing CSF genes (Lang *et al.*, 1987; Johnson, 1988). In these animals pathogenic changes can be apparent, particularly with GMCSF or IL-3. The pathology seen is consistent with the release of toxic cytokines such as IL-1 in response to the GMCSF or IL-3 (Gearing *et al.*, 1988, submitted for publication). GCSF does not seem to cause the release of IL-1 and is also not associated with generalised toxicity. Such secondary toxicity should be borne in mind in the monitoring of clinical trials.

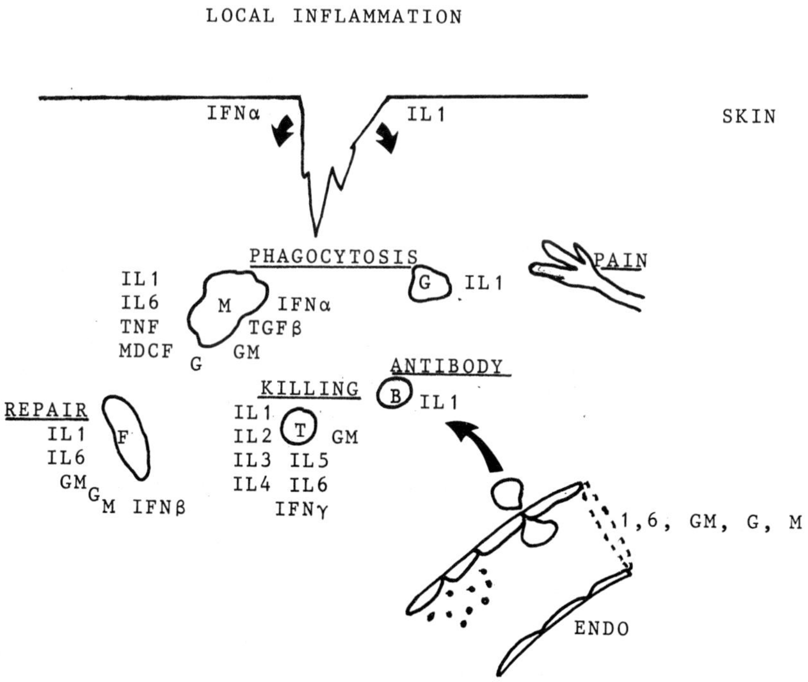

Figure 5. Diagram of cytokine production in localised inflammatory responses.

Local Inflammation

In any localised challenge, such as a cut in the skin, a number of steps can be recognised to be associated with particular groups of cytokines (Fig. 5). The initial trauma can release stored IL-1 from the epidermis (Fincham, 1988), which acts to enhance the subsequent activation of macrophages and T lymphocytes in response to introduced microbial antigens and mitogens. IL-1 can also stimulate nerves to enhance pain (Ferreira *et al.*, 1988). Activated T cells and macrophages can then release their full spectrum of cytokines which act in a number of ways. IL-1, Il-6, and the CSFs are all active on endothelial cells making the surface more sticky and procoagulant. A number of cytokines such as monocyte-derived chemokinetic factor, TGFβ, IL-2, γ IFN and macrophage inflammatory protein are chemoattractants for neutrophils, monocytes or lymphocytes and can attract these cell types from the circulation to the inflammed site. Phagocytosis of microbes or damaged cells is stimualted by IFNγ, IL-2, IL-4 and CSFs (Belosevic *et al.*, 1988), the various IFNs stimulate antiviral activity in cells, and specific T and B cell responses are maintained by IL-1, IL-2, IL-4, IL-6 and IFNγ (O'Garra *et al.*, 1988). Many cytokines are also though to be involved in the remodelling of tissue following damage. Cells such as fibroblasts, osteoclasts, muscle fibres, chondrocytes and synovial cells can be stimulated to proliferate or to secrete lytic enzymes in response to IL-1, IL-6, TNF, TGFβ, CSFs or IFNβ (Powanda *et al.*, 1988; Oppenheim, 1986).

It is less clear how the inflammatory response is down regulated. Prostaglandins are potent inhibitors of much cytokine action, as is TGFβ for T lymphocyte activation; IFNs are also antiproliferative for lymphocytes (Kehrl *et al.*, 1986; Tsunawaki *et al.*, 1988). The release of soluble IL-2 receptors by activated T cells may also act to mop up IL-2 and hence reduce T cell proliferation (Wagner *et al.*, 1986). It is likely that reduced antigen levels and the cumulative effects of inhibitory molecules, along with destruction of cytokines by proteolytic enzymes in tissue exudates will eventually shut off the response. In situations where local chronic stimulation occurs, for example in rheumatoid joints, chronically elevated cytokine levels, particularly IL-1 or TNF, can contribute to the degenerative pathology (Saklatvala, 1986).

Systemic Inflammation

When infections are severe, activated macrophages or lymphocytes can release cytokines into the circulation. The most obvious manifestations of this are the febrile and acute phase responses (Dinarello, 1987; Gitlin and Colten, 1987). Both responses are now known to be mediated by cytokines (Dowton and Colten, 1988). The febrile response can be mimicked by injections of IL-1, TNF, IL-6 or IFNγ causing resetting of the thermoregulatory centre in the hypothalamus. The acute phase response can be mimicked by injection of IL-1, TNF or IL-6 causing release of acute phase proteins from the liver. In addition to these effects, systemic administration of IL-1 can cause ACTH secretion and the induction of slow wave sleep, and in high doses can be analgesic. Systemic IL-1 or TNF can cause increased proliferation in bone marrow, release of calcium from bone by activation of osteoclasts, and release of amino acids by proteolysis of muscle. They can also cause breakdown of fat resulting in the cachectia often seen accompanying many tumours (Cerami and Beutler, 1988) (Fig. 6). Under conditions of moderate infection all these changes are beneficial for the host. Elevated temperatures enhance the antimicrobial actions of leucocytes, many acute phase proteins such as C reactive protein are directly antibacterial, increased sleep allows more energy to be directed to fighting infection, while the catabolism of muscle and fat increases the supply of nutrients to support the function of the immune system. When infection is severe, as in septic shock, toxic quantities of cytokines can be released in response to endotoxin, and the effects can be fatal (Girardin, 1988; Waage and Espevik, 1988). Excess TNF and IL-1 can cause wasting, haemorrhage, lung congestion and eventually death. It has been demonstrated that neutralising antibodies to TNF given immediately prior to a lethal dose of endotoxin can prevent death (Cerami and Beutler, 1988).

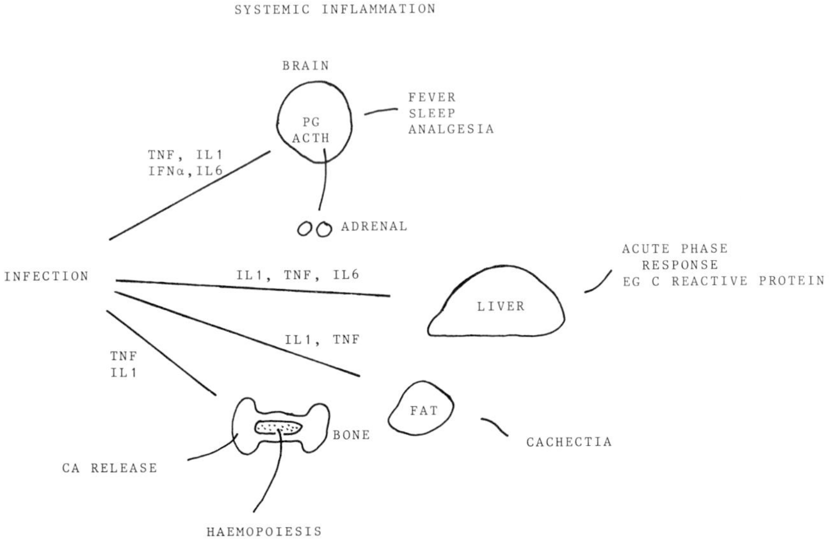

Figure 6. Diagram of cytokine involvement in systemic inflammation.

Current directions in cytokine research

The ultimate aim for the majority of cytokine research is to gain sufficient understanding of the way in which cytokines affect their target cells, either alone or in combination with other cytokines, hormones or conventional drugs, to enable agonists or antagonists to be used therapeutically. The crystal structures of some cytokines, for example IL-1β and IL-2, have been determined (Priestle *et al.*, 1988; Brandhuber *et al.*, 1987) and several cytokine receptors have now been identified or cloned (Sheehan *et al.*, 1988; Pfizenmaier *et al.*, 1988; Ambrus *et al.*, 1988; Park *et al.*, 1987; Smith *et al.*, 1987; Sims *et al.*, 1988; Yamasaki *et al.*, 1988) (Table VII). Analysis of post receptor binding events ranging from generation of second messengers to the control of gene expression is currently being undertaken. Many groups are also studying the regulation of cytokine gene expression (Webb and Goeddel, 1987). Continuing work on the interaction of hormones and neurotransmitters with cytokine mediated events, and the role of cytokines in neuroendocrine function also looks extremely promising (Lotz *et al.*, 1988; Robertson *et al.*, 1988; Lemaire *et al.*, 1988; Breder *et al.*, 1988; Giulian and Lachman, 1985). To date, however, there are no convincing receptor-agonists other than the recombinant cytokines themselves, nor are there antagonist drugs other than neutralising monoclonal antibodies (anti-cytokine antibodies) or anti-receptor antibodies, such as anti-TAC for the IL-2 receptor and the recently described MR6 that may recognise the human IL-4 receptor (Larché *et al.*, 1988; Larché and Ritter, Chapter 5 this volume). The next two to three years should see an explosion of information in all of these areas.

BIBLIOGRAPHY - Useful texts.

Balkwill, F.R. (1988) *Cytokines in Tumour Therapy*. Oxford University Press.

Clemens, M.J., Morris, A.G. and Gearing, A.J.H. (1987) *Lymphokines and Interferons: a Practical Approach*. IRL Press.

Gearing, A.J.H. and Hennessen, W. (1988) *Developments in Biological Standardisation, Volume 69, Cytokines: Laboratory and Clinical Evaluation*. Karger.

Gillis, S. (1987) *Recombinant Lymphokines and their Receptors*. Marcel Dekker.

Hamblin, A.S. (1988) *Lymphokines in Focus*. IRL Press.

Metcalf, D. (1984) *The Hemopoietic Colony Stimulating Factors*. Elsevier.

Moller, G. (1984) *Immunological Reviews, Volume 78. B cell Growth and Differentiation Factors*. Munksgaard.

Moller, G. (1986) *Immunological Reviews, Volume 92. IL-2: Receptors and Genes*. Munksgaard.

Moller, G. (1987) *Immunological Reviews, Volume 97. Gamma Interferon.*

Moller, G. (1988) *Immunological Reviews, Volume 102. IL-4 and IL-5: Biology and Genetics*. Munksgaard.

Pestka, S. (1986) *Methods in Enzymology, Volume 119. Interferons Part C*. Academic Press.

Powanda, M.C., Oppenheim, J.J., Kluger, M.J., Dinarello, C.A. (1988) *Monokines and Other Non-Lymphocytic Cytokines*. Alan R. Liss.

Schrader, J. (1988) *Lymphokines, Volume 16, Interleukin 3: The Pan-Specific Haemopoietin*. Academic Press.

Webb, D.R. and Goeddel, D.V. (1987) *Lymphokines, Volume 13 Molecular Cloning and Analysis of Lymphokines*. Academic Press.

REFERENCES

Ambrus, J.L., Jurgensen, C.H., Brown, E.G., McFarland, P. and Fauci, A.S. (1988) Identification of a Receptor for High MW BCGF. *J. Immunol.* **141**, 861-869.

Belosevic, M., Davis, C.E., Meltzer, M.S., Nacy, C.A. (1988) Regulation of Activated Macrophage Antimicrobial Activities. *J. Immunol.* **141**, 890-896.

Bird, T.A., Gearing, A.J., Saklatvala, J. (1988) Murine IL-1 Receptor. *J. Biol. Chem.* **263**, 12063-12069.

Brandhuber, B.J., Boone, T., Kenney, W.C. and McKay, D.B. (1987) 3-D Structure of IL-2. *Science* **238**, 1707-1709.

Breder, C.D., Dinarello, C.A. and Saper, C.B. (1988) IL-1 Immunoreactive Innervation of the Human Hypothalamus. *Science* **240**, 321-323.

Burke, D.C. (1985) The Interferons. *British Medical Bull.* **41**, 333-338.

Campbell, H.D., Tucker, W.Q.J., Hort, Y., Martinson, M.E., Mayo, G., Clutterbuck, E.J. *et al.* (1987) Molecular Cloning, Nucleotide Sequence and Expression of the Gene Encoding Human EDF (IL-5). *Proc Natl Acad Sci., U.S.A.*, **84**, 6629-6633.

Cerami, A. and Beutler, B. (1988) The Role of Cachectin/TNF in Endotoxic Shock and Cachexia. *Immunology Today* **9**, 28-31.

Clark, S.C. and Kamen, R. (1987) The Human Haematopoietic Colony- Stimulating Factors. *Science* **236**, 1129-1237.

Davatelis, G., Tekamp-Olson, P., Wolpe, S.D., Hermsen, K., Luedke, C., Gallegos, C. *et al.* (1988) Cloning and Characterisation of a cDNA for Murine Macrophage Inflammatory Protein (MIP). *J. Exp. Med.* **167**, 1939-1944.

Dinarello, C.A. (1984) Interleukin 1. *Reviews of Infectious Diseases* **6**, 51-95.

Dinarello, C.A. (1987a) Interleukins, TNFs and Interferons as Endogenous Pyrogens and Mediators of Fever. In *Lymphokines 14*, 1-31. Academic Press.

Dinarello, C.A., (1987b) The Biology of IL-1 and Comparison to TNF. *Immunol. Letts.* **16**, 227-232.

Dowton, S.B. and Colten, H.R. (1988) Acute Phase Reactants in Inflammation and Infection. *Seminars in Haematology* **25**, 84-90.

Ferreira, S.H., Lorenzetti, B.B., Bristow, A.F., and Poole S. (1988). IL-1β as a Potent Hyperalgesic Agent Antagonized by a Tripeptide Analogue. *Nature* **334**, 698-700.

Fibbe, W.E., Van Damme, J., Billiau, A., Goselink, H.M., Voogt, P.J., Van Eeden, G. *et al.* (1988) IL-1 Induces Human Marrow Stromal Cells in Long-Term Culture to Produce GCSF and MCSF. *Blood* **71**, 430-435.

Fincham, N.J., Camp, R.D.R., Gearing, A.J., Bird, C.R. and Cunningham, F.M. (1988) Neutrophil Chemoattractant and IL-1-Like Activity in Samples from Psoriatic Skin Lesions. *J. Immunol.* 140, 4294-4299.

Gearing, A.J.H., Bird, C.R., Thorpe, R.C., Bristow, A. and Poole, S. (1987) A Simple Bioassay for IL-1 which is unresponsive to IL- 2. *J. Immunol. Meths.* **99**, 7-11.

Gearing, D.P., Gough, N.M., King, J.A., Hilton, D.J., Nicola, N.A., Simpson, R.J. *et al.* (1987) Molecular Cloning and Expression of cDNA Encoding a Murine Myeloid Leukaemia Inhibitory Factor. *EMBO Journal* **6**, 3995-4002.

Ghezzi, P. and Dinarello, C.A. (1988) IL-1 Induces IL-1. *J. Immunol.* **140**, 4238-4244.

Girardin, E., Grau, G.E., Dayer, J., Lombard, P. *et al.* (1988) TNF and IL-1 in the Serum of Children with Severe Infectious Purpura. *New Eng. J. Med.* **319**, 397-400.

Gitlin, J.D. and Colten H.R. (1987) Molecular Biology of the Acute Phase Plasma Proteins. In: *Lymphokines Volume 14*, 123-153 Academic Press.

Guilian, D. and Lachman, L.B. (1985) IL-1 Stimulation of Astroglial Proliferation After Brain Injury. *Science* **228, 497- 499.**

Hirano, T., Yasukawa, K., Harada, J., Taga, T., Watanabe, Y., Matsuda, T. *et al.* (1986) cDNA for a Novel Human Interleukin (BSF-2) that Induces B Lymphocytes to Produce Immunogloublin. *Nature* **324**, 73-76.

Hopkins, S.J., Humphreys, M., Jayson, M.I.V. (1988) Cytokines in Synovial Fluid: *I. Clin. Exp. Immunol.* **72**, 422-427.

Johnson, G.R. (1988) Effects of Excess GMCSF in Mice Infected with a GMCSF Retrovirus. In: *Developments in Biological Standardisation* Volume 69, 1-8. Karger.

Kehrl, J.H., Wakefield, L.M., Roberts, A.B., Jakowlew, S., Mon, M., Derynck, R. *et al.* (1986) Production of TGFβ by Human T Lymphocytes and its Potential Role in the Regulation of T Cell Growth. *J. Exp. Med.* **163**, 1037-1050.

Keller, J.R., Mantel, C., Sing, G.K., Ellingsworth, L.R., Ruscetti, S.K. and Ruscetti, F.W. (1988) TGFβ Selectively Regulates Early Murine Haematopoietic Progenitors and Inhibits the Growth of IL-3 Dependent Lines. *J. Exp. Med.* **168**, 737-750.

Lang, R.A., Metcalf, D., Cuthbertson, R.A., Lyons, I., Stanley, E., Kelso, A., Kannourakis, G. *et al.* (1987) Transgenic Mice Expressing a Haemopoietic Growth Factor Gene (GM-CSF) Develop Accumulations of Macrophages, Blindness and a Fatal Syndrome of Tissue Damage. *Cell* **51**, 575-586.

Larché, M., Lamb, J.R., O'Hehir, R.E., Imami-Shita, N., Zanders, E.D., Quint, D.E., Moqbel, R. and Ritter, M.A. (1988) Functional Evidence for a Monoclonal Antibody that Binds to the Human IL-4 Receptor. *Immunology*, in press.

Le, J. and Vilcek, J. (1987) Biology of Disease. TNF and IL-1: Cytokines with Multiple Overlapping Biological Activities. *Lab. Invest.* **56**, 234-248.

Lemaire, I. (1988) Neurotensin Enhances IL-1 Production by Activated Alveolar Macrophages. *J. Immunol.* **140**, 2983-2988.

Lin, F.K., Suggs, G., Lin, C.H., Browne, J.K., Smalling, R., Egrie, J.C. *et al.* (1985) Cloning and Expression of the Human Erythropoietin Gene. *Proc. Natl. Acad. Sci. U.S.A.* **82**, 7580-7584.

Lotz, M., Vaughan, J.H. and Carson, D.A. (1988) Effect of Neuropeptides on Production of Inflammatory Cytokines by Human Monocytes. *Science* **241**, 1218-1221.

March, C.J., Mosley, B., Larsen, A., Cerretti, D.P., Braedt, C., Price, V. *et al.* (1985) Cloning Sequence and Expression of Two Distinct Human IL-1 Complementary DNAs. *Nature* **315**, 641-647.

Morstyn, G., Campbell, I., Souza, L.M., Alton, N.K., Keech, J., Green, M. *et al.* (1988) Effect of GCSF on Neutrophenia Induced by Cytotoxic Chemotherapy. *Lancet* 667-672.

Namen, A.E., Lupton, S., Hjerrild, K., Wignall, J., Mochizuki, D.Y., Schmiere, A. *et al.* (1988) Stimulation of B Cell progenitors by Cloned Murine IL-7. *Nature* **339**, 571-573.

O'Garra, A., Umland, S., De France, T. and Christiansen, J. (1988) B Cell Factors are Pleiotropic. *Immunology Today* **9**, 45- 54.

Old, L.J. (1985) TNF. *Science* **230**, 630-633.

Oppenheim, J.J., Kovacs, E.J., Matsushima, K. and Durum, S.K. (1986) There is more than one IL-1. *Immunology* **7**, 45-56.

Oppenheim, J.J., Matsushima, K., Yoshimura, T. and Leonard, E.J. (1987) The Activities of Cytokines are Pleiotropic and Interdependent. *Immunol. Letts.* **16**, 179-184.

Park, L.S., Friend, D., Grabstein, K., Urdal, D.L. (1987) Characterisation of the High-Affinity Cell-Surface Receptor for Murine BSF1. *Proc. Natl. Acad. Sci. U.S.A.* **84**, 1669-1673.

Pfizenmaier, K., Wiegmann, K., Scheurich, P., Kronke, M., Merlin, G., Aguet, M. *et al.* (1988) High Affinity Human IFN gamma binding Capacity is Encoded by a Single Receptor Gene in proximity to c-ros on Human Chromosome Region 6q16 to 6q22. *J. Immunol.* **141**, 856-860.

Priestle, J.P., Schar, H. and Grutter, M.G. (1988) Crystal Structure of IL-1β. *EMBO Journal* **7**, 339-343.

Robertson, B., Dostal, K. and Kaynes, R.A. (1988) Neuropeptide Regulation of Inflammatory and Immunologic Responses. *J. Immunol.* **140**, 4300-4307.

Robb, R.J. (1984) IL-2: The Molecule and its Function. *Immunol. Today* **5**, 203-209.

Saklatvala, J. (1986) TNF Stimulates Resorption and Inhibits Synthesis of Proteoglycan in Cartilage. *Nature* **322**, 547-550.

Sharma, S., Mehta, S., Morgan, J., Maizel, A. (1987) Molecular Cloning and Expression of a Human B-Cell Growth Factor Gene in E.coli. *Science* **235**, 1489-1492.

Sheehan, K.C.F., Claderon, J., and Schreiber, R.D. (1988) Generation and Characterization of Monoclonal Antibodies Specific for the Human IFN gamma Receptor. *J. Immunol.* **140**, 4231-4237.

Sims, J.E., March, C.J., Cosman, D., Widmer, M.B., McDonald, H.R., McMahan, C.J. *et al.* (1988) cDNA Expression Cloning of the IL-1 Receptor A Member of the Immunogloublin Superfamily. *Science* **241**, 585-589.

Smith, K.A. (1987) The Two Chain Structure of High-Affinity IL-2 Receptors. *Immunol Today* **8**, 11-13.

Sporn, M.B., Roberts, A.B., Wakefield, L.M., Assoian, R.K. (1986) TGFβ: Biological Function and Chemical Structure. *Science* **233**, 532-534.

Sporn, M.B. and Roberts, A.B. (1988) Peptide Growth Factors are Multifunctional. *Nature* **332**, 217-219.

Symons, J.A., Dickens, E.M., Di Giovine, F. and Duff, G.W. (1987) Measurement of IL-1 Activity. In *Lymphokines and Interferons A Practical Approach.* Eds. Clemens, M.J., Morris, A.G. and Gearing A.J.H. IRL Press.

Trinchieri, G., Kobayashi, M., Murphy, M. and Perussia, B. (1987) Human Interferons and Cytotoxins: Regulatory Effects on Myeloid Cells. *Lymphokines* Volume 14, 267-305. Academic Press.

Tsunawaki, S., Sporn, M., Ding, A. and Nathan, C. (1988) Deactivation of macrophages by TGFβ. *Nature* **334**, 260-263.

Vadhan-Raj, S., Keating, M., Le Maistre, A., Hittelman, W.N., McCredie, K., Trujillo, J.M. *et al.* (1987) Effects of rGMCSF in Patients with Myelodysplastic Syndromes. *New England J. Med.* **317**, 1545-1552.

Waage, A. and Espevik, T. (1988) IL-1 Potentiates the Lethal Effect of TNF in mice. *J. Exp. Med.* **167**, 1987-1992.

Wagner, D.K., Jolley, J., Malek, T.R., Berzofsky, J.A. and Nelson D.L. (1986) Antigen Specific Murine T Cell Clones Produce Soluble IL-2 Receptor on Stimulation with Specific Antigens. *J. Immunol.* **137**, 592-596.

Yamasaki, K., Taga, T., Hirata, Y., Yawata, H., Kawanishi, Y., Seed, B. *et al.* (1988) Cloning and Expression of the Human IL-6 Receptor. *Science* **241**, 825-828.

Yang, Y., Ciarletta, A.B., Temple, P.A., Chung, M.P., Kovacic, S., Giannotti, J.S. *et al.*, (1986) Human IL-3: Identification by Expression Cloning of a Novel Haematopoietic Growth Factor Related to Murine IL-3. *Cell* **47**, 3-10.

Yokota, T., Coffman, R., Hagiware, H., Rennick, D., Takebe, Y., Yokota, K. *et al.* (1987) Isolation and Characterisation of cDNA Clone Encoding Mouse and Human IgA-Enhancing Factor and eosinophil CSF: Relationship to IL-5. *Proc. Natl. Assoc. Sci., U.S.A.* **84**, 7388-7392.

Yoshimura, T., Matsushima, K., Tanaka, S., Robinson, E.A., Appella, E. *et al.* (1987). Purification of a Human Monocyte Derived Neutrophil Chemotactic Factor that has Peptide Sequence Similarity to other Host Defence Cytokines. *Proc. Natl. Acad. Sci., U.S.A.*, 84, 9233-9237.

CONFERENCE REPORTS

THE THYMUS: HISTOPHYSIOLOGY AND DYNAMICS IN THE IMMUNE SYSTEM Conference held at Centrum Rolduc, Kerkrade, The Netherlands, April 17-20, 1988.

INTRODUCTION

The beautiful and peaceful environment of the 12th century monastery in the south of the Netherlands was the setting for approximately 100 researchers, mainly from Europe, to gather and discuss informally their recent work in the field of thymology.

The meeting was opened by a stimulating plenary lecture from **Bruce Roser** (Cambridge, UK) who reminded us that all organisms are built from the same building blocks, for example the same 20 or so amino acids, yet we expect the lymphocytes to recognise something odd about invaders and to be able to discriminate self from non-self and destroy unwanted elements. Quite remarkable when thought of from this angle. Polymorphism is a necessity for this phenomenon and diversity must be expressed before the destructive events as correct recognition is the key to good health. A clue to the mechanism of self tolerance was provided recently by the elegant work of Kappler and Marrack who showed that T cells reactive with a self antigen (in this case MHC class II) are deleted from the repertoire at an early stage of thymic development, soon after the acquisition of surface receptors for antigen. This negative selection is thought to take place in the thymic cortex. Studies on grafting have shown that the prevention of rejection of a graft can be transferred from a tolerant to a non-tolerant individual, indicating that some tolerance involves an active process ie. suppression. Roser cited his own experiments showing that clonal deletion tolerance can be induced in the thymus in the first cohort of T cells (CD 17 positive) if antigen is present before birth. He proposed that this active rejection enacts tolerance. In other words, if there is clonal suppression, there should be no need for deletion.

SESSION 1

Physiology and Embryology of the Microenvironment. Chairmen: ***Marion D. Kendall*** *(London, UK) and* ***Catherine Corbel*** *(Paris, France).*

The first session started with **A. Moraux** (Brussels, Belgium) who studied thymus histogenesis in C3H mice from the twelfth embryonic day to the seventh postnatal day using light and electron microscopy. At day 12, each thymus lobe consisted of a nonvascularized epithelial bud of undifferentiated cells containing a few lymphoblasts and surrounded by mesenchymal cells and blood vessels. During the following days, all epithelial cell types seemed to differentiate from this unique primitive cell population. In the developing cortical regions, lymphoblasts accumulated rapidly, stretching the epithelial cells which became, from day 15 on, stellate in shape. At day 17, the deep cortex contained medium-size thymocytes. After day 18 the cortex was well developed and essentially populated by small thymocytes. Most lymphoblasts, incorporating tritiated thymidine, were located in the subcapsular region.

Medullary regions were clearly visible on day 18. They were observed as early as day 13 in areas where epithelial cells were joined by numerous desmosomes. At day 15, these regions consisted almost exclusively of spindle-shaped epithelial cells that were the precursors of medullary stellate epithelial cells which developed either vesicles and primary lysosome or formed intra- or extra-cellular cysts or Hassall's corpuscles. On day 17, the presence of interdigitating cells in the medullary areas completed medullary differentiation. Thus, the full differentiation of epithelial and non-epithelial stromal cells preceeds the appearance of the small cortical thymocyte population and the histogenesis of the mouse thymus is achieved at birth.

Furthermore, peculiar multivesicular bodies were observed in epithelial cells. At day 12 or 13, these bodies were completely filled with very small vesicles and were near the Golgi area. At day 15, they were more numerous, sometimes heterogeneous, and had empty spaces between clusters of small vesicles that became larger by day 16. These bodies resemble the "clear-vacuoles" described in the adult thymus. These vacuoles seem to belong to the lysosomal compartment and could be involved in the storage of a secretion product.

Jaap Kampinga (Groningen, The Netherlands) presented his work on the relationship between different monoclonal antibodies and the development of thymocytes from the prothymocyte to the mature medullary form. Figure 1 summarises his findings for thymocytes, non-thymocytes and thymic epithelium in the rat. On the first day that thymocytes are seen within the anlage (day 14) these cells can be stained with HIS 44 and they rapidly come to express

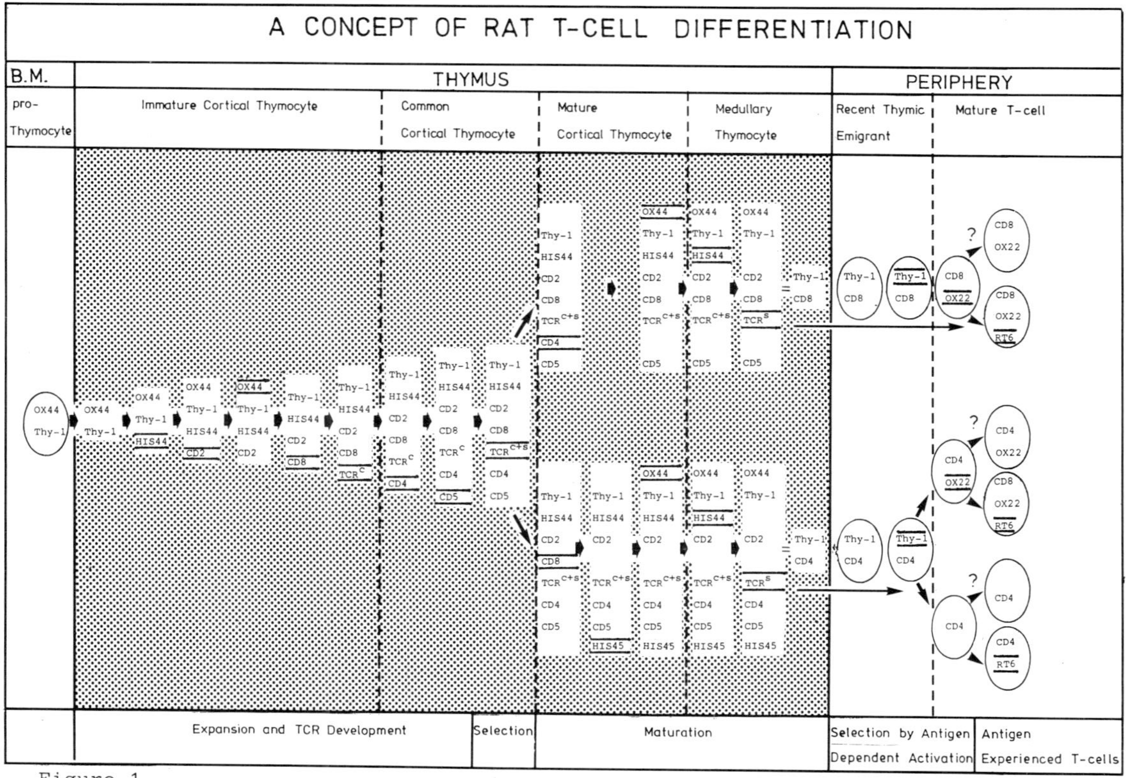

Figure 1

A concept of rat T cell differentiation according to J.Kampinga, H.Groen and R.Aspinal. This concept is based on immunohistology studies of the thymus during ontogeny and immunofluorescence studies of peripheral T cells in vascular thymus grafted, peripheral T cell transfered or thymectomized RT7 congenic rats and diabetes-prone BB rats. Changes in phenotype between two successive T cell differentiation stages are outlined. The different compartments are not to scale.

CD2. By day 17, CD8 is expressed and the next day positive staining for CD4 and HIS 42 indicates the presence of the T cell receptor (TcR). Only after this, can CD5 be visualised with monoclonal antibodies. Interestingly there is no expression of CD25, (the IL-2 receptor) and it is thought that the expression of this receptor is not needed for thymocyte development.

Dendritic cells cannot be visualised with monoclonal antibodies (IF119) until quite late in ontogeny (day 20) and some of the monoclonal antibodies raised against the thymic epithelium do not stain during ontogeny, but this may be related to the development of the epithelium itself. This powerful set of monoclonal antibodies are proving to be most useful for developmental studies.

Brita von Gaudecker (Kiel, West Germany) presented data from studies on the localization and characteristics of IL-2-positive cells in the human thymus. Monoclonal antibodies against IL-2 and against its receptor were used to demonstrate interleukin-2- positive cells in the human thymus and to compare the reaction pattern with T-cell lines of known chracteristics.

Light microscopy of Jurkat cells, a human leukaemia cell line which produces IL-2, showed intensive homogeneous intracytoplasmatic reaction with anti-IL-2, by indirect immunoperoxidase staining. Cells of the IL-2 dependent murine cytotoxic T-lymphocyte line (CTLL) revealed surface staining and capping phenomena. Incubation of human thymic tissue with anti- IL-2 showed many positive round cells in the cortex with weaker staining of the medulla.

This could be confirmed by immunoelectron microscopy using an immunogold-silver enhancement technique before embedding the tissue. Many of the cortical thymocytes showed distinct surface reactivity. Frequently the gold particles were arranged in small groups. Similar results were also obtained after incubation with a monoclonal antibody against the IL-2 receptor. To investigate if this labelling was related to receptor mediated endocytosis of IL-2, specimens were prepared and examined for coated pits and coated vesicles. CTLL cells were incubated with the antibodies before fixation and embedding. The cells showed heavy surface labelling. Frequently the reaction product was within clear cytoplasmic vesicles formed by an internalization process during the incubation time with the secondary gold coupled antibody. Coated pits were recognized at the periphery of these highly active cells, but surprisingly these structures were not associated with labelled compound.

In the medulla two types of IL-2 positve cells were recognised according to their reaction patterns: one with intracytoplasmatic labelling and the other with surface labelling. In the preparations for immunoelectron microscopy the reaction pattern was different depending on the kind of fixation. After fixation with glutaraldehyde and formaldehyde only a few positive cells were recognised in the medulla, and these showed the surface reaction. After fixation with formaldehyde alone, medium sized thymocytes showed a cytoplasmic reaction.

One interpretion of these preliminary results would be that medullary thymocytes can produce interleukin-2, and cortical thymocytes may be stimulated to express receptors and thus become IL-2 responder cells. Receptor-ligand binding and internalization of IL-2 then could trigger the proliferation of thymocytes.

Annick Tournefier (Paris, France) introduced us to the thymus of the Axolotl *Ambystoma mexicanum* which is a very primitive neotenic Urodele. In this amphibian the thymus is formed by 3 bilateral nodules from the 3rd, 4th and 5th pharyngeal pouches. Ten to 12 days after hatching the epithelial component of the thymic anlagen is colonized by stem cells originating from the ventral blood islands and migrating through the lateral and dorsal mesoderm toward the organ. The thymic development reaches its maximum before sexual maturity around 8 to 9 months. No cortico-medullary differentiation can be observed as in mammals, but some epithelial clumps are scattered among densely packed thymocytes.

The thymic microenvironment was studied by immunofluorescence using mouse monoclonal anti-human antibodies directed against cytokeratins (KL1 and KL4) and rat monoclonal antibodies directed aginst stromal cells (Willem van Ewijk's ER-TR series). KL1 gives a fine reticular staining pattern in the whole thymus with a higher degree of reactivity in the epithelial areas whereas KL4 reacts with some epithelial cells in the subcapsular region and some cells sparsely scattered in the organ. No staining was observed when ER-TR 1, 2 and 3 (anti-H-2) were used. ER-TR4 labels a few cells in the subcapsular region and in the epithelial areas and ER-TR5 labels a fine reticular network in the whole thymus and also the walls of cystic structures in a similar manner to the staining pattern observed with KL1. ER-TR6 stains some cells at the periphery of the epithelial areas but does not stain the cyst walls. Her scanning electron micrographs illustrated the very close assoications between thymocytes and epithelial cells with long processes. Some macrophages and possible dendritic cells with cytoplasmic vacuoles were distinguishable.

The problem of the thymic-blood barrier has fascinated and puzzled thymus workers since it was first proposed by Marshall and White in 1961 to account for the fact that following parenteral administration of antigens no antibody formation occurred within the thymus despite thymocytes having full immunocompetence to induce lethal Graft *versus* Host (GvH) disease. The cortex, as a privileged site, must allow the differentiation of immature T cells without interference from the outside world, yet at the same time, these maturing cells acquire self tolerance against circulating self-proteins and cell surface molecules. Raviola and Karnovsky using morphological techniques concluded that although the brain blood barrier has a tighter control of the passage of materials, several mechanisms concur in protecting the proliferating lymphocytes of the cortex from the influence of circulating antigen and that medullary lymphocytes are freely exposed to blood-borne substances. Thus the thymus-blood barrier was accepted but limited to the cortex. Evidence for antigens reaching the thymic parenchyma after intraperitoneal or intramedastinal injections existed since the early 1960s and has led to the conclusion that the capsule-thymus barrier is less efficient than the blood-thymus barrier, and that peripheral stimulation, especially via the peritoneal cavity might play a role in the overall process of differentiation within the thymic cortex.

Paul Nieuwenhuis' group (Grongingen, The Netherlands) interested in the cells and mechanisms involved in the development of GvH disease in the P-1/2F1 model in rats, injected (*i.v.* or *i.p.*) anti Ia (MHC Class II) monoclonals to suppress it, and HIS 19 (directed at the class II antigen of the m haplotype). Surprisingly, it was found that HIS 19 had penetrated into the thymic parenchyma. These experiments were extended and included other monoclonal antibodies. One hour after an *i.p.* injection of HIS 19 it was bound to small numbers of cortical epithelial cells in the immediate subcapsular area especially adjacent to the parathymic lymph node, which itself exhibited strong HIS 19 binding. Between 1 and 12 hrs after this a sharply defined front of antibody binding to epithelial cells gradually moved centripetally towards the corticomedullary junction, presumably moving with the flux of interstitial fluid. There was no primary association with the parathymic lymph node when it was injected *i.v.*, and HIS 17 either *i.v.* or *i.p.* behaved as HIS 19. A human melanoma antigen that does not cross-react with rat antigens (MEL) was not detectable in the thymus. EM studies to detect *i.v.* injected HIS 17 showed that 1 hour after injection it was bound to the cell surface of subcapsular thymocytes, but thymocytes surrounding cortical capillaries did not show HIS 17 binding.

These experiments showed that antibodies specific for antigenic determinants present in that particular area, entered the thymus cortex via the capsule and were absorbed out, binding until saturation of these determinants. This allowed them to be visualised as the levels rose about those normally present in the interstitial fluid. Thus it can be envisaged that self-proteins carried in the circulation or present outside the thymus can penetrate the microenvironment where they could be instrumental in influencing locally maturing T cells at the time of T cell receptor expression resulting in self-tolerance induction either by clonal deletion or by the induction of self-antigen specific suppressor cells.

The early work of Hammar indicated that the thymus has both a parasympathetic and sympathetic innervation, and it has now been shown from many laboratories that the nervous system can influence the immune system. The noradrenergic innervation originating from the superior cervical ganglion is a likely efferent system. Lymphocytes bear adrenergic receptors and the effects of adrenergic compounds reveal inhibitory influences on cell proliferation or antibody secretion. It is possible that the cholinergic innervation is associated with increased lymphopoiesis.

At this workshop, three papers were given on aspects of innervation of the thymus. **Ali Al-Shawaf** (London, UK) presented work on protein gene product 9.5 (PGP 9.5), RT97, sucrose- phosphate-glycoxilic acid (SPG) and vasoactive intestinal peptide (VIP) staining of neural elements in the rat thymus. He found that PGP, RT97 and SPG stained more or less the same structures, i.e. the catacholaminergic element of thymic innervation that Felten had earlier described in the thymic cortex. VIP, which is supposed to be present in some acetylcholinergic neurones, did not co-localise with acetylcholinesterase (AChE) positive nerves in the medulla or corticomedullary junction (where previous work had suggested an AChE network of fibres) but it did show numerous very fine nerves of distinctinve distribution in the cortex. In addition a few groups of small VIP positive cells were observed that could be part of the previously described 2 populations of AChE positive cells.

Joop van Baarlen (Utrecht, The Netherlands) used an acetylcholinesterase method to show that AChE positive nerves can be observed in the human thymus, and that the distribution in children is comparable to that of the mouse, rat and rabbit thymus. He also observed some AChE positive cells in the medulla. The use of MOC-1 demonstrated nerve fibres in the medulla, sometimes in close contact with Hassall's corpuscles, and sporadic positive cells in the medulla.

Vincent Geenen (Liege, Belgium) presented data to support the view that neuroendocrine systems may affect the pathways of T cell differentiation as he found the epithelial cells of the subcapsule and medulla react with monoclonal antibodies against oxytocin (OT), neurophysin (NP) and to a lesser extent vasopressin (VP). Thymic nurse cells (but not the enclosed thymocytes) react strongly with A2B5 (a ganglioside of nervous and neuroendocrine cells), anti-OT, VP, NP and neurone-specific enolase antisera. High levels of mRNA for OT and VP were detected in human thymus extracts thus confirming the intrathymic synthesis of these peptides and the immunoreactivity patterns were identical in serial sections suggesting OT and VP gene expression by the same cells.

Thomas Nebe (Orpegen, West Germany) introduced his talk by pointing out that since the islolation and purification of thymic peptides in the 1970s very few were developed for clinical use and 2 were marketed as a drug: Thymopentin (TP-5, Timunox) a pentapeptide of Gideon Goldstein's laboratory and TP-1, a partially purified thymic extract. He reviewed the literature and discussed the results of his repetition and extension of *in vitro* assays in mice.

One of the problems in the evaluation of the action of thymic hormones is their measurement. Thymulin, thymopoietin and thymosin α1 are all hormones with well-defined peptides yet not all of the available immunoassays are well established and bioassays are still necessary.

Bared Safieh (London, UK) finished the first session with the news that he has recently developed a sensitive and specific radio-immunoassay (RIA) for the measurement of plasma thymulin in man. The data presented at the meeting were preliminary, being based on the values obtained from normal healthy children and adults. Young children had the highest levels of thymulin, yet after an age-related decline, thymulin was still measurable in the oldest patients (in the sixth decade). He also had preliminary data indicating lower thymulin levels in the first trimester of pregnancy. The results illustrated the sensitivity of the assay (important as most immunodeficient conditions would be expected to decrease serum thymulin levels), and we can now look forward to a better understanding of changes in thymulin with age and disease.

The first session ended late into the evening, but was an exciting introduction to the thymus workshop.

SESSION II

Thymocyte Subpopulations: from Bone Marrow Precursor to Thymic Emigrant Chairmen: ***Bruno Kyewski*** *(Heidelberg, Germany) and* ***Martine Papiernik*** *(Paris, France)*

The two first papers of this session were devoted to the bone marrow T cell precursors and their migration properties. T cell precursors in the bone marrow have up to now no characteristic antigenic profile, and the only way to identify a potential pre-T cell subset in the bone marrow is to use their thymus-homing capacity. **Sophie Ezine** (Paris, France) has used density gradient centrifugation to separate a low density fraction from the bone marrow that is highly enriched for pre-T cell activity. This fraction was intrathymically (IT) transferred into Thy-1 congenic irradiated hosts, and the kinetics of reconstitution were compared to those in whole bone marrow reconstituted hosts. The chimeric thymuses obtained after IT transfer of the pre-T cell enriched fraction were more rapidly reconstituted, for a longer time, and contained a higher percentage of donor type cells, but the absolute number of cells per thymus was not different in both groups. This may indicate that the input of stem cells in the thymus from the bone marrow is under the control of intra-thymic precursors: a high number of donor type precursors after IT injection of the precursor enriched fraction blocks the input of host type precursors.

The second point of this paper was the characterisation of cells in these IT constituted thymuses, which may for a long time retain their capacity to form CFU-S (colony forming unit in the spleen) when transferred IT into a second host.

The mechanism by which bone marrow precursors migrate to the thymus was addressed by the next paper. J.P. Thierry's group proposed that epithelial cells of the thymus are producing a chemotactic factor for hemopoietic precursors. **Marie-Auge Deugnier** (Paris, France) described the characteristics of this thymic peptide of 11 KD that was isolated from an epithelial cell line *in vitro*. This peptide called thymotaxin was tested in an *in vitro* assay where fresh bone marrow cells were selected for their capacity to be attracted by the peptide. Only 0.1% of the cells respond to thymotaxin and 70% of them were Thy-1^+, and all were $CD4^-$, $CD8^-$. They may acquire these two markers by co-culture with embryonic thymic tissue. On the contrary, they do not grow with IL-3, suggesting that BM cells migrating toward thymotaxin are specific pre-T cells, and not myelo-monocytic precursors.

The next papers were devoted to intra-thymic precursors, and tried, using new, or already defined markers, to analyse the different sub-populations of these precursors.

Philippe Naquet from the group of M. Pierres in Marseille, France, has defined a new molecule called THAM (thymocyte activating molecule) which is a pan-lymphoid marker of 128,000- 115,000 M.W. It is present on 15% BM cells, T and B cells, 100% thymocytes, on some non lymphoid tissues (liver, gut, kidney and lung) and absent from macrophages, dendritic cells, and thymic epithelial cells. THAM disappeares from activated T and B cells. Expression of THAM is particularly interesting in thymocyte precursors, as 70% of $CD4^-$ $CD8^-$, $J11D^+$ and day 14-16 foetal thymocytes express a high level of THAM. High-$THAM^+$ cells are localized in the sub-capsular area at birth. Cross linked anti- THAM antibody can induce thymocyte proliferation in the presence of submitogenic doses of PMA, and may be defined as a thymocyte activating molecule.

Rawleigh C. Howe (Epalignes, Switzerland) attempted to identify functionally distinct subsets within $CD4^-$ and $CD8^-$ thymocytes, using mAb specific for antigens present on activated lymphocytes. Four discrete subsets differing in their *in situ* proliferation status and cell size have been identified. When these subsets were analysed for their *in vitro* proliferation and lymphokine production, distinct differences were noted. Two of these populations appeared early in ontogeny, were actively cycling and non responsive to T cell mitogens (Con A or anti-Thy-1). In contrast, the two populations which appeared later in ontogeny were responsive to Con A and anti-Thy-1, and express CD3, some in almost exclusive association with γ/δ T cell receptor molecule, and the other (B2A2$^-$) primarily with the α/β T cell receptor. Finally, all the populations responded to IL1 and specific high affinity receptors were detected on unfractionated CD4$^-$ CD8$^-$ thymocytes, consistent with a role for IL1 in thymic ontogeny.

The same type of approach was used by **Nick Crispe** (London, UK) using Thy-1, J11D, IL-2R and Pgp-1 as markers to identify different sub-populations within CD4$^-$ CD8$^-$ prothymocytes. The J11D marker seems to make a fundamental distinctions as it separates cells with CD3-associated γ/δ receptor (J11D$^+$) from those with CD3-associated α/β receptor (J11D$^-$), and precursor cells from cells without precursor activity. Other workers have suggested that precursor thymocytes express Pgp-1, an apparent paradox since most J11D$^+$ cells are Pgp-1$^-$. In the early foetal thymus, however, all cells are J11D$^+$, Pgp-1$^+$. Cells with this phenotype are also present in the adult CD4$^-$ CD8$^-$ thymic subpopulation, and may be the earliest cells in the adult thymus as well.

New sets of antibodies defining rat antigens were described by **F. Koch** from the laboratory of **Gunther Thiele** (Hamburg, West Germany) and **Jaap Kampinga** (Groningen, The Netherlands) analysing the expression of RT6 and Thy-1 on thymic and peripheral T cells. Rat thymocytes are Thy-1$^+$ RT6$^-$, as well as the recent thymic migrants in the periphery. The ratio of Thy-1$^+$/RT6$^+$ cells slowly decreases during postnatal ontogeny. Adult peripheral T cells are Thy-1$^-$RT6$^+$. RT6 is a useful marker, able to distinguish recent thymic migrants from mature long living peripheral T cells.

Jaap Kampinga also described two potentially interesting antibodies in rat: HIS 44 and HIS 45. Antigen recognised by HIS 44 is present on 85% of thymocytes but not on peripheral T cells, and may be the equivalent of CD1. HIS 44 also recognises determinants on non lymphoid organs. HIS 45 recognises a 85-90 Kd molecule, and labels 12% of thymocytes in the medulla, all peripheral T cells and a low percentage of non T cells in the spleen. This antibody may be the equivalent of CD6.

Enzymes may also be interesting tools. **Jean Plum** (Ghent, Belgium) is studying lactate dehydrogenase (LDH) which is a key enzyme of carbohydrate metabolism, as a marker for T cell differentiation. Five isoenzymes are described which may be differently expressed depending on cell maturity. LDH isoenzyme always evolves in the same way during development from LDH5 to LDH1, from an anaerobic to an aerobic pattern. Prothymocytes have the highest activity of LDH-5. **Claude Penit** (Paris, France) reported studies on thymic population dynamics and developmental interrelationships between thymocyte subsets. Using a combination of bromodeoxyuridine to label cycling cells and anti- mitotic drugs to specifically eliminate cycling cells *in situ* he derived the following results. Double negative (CD4$^-$/CD8$^-$) cells enter the thymus as non-cycling cells and resume proliferation within the thymus. They give rise to double-positive cells via $CD8^+$ intermediates. These events occur during migration from the inner cortex to the subcapsular region. When entering the pool of double-positive thymocytes they cease to proliferate and either die *in situ* or are selected to differentiate into single- positive mature T cells. The transition from double-positive to single-positive cells seems to be independent of cell division and occurs after or during migration from the cortex to the medulla. The studies support a single lineage module of T cell differentation characterized by a strict correlation of cell division, migration and acquisition of differentation antigens.

Gosia Murawska (Groningen, The Netherlands) addressed the question whether intrathymic T cells originate from an intrathymic stem cell pool (seeded once during embryonic life) or from the continuous influx of bone marrow-derived pre-T cells with limited self-renewal capacity. Thymus repopulation was studied after depletion of cycling cells with adriamycin in combination with thymus grafting. The results indicate that thymus repopulation is strictly dependent on the recruitment of bone marrow-derived pre-T cells whereas intrathymic progenitor cells have only a limited self-renewal capacity.

Benedetta Rocha (Paris, France) studied the population dynamics of post thymic T cells in peripheral lymphoid organs by transferring mature T cells into Thy-1 congenic nude mice. Transferred T cells are able to expand by a factor of 40,000 corresponding to a minimum of 15 productive cell divisions *in vivo*. The initial $CD4^+/CD8^+$ ratio of inoculated T cells is maintained during this expansion process indicating a similar proliferation capacity among both subsets and possibly a peripheral mechanism controlling homeostasis among both subsets. Surprisingly, the post thymic T cell expansion seems to be completely independent of IL-2 as indicated by (1) the absence of IL-2 receptors on these cycling T cells, (2) the lack of IL-2 production by these T cells *in vitro*, (3) the absence of m-RNA for IL-2 and (4) the lack of interference of T cell expansion by anti-IL-2 receptor antibodies *in vivo*. A similar lack of IL-2 dependence pertains to T cell division in draining lymph nodes in vivo in response to administration of specific antigens. These *in vivo* data contrast with the known role of IL-2 in T cell activation *in vitro* and thus caution us against an extrapolation of *in vitro* results to the *in vivo* environment.

The same issue was studied in nude rats by **Eric Bell** (Manchester, UK) using thoracic duct-derived cells as a donor population. Again a substantial increase (10-15 fold) of the transferred donor population was observed. In contrast to the previous study, however, only $CD4^+$ T cells comprised the donor cell pool. The reconstituted nude rats performed normal T cell functions and survived for more than 2 years. The T cell immunity of the nude rats was completely donor cell-derived without evidence for a "bystander" maturation of host T cells. Thus post-thymic T cells can restore T cell competence.

Richard Aspinall (Cambridge, UK) described a new rat antibody, HIS 42, which stains a minor fraction of peripheral T cells amongst both $CD4^+$ and $CD8^+$ subsets. The immobilized antibody induced T cell proliferation in the presence of Il-2. The antibody stains a minority of thymocytes (4%) in the cytoplasm as well as membrane determinants on 1-2% of mature thymocytes. Thus, the antibody is possibly directed against a T cell receptor determinant present on different T cell clones.

Robert Dawson (London, UK) presented studies on Thy-1.1 transgenic mice. Thy-1 transgenic mice were constructed in order to study the molecular control of the developmental and tissue- specific expression and function of this molecule. By exploiting the different tissue expression of Thy-1 in mice and man, two types of constructs were made and injected into Thy-1.2 mice; (1) mouse-derived Thy-1.1 coding and 5' and 3' flanking regions with human 3' segments and (2) mouse derived 5' and 3' flanking regions with human coding regions. The first construct showed a typical "human" tissue expression of the transgene with Thy-1.1 expression in brain and kidney but not on peripheral T cells. The kidney expression led to proteinuria probably due to destruction of podocytes which express the transgene. The second construct surprisingly induced tumours in proximal kidney epithelial cells again expressing the transgene. Many tumour lines could be derived from these transformed kidney epithelial cells. These renal tumour-derived cell lines may prove to be valuable tools in establishing thymic/renal epithelial hybridomas. The different constructs will allow a dissection of the coding and flanking regions responsible for the species-specific differences in Thy-1 expression and a possible role of aberrant Thy-1 expression in tumourgenesis.

SESSION III

Thymus Stromal Cells: Subset Identification, Isolation and Functional Roles. Chairmen: ***Mary A. Ritter*** *(London, UK) and* ***Willem van Ewijk*** *(Rotterdam, The Netherlands).*

The thymic stroma plays an essential role in the differentiation of T lymphocytes. It is now generally accepted that MHC restriction and tolerance are induced by stromal elements within the thymus. As can already be observed at the morphological level, the thymic stroma forms a complicated network (Fig. 2), and close interrelationships between the thymic stroma and developing lymphocytes demonstrate the importance of cell-cell interaction during T cell development (Fig. 3).

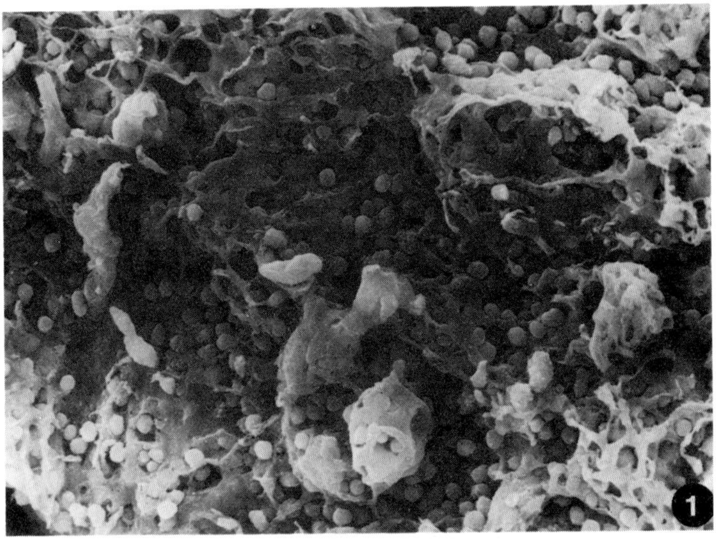

Figure 2 Epithelial-reticular cells and thymocytes in the cortex of the mouse thymus (x650).

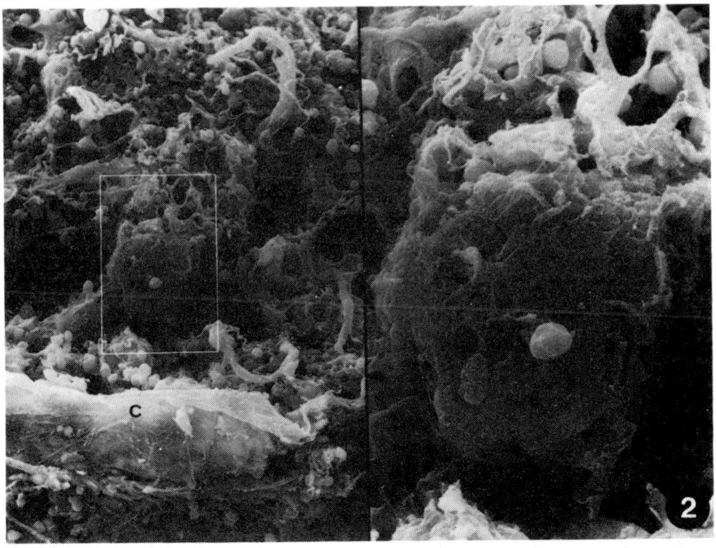

Figure 3 Mouse thymus, subcapsular cortex. Thymocytes associate with thymic epithelial cells in structures which resemble thymic nurse cells. c=thymic capsule. (left panel: x310; right panel: x1050).

A first step in the unravelling of the precise role of the various stromal cell types is the exact identification of thymic stromal cell subsets. This was a major item at the stromal cell session of the Rolduc meeting. Today, more than eight laboratories have developed monoclonal antibodies directed to various stromal elements in the thymic network (see also Table 1).

Richard Boyd (Melbourne, Australia) presented extensive panels of monoclonal antibodies to stromal cells of the mouse, rat, and chicken thymus, recognising at least 11 distinct microenvironments. He furthermore showed that the stromal elements develop independently during embryogenesis. Moreover, pregnancy disturbs the stromal architecture of the mother, indicating a hormonal influence on thymic microenvironments. Surprisingly, antibodies to lymphoid markers (Thy-1, L3T4) do also bind to stromal cells. This phenomenon was particularly manifest *in vitro,* and may represent cellular communication networks *in vivo.*

Jaap Kampinga (University of Groningen, The Netherlands) has developed a set of monoclonal antibodies reactive with stromal cells of the rat thymus. Also in this species, stromal cells detected by "HIS" monoclonal antibodies follow a strict developmental pattern, indicating an ordered sequence in the development of thymic microenvironments. In the mouse, however, it has been shown previously that stromal cells in the two major compartments, i.e. cortex and medulla, are already detectable by day 14 (van Vliet *et al., Eur. J. Immunol.* **15**, 675, 1985).

In the same context, **Miodrag Colic** (Belgrade, Yugoslavia) presented a panel of monoclonal antibodies directed to stromal subsets in the rat thymus.

From the above presentations and with the help of **Pieter Breckelmans** (Rotterdam, The Netherlands) we have selected the most relevant monoclonal antibodies and we present these together with other published antibodies in Table 1. Obviously the information in Table 1 is not complete and a major conclusion from this Rolduc session was a clear need to exchange monoclonal antibodies and to compare their reactivity in different systems.

Another important topic of the stromal cell session considered the isolation of stromal cells from the thymus. **D. Toussaint- Demylle** (Brussels, Belgium) examined the various isolation steps of "thymic nurse cells" (TNC) with the electron microscope. A major conclusion from her work was that TNC develop during the preparation steps, indicating the pleiomorphic character of the thymic stroma. Most TNC however are of cortical origin, but some small TNC are derived from the medulla. It may be noted here that we have found morphological evidence for the presence of intact TNC-like structures in the subcapsular region of the thymus (van Ewijk, Lab. Invest., 1988, in press; see also Fig. 3).

TABLE I. Binding patterns of anti-thymic stromal cell antibodies

		Species	Thymic epithelium		
Code nr.	Source	(target)	subcapsule	cortex	medulla
ER-TR1	1	mouse	+	+	+
ER-TR2	1	"	+	+	+
ER-TR3	1	"	+	+	+
ER-TR4	1	"	+	+	-
ER-TR5	1	"	-	-	+
ER-TR6	1	"	-	-	+
ER-TR7	1a	"	-	±	±
CDR-1	2	mouse	-	+	-
MD-1	2	"	±	-	+
MD-2	2	"	-	-	+
B6TS-1	3	mouse	+	-	+
AKTS-1	3	"	+	+	-
NLDC-145	4	mouse	?	+	-
RCK102	5*	mouse	?	+	-
RCK103	5	"	-	-	+
4F1E4	6	mouse	-	+	-
IVC4	6	"	+	-	+
HIS46	7	rat	-	+	+
HIS37	7	"	-	+	±
HIS38	7	"	-	+	-
HIS39	7	"	+	±	+
R-MC13,14,15	8	rat	-	+	-
R-MC18,19	8	"	+	-	+
R-MC20	8	"	+	—	±
R-MC22	8	"	-	-	±
R-MC23	8	"	-	-	-
MTS1,2,3,4,5	9	mouse	+	+	+
MTS6,7,8,9	9	"	+	+	+
MTS10,11	9	"	+	-	+
MTS17,19	9	"	+**	-	-
Mui 53	9	chicken	+	-	-
Mui 63	9	"	+	+	-
Mui 68	9	"	-	-	+
Mui 67	9	"	-	-	-
Mui 58	9	"	-	-	-

* mouse anti-human cytokeratins crossreactive with mouse thymus. Ref. 5 shows specificities of these two antibodies.

** small subcapsular clusters.

Code nr.	Source	Species (target)	Non-epithelial cells: fibroblasts	smooth muscle cells	macro-phages	neutrophils	IDC
ER-TR1	1	mouse	-	-	-	-	-
ER-TR2	1	"	-	-	-	-	-
ER-TR3	1	"	-	-	-	-	-
ER-TR4	1	"	-	-	-	-	-
ER-TR5	1	"	-	-	-	-	-
ER-TR6	1	"	-	-	-	-	+
ER-TR7	1a	"	+	-	-	-	-
CDR-1	2	mouse	-	-	-	-	-
MD-1	2	"	-	-	-	-	-
MD-2	2	"	-	-	-	-	-
B6TS-1	3	mouse	-	?	-	-	-
AKTS-1	3	"	-	?	-	-	-
NLDC-145	4	mouse	-	?	-	-	+
RCK102	5	mouse	-	?	-	-	-
RCK103	5	"	-	?	-	-	-
4F1E4	6	mouse	-	?	-	-	-
IVC4	6	"	-	?	-	-	-
HIS46	7	rat	-	+	-	-	-
HIS37	7	"	-	-	-	-	-
HIS38	7	"	-	-	-	-	-
HIS39	7	"	-	+	-	-	-
R-MC13,14,15	8	rat	-	-	-	-	-
R-MC18,19	8	"	-	-	-	-	-
R-MC20	8	"	-	-	+	-	+
R-MC22	8	"	±	-	+	+	-
R-MC23	8	"	+	-	-	-	-
MTS1,2,3,4,5	9	mouse	?	?	?	?	?
MTS6,7,8,9	9	"	?	?	?	?	?
MTS10,11	9	"	?	?	?	?	?
MTS17,19	9	"	?	?	?	?	?
Mui 53	9	chicken	?	?	?	?	?
Mui 63	9	"	?	?	?	?	?
Mui 68	9	"	?	?	?	?	?
Mui 67	9	"	-	+	?	?	?
Mui 58	9	"	+	-	-	-	-

code nr.	Source	Species target	Antigen detected	Ig-class
ER-TR1	1	mouse	Ia kdqr	IgG2c
ER-TR2	1	"	Ia ksr	IgG2b
ER-TR3	1	"	Ia kbdqr	IgG2b
ER-TR4	1	"	cytokeratin?	IgM
ER-TR5	1	"	cytokeratin?	IgM
ER-TR6	1	"	?	IgM
ER-TR7	1a	"	?	IgG2a
CDR-1	2	mouse	analogous to GC3; MW 130-160 kD	IgG2a
MD-1	2	"	?	IgM
MD-2	2	"	?	IgM
B6TS-1	3	mouse	?	?
AKTS-1	3	"	?	?
NLDC-145	4	mouse	?	?
RCK102	5	mouse	cytokeratin 5,8	IgG1
RCK103	5	"	at least cytoker.5	IgG1
4F1E4	6	mouse	?	IgM
IVC4	6	"	?	IgM
HIS46	7	rat	?	IgM
HIS37	7	"	?	IgM
HIS38	7	"	?	IgM
HIS39	7	"	?	IgM
R-MC13,14,15	8	rat	?	IgGl
R-MC18,19	8	"	?	IgM
R-MC20	8	"	?	?
R-MC22	8	"	?	IgGl
R-MC23	8	"	?	IgGl
MTS1,2,3,4,5	9	mouse	?	?
MTS6,7,8,9	9	"	?	?
MTS10,11	9	"	?	?
MTS17,19	9	"	?	?
Mui 53	9	chicken	?	?
Mui 63	9	"	?	?
Mui 68	9	"	?	?
Mui 67	9	"	?	?
Mui 58	9	"	?	?

Source:

1. Van Vliet E., Melis M., van Ewijk, W. *Eur. J. Immunol.*, **14**, 524, 1984.
1a. Van Vliet E., Foidart J.M., van Ewijk W. *J. Histochem. Cytochem,.* **34**, 883, 1986.
2. Rouse R.V., Bolin L.M, Bender J.R, Kyewski B.A. *J. Histochem. Cytochem.*, 1988 in press.
3. Kaneshima H., Ito M., Asai J., Taguchi O., Hiai H. *Lab. Invest.*, **56**, 375, 1987.
4. Kraal G., Breel M., Janse M., Bruin G. *J. Exp. Med.*, **163**, 981, 1986.
5. Ramaekers F., Huysmans A., Schoort G., Moesker O. and Vooys P. *Exp. Cell Res.* **170**, 235, 1987.
6. Ritter M.A., submitted for publication.
7. Kampinga J., Kroese F.G.M, Duijvesteijn A.M., Murawska A.M., Pol G.H., Nieuwenhuis P. *Transpl. Proc.*, **XIX(3)**, 3171, 1987.
8. Colic M., Matanovic D., Hegedis L. and Dujic A. *Immunology*, in press.
9. Boyd R. Dept. Pathology/Immunology, Monash Univesity, Commercial Road, Prahrau, Victoria 3181, Australia.

Ed Kamperdijk (Amsterdam, The Netherlands) has isolated interdigitating cells (IDC) from the thymus. *In situ,* these cells are located at the cortico-medullary junction and in the medulla. IDC are characterized by a strong expression of class II markers and by their voluminous pale cytoplasm, containing a central "spot" with acid phosphatase activity. About 60 per cent of IDC show Birbeck granules. Although IDC do not phagocytize to a high extent, they can develop this capacity *in vitro.*

TABLE II. Mouse and rat thymic stromal cell lines

	Species	Origin	Reference
B6TE	mouse	cortex(?)	
B6TE/A	"	cortex	1
A2T	"	?	2
A17A	"	?	
A89A	"	?	
A26A	"	?	
TEC	"	?	3
TEPI	"	?	4
E5	"	medulla	5
IT76M	"	?	
IT26R21	rat	?	6
IT45R1	rat	?	7

Source
1. Miyazawa T, Sato C, Hiai H, Nishi Y, Matsuyama M. *Cell Struct. Function* **5**, 305, 1980.
2. Yanagihara K, Kajitani T, Kamiya K, Yokoro K. *Leukemia Res.* **5**, 321, 1981.
3. Glimcher L.H., Kruisbeek A.M., Paul W.E., Green I. Scand. *J. Immunol.* **17**, 1, 1983.
4. Beardsley T.R., Piersbacher M, Wetzel G.D., Hayes E. *Proc. Natl. Acad. Sci.*, **80**, 6005, 1983.
5. Potworowski E.F., Turcotte F, Beauchemin C, Hugo P, Zelechowska M.F. *In vitro* **22**, 557, 1986.
6. Itoh T, Kasahara S, Aizu S, Kato K, Takenchi M, Mori T. *Cell Tissue Res.*, **226**, 469, 1982.
7. Itoh T, Aizu S, Kasahara S, Mori T. *Biomedical Res.*, **2**, 11, 1981.

A major task in thymology is to try to culture the thymic stroma in order to investigate the role of the various thymic microenvironments *in vitro.* **Myra Small** (Tel-Aviv, Israel) cultured lobes of the thymus after giving mice repeated injections of hydrocortisone. She obtained considerable growth of epithelial cells from control mice (over 16 months in culture!) both of cortical and medullary origin. However, a difference in growth patterns was observed between cultured stromal cells from male and female mice. Most "male" cultures developed a normal growth pattern (cortical and medullary stromal cells), whereas the "female" cultures only developed medullary stromal cells. This "sex difference" between stromal cell cultures of male and female is interesting, but further work is needed to clarify the meaning of this phenomenon.

There is an obvious need to clone out the various stromal elements from the thymus. So far, for the mouse this appears to be a difficult task. Only a few stromal cell lines (Table II) are now available and a major problem is their identification after prolonged culture. Only after isolation of well defined cloned stromal cell populations, can the precise role of the various thymic microenivonments in T cell differentiation be unravelled.

The hetereogeneity of the thymus microenvironment, as discussed above, implies that each microenvironmental niche will be responsible for the maturation and migration of discrete phases of T lymphocyte differentiation. The first stage in thymopoiesis must be the attraction of lymphoid prescursor cells from the blood stream into the thymus. **Beat Imhof** (Paris, France) used the rat thymic epithelial cell line IT45 to isolate and characterise an 11kD molecule - "thymotaxin" - that is chemotactic for the precursors of T lymphocytes. A mouse monoclonal antibody, PA-1, was raised to thymotaxin and used in immunohistochemistry of thymus sections to show that the microenvironmental cells which *in vivo* produce this chemoattractant are the epithelium of the subcapsular and perivascular regions. These data therefore fit well with the known routes of precursor cell entry.

Soon after this, the early thymocytes associate preferentially with a specialised subset of thymic macrophages, shown by *in vitro* cluster formation. These stromal cells, distingushed by immunolabelling with the monoclonal antibody ED2, are located only in the thymic cortex and have a relatively slow turnover rate (**Gosia Murawska**, Groningen, The Netherlands). The lymphocytes that form clusters with the $ED2^+$ macrophages are of the double negative and immature $CD8^+$ phenotype. Other macrophages in the thymic cortex may be involved in phagocytosing the debris resulting from intrathymic deletion of self reactive cells (**Nicholas Dourov**, Bruxelles, Belgium).

A variety of antigen-presenting cells (APC) are thought to reside within the thymus and to play a critical role in positive and negative selection for the development of MHC restriction and self tolerance respectively. Cultured thymic epithelial cells respond to γ interferon by upregulating surface expression of MHC class II molecules - a response typical of APC. The supernatant medium from cultures of these stromal cells has been shown to contain a thymocyte-stimulating activity which was mediated by an IL-1-like molecule which by molecular weight analysis and antigenic charcteristics was indistinguishable from typical monocyte/APC IL-1 (**Sylvia Cohen-Kaminsky**, Paris, France). Bacterial agents such as LPS enhanced the production of this IL-1-like molecule. It would therefore be interesting to identify the intrathymic stimulus for IL-1 production and to determine whether it was regulated by an interaction between thymic epithelium and developing lymphocytes. However, many other cell types within the thymus appear to have the characteristics of APC, and **Nayef Aqel** (London, UK) demonstated that there were probably even more than previously suspected since in the medulla $CD1^+$ Langerhans cells could be distinguished from other interdigitating cells and medullary macrophages.

Thus in conclusion, thymologists have now developed a wide range of monoclonal antibodies that can be used as tools with which to identify, separate and ultimately clone the various cellular components of the thymic microenvironment. The next stage is to elucidate the function of these and other microenvironmental molecules in the complex series of maturation, selection and migration events that leads to the emergence of a fully functional T cell repertoire.

SESSION IV

Artificial and natural involution of the thymus: models for physiology?
Chairmen: ***Jan Rozing*** *(Rijswijk, The Netherlands) and* ***Willem Seinen*** *(Utrecht, The Netherlands)*

In this session data were presented which were obtained in a variety of experimental systems. All of the models presented, however, had a common theme: the influence of certain stressors on the physiology of the thymus. These stressors included industrial chemicals, hormones, cyclopsorin A and "aging".

Whereas most of the substances induced atrophy of the thymus, exciting data were also presented on treatments which seem to induce a regeneration of an already involuted thymus.

Nick J. Snoeij (Utrecht, The Netherlands) showed that the organotin compound dibutyltin (DBTC) causes a rapid reduction in numbers of thymic lymphoblasts when given as a single oral dose to rats. Within one day the number and proliferative activity of the lymphoblasts are reduced. As a consequence the number of small thymocytes decreased at days 3 and 4 after a single dose of DBTC. After nine days proliferative activity and numbers of cells in the various fractions are again at a normal level.

André H. Penninks (Utrecht, The Netherlands) presented data on the immune histopathology of organotin induced thymus atrophy using monoclonal antibodies against thymocytes at various maturation stages as well as cells of the stromal compartment. He showed that organotins mainly effect the cortical area of the thymus, due to a reduction in the number of cortical thymocytes. Furthermore a clear relation exists between the reduction of the absolute thymus weight and the reduction of the cortical area after treatment. No obvious changes were observed with the monoclonal antibodies against macrophages (ED_1, ED_2) and epithelial cells (HIS 19, HIS 37, HIS 38, HIS 39), except that due to the reduction of cortical thymocytes a relative increase of cortical macrophages as well as epithelial cells was observed. As far as effects of organotin treatment on thymocytes were concerned, a clear time dependent increase of OX 44^+ cells was observed, with maximum levels at day 4 and returning to normal levels at day 9. Moreover a reduction of cortical thymocytes was observed starting with thymocytes with early differentiation markers (CD2) followed by cells with more mature markers (CD4, CD8, CD5). These data indicate that DBTC might disturb the onset of an early blast transformation of OX 44^+ cells. Since the small cortical thymocytes will follow their normal maturation and differentiation process in which most of them are supposed to die, thymus atrophy will occur.

A completely different effect seems to underlie the thymus atrophy as caused by dioxin (TCDD). **Johan Garssen** (Bilthoven, The Netherlands) presented elegant experiments indicating that TCDD binds primarily through specific receptors in a variety of cell types. In particular, thymus epithelial cells express large numbers of these cytoplasmic TCDD receptors. It was hypothesized that TCDD indirectly through interaction with the thymic epithelial cells influences T cell in the thymus. Indeed, the proliferation of thymocytes co-cultured on TCDD-treated thymus epithelial monolayers was reduced.

The thymic effects as found in these toxicology studies differed largely from those observed in the thymus of children in acute stress situations. **Joop van Baarlen** (Utrecht, The Netherlands) presented data on a large series of such pediatric thymuses at varying degrees of involution. From these studies it became clear that there exists a strong correlation between the grade of involution of the thymus and the duration of the illness. Grading of accidental involution of such a thymus can therefore be of use in estimating the duration of illness when unknown. Furthermore he examined the expression of various stromal markers as well as of thymosin-α1 and thymosin-β4 in these thymuses. He could show that with advanced accidental involution the markers MR10 and MR19 appear in the thymus cortex without significant changes in the expression of MR3 and HLA-DR. There is however, with advanced accidental involution a loss of HLA-DR expression in the medulla of the thymus. Also a loss of expression of thymosin-α1 and thymosin-β4 is observed. predominantly in the subcapsular area of the thymus. Thus, due to the acute stress situation major changes in stromal cell populations occur in the thymus, leading to severe thymus atrophy.

Cyclosporin A (CSA) treatment also causes significant changes to occur in the thymus. **Heather Ladyman** (London, UK) analysed the changes that occur within both the thymic microenvironmental as well as the thymocyte subpopulations in mice treated with CSA using monoclonal antibodies to epithelial cell subsets, lymphocyte subpopulations, macrophages and MHC class II antigens. The results showed that CSA is selective, i.e. gross changes were observed only in the medulla, whereas the cortex was left unimpaired. This reduction in the medulla involved all cell populations which are normally present in the medulla: $CD4^+$ and $CD8^+$ lymphocytes, epithelial and dendritic cells. Only macrophages were relatively increased in number, located around blood vessels and in the septa. These alterations increased with the duration of CSA therapy, but seemed to level off after approximately 5 weeks. At that stage only a small medullary remnant remained. Total ablation of the medulla was not observed. When CSA treatment was stopped the architecture of the medulla returned to normal within 2 weeks.

Nick A. Willcox (London, UK) investigated the corticosteroid sensitivity of human thymus in pre-treated myasthenia gravis patients. As a result of the corticosteroid treatment the total cell yield of the thymus is greatly reduced (10 times or more) in the majority of patients. However, some patients showed only partial or no depletion at all. This phenomenon is probably due to differences in steroid resistance between individuals. The same conclusions can be drawn from the analysis of CD1 positive and CD1 negative thymic cells. When tested for function in various assays, such as the spontaneous production of anti-acetylcholine receptor (AChR) antibodies, spontaneous and pokeweed mitogen induced IgG production and AChR induced T cell stimulation, all of these responses in these patients are close to normal when assessed on a per cell basis. They only show a reduction when the total cell yield is taken into account. Also the results of **Naglaa Abou-Rabia** (London, UK) illustrated the strong interactions between the endocrine system and the thymus. In these experiments rats were treated with antithyroid drugs. This resulted in a state of hypothyroidism as shown by significantly reduced T4 levels in the serum. In these animals a vast decrease in thymus weight was observed throughout the experiment. When studied morphologically these thymuses showed depopulation of the cortex and an apparent increase in epithelial cells, mast cells and connective tissue at 4 weeks. At 6 and 7 weeks there were more significant changes resulting in a marked loss of architecture and decrease of mitosis. Hassall's corpuscles were frequently observed in the medullary regions with many cystic vacuoles in the epithelial cells. An interesting feature at this stage is the appearance of the blood vessels with the endothelial cells bulging into the lumen due to the presence of high numbers of lymphocytes in the walls. This suggests an active migration of such lymphocytes, thereby possibly adding to the thymus involution caused by cell death. Also plasma cells were frequently observed within the thymic parenchyma in the experimental groups at 6-7 weeks of treatment. The origin of these cells awaits further clarification. Indications for hormonal influences on the thymus can also be derived from studies during pregnancy as presented by **Ann G. Clarke** (Cambridge, UK). Dramatic changes occur in the thymus of all pregnant mammals. In the mouse, thymus weight and cell number increase at the time of implantation before falling by 75% in late pregnancy. Both responses occur in pregnancy and pseudopregnancy and are greater in allogeneic than syngeneic pregnancies. The subpopulations of thymocytes carrying the CD4 and CD8 markers and the size of the cells are also altered after mating. The percentage of $CD4^+$ and $CD8^+$ cells both decrease by day 3 of pregnancy. Two colour analysis showed that $CD4^+CD8^-$, $CD4^-CD8^+$ and $CD4^-CD8^-$ cells all increase in early pregnancy. Levels of double negative cells remained high over the entire period. There were no changes in the double positive subset. These results suggest that the most immature and mature cells of the thymus are involved in the thymic changes during pregnancy. The significance of these changes remains to be resolved. it seems likely however that such radical alterations in the thymus will influence the changes known to occur in the mother's peripheral immune system during pregnancy. Further evidence for a direct interation between the thymus and the endocrine system came from studies presented by **Fenella Fitzpatrick** (London, UK and Paris, France). In these experiments old male rats were treated with a stable analogue of LHRH, goserelin. This treatment results in chemical castration after 28 days as shown by a dramatic reduction in accessory sex organ weight, their non-secretory histological appearance and the virtual absence of serum testosterone in these rats. Strikingly, the thymus weight of treated animals was significantly higher than of control animals. Histologically, these thymuses resembled those of young adult rats, the tissue being packed with thymocytes and clearly divided into medullary and cortical areas. These findings were confirmed in immunohistochemical analysis with anti-T cell and anti-MHC class II reagents. Also the white blood cell count was significantly higher in treated rats as compared with control animals. The phenotype of such cells however still needs to be determined. Nevertheless, these studies indicate that in contrast to most (experimental) manipulations the treatment with this LHRH analogue leading to chemical castration seems to result in a recovery of an already involuted thymus. This opens new perspectives in the field of spontaneous or induced thymus involution, although the type of treatment at this moment is rather drastic.

One of the most outstanding areas of thymus involution is obviously the age-related phenomenon. **Nicholas Dourov** (Brussels, Belgium) presented data on a large series of thymic remnants in aged people. From these studies it can be concluded that the systematical investigation of these thymic remnants at various ages will be very useful to describe and investigate the role and function of the large thymic lobules, with a well preserved cortico-medullary delimitation persisting in many patients up to high ages. Also the increased presence of plasma cells in these remnants as well as their function and signficance are of interest. Finally, thymic structures in aged people offer a good model to study the importance of epithelial nests and their trabecular or micronodular configuration. Once more it became clear from these studies that the dogma of a complete absence of the thymus in old people is not as absolute as stated in most text books. Quite large thymuses or thymic structures can be found in adults at autopsy, whereas on the other hand one has to be aware of the fact that involution in some other samples may be the result of a period of stress prior to death. More detailed and defined studies are therefore needed to address the significance of the thymus in adult and old life.

In conclusion, this session brought together people working on a variety of aspects dealing with articifical and natural involution of the thymus. From the presentations, and perhaps even more, the lively discussions afterwards it became obvious that such an integral approach will lead to a better understanding by non-thymologists, such as toxicologists or endocrinologists, for the target organ they are studying, but perhaps even more important it opened a new area for true thymologists with exciting models of experimental and "clinical" manipulation, which will definitely add significantly to the understanding of the very complex role and function of the thymus under normal physiological conditions.

SESSION V

The Thymus in Pathology: Lessons for Histophysiology. Chairmen: ***Mireille Dardenne*** *(Paris, France) and* ***Henk-Jan Schuurman*** *(Utrecht, The Netherlands).*

The study of the thymus in pathological conditions not only emphasises our knowledge of thymus histophysiology, but may also allow a new look at structural/functional relationships. The first stages require a detailed (immuno)histologic characterisation of thymic subcomponents. A number of monoclonal antibodies are available to identify stromal cell types and lymphocyte subsets. Based on their definition in the normal state (viz., ontogeny) the thymic microenvironment in a pathological situation can be described which in turn affects the understanding of subcomponents in physiology. **Mireille Dardenne** (Paris, France), in her introduction pointed to a major drawback of this approach - the absence of firmly based functions for antigens recognised by monoclonal antibodies to distinct stromal components.

The reports presented in this session can be roughly divided into three main topics: neoplasia, autoimmunity and immunodeficiency/infection.

Neoplasia

Neoplastic conditions can occur in both the lymphocytic compartment and in the stromal component. Considering the former, the group of **Jacques Boniver** (Chantal Humblet, Anne Rongy) from Liege, Belgium, has developed an interesting model of thymic lymphoma development that occurs several months after fractionated irradiation of C57Bl/Ka mice. During the latency period "preleukemic cells" are detectable in the thymus. Their presence is associated with changes in phenotypicaly defined thymocyte subsets. Suspensions containing "preleukemic" cells can yield donor type lymphomas after intrathymic injection in congenic irradiated recipients. The presence of preleukemic cells in the thymus is associated with changes in phenotypically defined thymocyte subsets (especially a reduction in cortical-style thymocytes). But, also other changes occur in mice during the latency period. There is impaired prothymocyte activity in the bone marrow, and in the thymus there was a decreased number of thymic nurse cells as well as a decreased capacity of epithelial cells to interact *in vitro* with foetal thymocytes.

These alterations culminating in lymphoma development, can be prevented by grafting the animals with bone marrow cells just after the last irradiation. This procedure restores the thymocyte subpopulations and the microenviroment (nurse cells) and inhibits lymphoma development. A bone marrow graft given later on after the last irradiation has less or no effect. These studies lead to the conclusion that alterations in the thymic microenviornment due to the fractionated irradiation play a role in lymphoma development, i.e. the progression of "preleukaemic cells" into neoplastic cells. This elegant prothymocyte-microenvironment interaction in lymphopoiesis and lymphoma development has relevance for other disease conditions, e.g. bone marrow transplantation in man, where after grafting, newly generated T lymphocytes have to pass through the (radiation-damaged) thymic microenvironment.

The various steps in intrathymic T-cell differentiation are defined by the immunophentoype and, more recently molecular biology characteristics. **Jacques van Dongen** (Rotterdam, The Netherlands) gave an excellent review of the presence of the cytoplasmic CD3 molecule and T-cell receptor γ and δ chains during normal T-cell differentiation and in T cell leukemias. It was stressed that the sole analysis of rearranged γ and δ chain genes does not give useful information on the maturation stage or subset lineage of the T lymphocyte, but that such analyses should be accompained by the assessment of gene function, i.e. expression of the protein receptor chains (see also report from the final session). An interesting observation for the assessment of the cytoplasmic CD3 molecule was more intense staining when antibodies to the denaturated T3 molecule were used instead of reagents to the intact (transmembrane) molecule. Apparently, the CD3 molecule is present in incomplete form in the cytoplasm of immature T cells.

Concerning tumours of the microenvironment, **Thomas Kirchner** (Wurzburg, Germany) presented a classification of thymic epithelial tumours (Table III). This classification is purely based on the histomorphology of the tumour cells. It is well-known that immunophenotypical characterisation using antibodies to distinct epithelial subtypes does not fit with the histopathological classification. Further analysis of clinico-pathological correlations yielded a clear association between histomorphology and growth behaviour and the risk of development of paraneoplastic myasthenia gravis. The highest prevalence of myasthenia gravis was found in cortical thymoma and well-differentiated thymic carcinoma, being tumours with immature type or with partial loss of differentiation, respectively. In addition, neoplastic epithelial cells can express acetycholine receptors, and their presence correlates with development of

TABLE III Histogenetic concept for the classification of thymic epithelial tumours (**Kirchner** *et al.*, Wurzburg, West Germany)

Tumour type	Characterization
Medullary thymoma	Monophasic medullary differentiation. Medullary type spindle cells and sometimes gland-like structures
Mixed thymoma	Composite cortico-medullary differentition. Areas with medullary type spindle cells and organoid cortical lobules
Predominantly cortical thymoma	Organoid cortical differentiation. Large cortical lobules. Sometimes small medullary islands with Hassall's corpuscles.
Cortical thymoma	Immature cortical differentition. Features of the outer thymic cortex
Thymic carcinoma	Obvious loss of organo-specific differentiation

paraneoplastic myasthenia gravis. From the same group, **Alex Marx** further elaborated on the characteristics of acetylcholine receptors on thymic epithelial cells. Immunoprecipitation of membrane fractions of epithelial cell cultures from thymuses/thymomas of myasthenia gravis patients and from normal thymus yielded a 44 kD single polypeptide chain. This was obtained using an antibody which recognizes the cytoplasmic side of the a-subunit of human acetylcholine receptor, but not with a reagent to the extracellular part of the receptor. It thus seems plausible that thymic epithelial cells in thymomas and myasthenia gravis (probably also in the normal thymus) express a protein similar to the a-subunit of human acetylcholine receptor, without expression of other subunits of the muscular acethylcholine receptor (e.g. those present in medullobastoma cells).

Autoimmunity

Considering the thymus in autoimmune conditions, myasthenia gravis still is one of the most interesting topics of research. **Norbert Sommer** (Oxford, UK) and **Nick Willcox** (Oxford, UK) presented data on acetylcholine receptor antibody synthesis and *in vitro* T-cell responsiveness to this antigen by lymphocytes isolated from thymectomy tissue and peripheral blood of myasthenia gravis patients. Spontaneuous anti-acetylcholine receptor antibody synthesis by thymus cell suspensions correlated with the extent of thymic hyperplasia and serum antibody titres, but not with the frequency of germinal centres in the hyperplastic thymus. In contrast, IgG production correlated with the frequency of germinal centres, the percentage of B cells and extent of hyperplasia, but not with serum antibody titre. Sommer noted the presence of T cells in hyperplastic thymuses and thymomas from myasthenia gravis patients, which responded *in vitro* to acetylcholine receptor stimulation. The vast majority (18 of 25 samples) of myasthenia gravis patients had such cells in their thymic tissue, whereas their presence in blood was found in only 35% of the patients. T cells responding *in vitro* to PPD-tuberculin were found in about half of the thymectomy samples, and their presence correlated with responses by peripheral blood T cells. Thus, hyperplastic thymuses and thymoma in myasthenia gravis comprises lymphocytes of both T and B cell lineage with specificity to the acetylcholine receptor, and these cells are enriched at this location with respect to their occurrence in the recirculating blood lymphocyte pool

It is plausible to relate the presence of these acetylcholine receptor-specific autoreactive lymphocytes to autosensitization inside the gland. Myoid cells have been proposed as a likely candidate involved in antigen presentation (also epithelial cells? see above). Such cells were demonstrated by both Thomas Kirchner and Nick Willcox but at slightly different locations. Willcox showed their presence especially in T-zone areas in hyperplastic thymuses, whereas Kirchner found these especially in the medullary epithelial reticular meshwork, sometimes aggregated around Hassall's corpuscles. The presumed location of autosensitization that results in autoreactive T and B cells inside the thymus (epithelial) microenvironment is an interesting phenomenon. **Bruce Roser** (Cambridge, UK) stressed this point from the view of tolerance induction, by stating that this autosensitization contradicts the major role of the thymus in tolerance induction. In his view, the acetylcholine receptor is the very first antigen to which an immune response is elicitated inside the microenvironment proposed to induce tolerance.

Immunodeficiency/infection

Because of the lack of thymic function, congenitally athymic nude animals are an interesting tool for thymologists, for example they can be used to distinguish thymus-dependent phenomena from thymus-independent ones in T-cell lineage processes. **Jan Rozing** (Rijswijk, The Netherlands) gave his data on "T-cell" competence in nude rats. The occurrence of cells with T-phenotype expression in these animals is well established, especially their increase with ageing. These cells are mitogen-responsive *in vitro* and even produce lymphokines like interleukin 2. More recent work indicated the presence of cells stained by the antibody HIS-42 presumed to detect a T-cell receptor molecule (see report from session II). But, functional antigen-specific immune responses can not be elicited. This indicates the privileged role of the thymus in the generation of the full antigen-recognition repertoire. These data, in combination with other reports, yielded a lively discussion on the definition of "T-cells" occurring in nude animals. When these cells have developed without a functional thymus microenvironmental influence, the definition "Thymus-dependent" lymphocytes is not strictly applicable (see report from session VII).

Two presentations dealt with the thymus and HIV-infection. **Martine Papiernik** (Paris, France) used her skills as a former pathologist in a study on thymus samples from 24 foetuses of HIV-infected women. Foetuses were 13 to 30 weeks of age. Thymic foetopathy was found in two of them, at 20 and 26 weeks. Focal depletion of the $CD1^+$, $CD4^+$ $CD8^+$ lymphocytes was found in the cortical area, while the rest of the cortex was normal. In these zones, the epithelial stroma was deeply modified and infiltrated by $CD4^+$ macrophages. The medulla was nearly normal, considering both the lymphoid and the stromal compartment. As no viral antigen was found by the classical immunofluorescence technique, the direct involvement of HIV in this thymic foetopathy remains to be proven. On the other hand, **Jonne Huber** (Utrecht, The Netherlands) presented data on the histology of the thymus at autopsy of HIV-infected children and adults. He was unable to demonstrate any difference between these hyperinvoluted thymuses and control tissue from other patients (congenital immune deficiency, bone marrow receipients) after long and debilitating disease. In the analysis of HIV components, it was shown that thymus epithelium in normal tissue and in the involuted state, irrespective of HIV-infection, is stained by some antibodies to the p17 *gag*-protein of HIV. Apparently, subtypes of thymus epithelium expresses epitopes of HIV (*gag* protein), similar to the expression of epitopes of HLTV-1.

Related to immunodeficiency is the thymic atrophy observed in osteopetrotic rats. The ultrastructural characteristics of the involuted state were presented by **Nicholas Dourov** (Bruxelles, Belgium). The atrophy is in part secondary to a stress or nutritional deficiency (wasting) in the animals. It is directly related to defects in haemopoiesis. A single intraperitoneal injection of bone marrow cells restores the involution.

Wilson Savino (Paris, France) reported preliminary data obtained with the group of Mireille Dardenne (F. Lepault, M. Leite de Moraes) on thymic abnormalities in acute experimental Chagas' disease. These abnormalities were summarized as follows: the injection of 10^5 *Trypanosoma cruzi* induces, within 7 to 14 days, thymic atrophy with a decrease in the numbers of viable thymocytes, a progressive decrease of Thy-1^+, $CD4^+$ $CD8^+$ cells and a relative increase of $CD3^+$ cells. The epithelial compartment is also affected as evidenced by the decrease in thymic hormone production. Circulating autoantibodies directed against thymic epithelial cells can be detected in the sera of injected animals. In addition, a factor able to kill normal thymocytes *in vitro* appears in serum of these animals. These findings lead to the hypothesis that the thymus can be considered as a target organ in Chagas' disease and that thymic alterations may have an influence at least in some pathophysiologic aspects of this disorder.

In summary, a number of topics in thymus pathology were highlighted during the session. The session was successful in achieving the exchange between pathology and physiology, thereby broadening insights into thymus function under normal and abnormal conditions.

SESSION VI

The Thymic Environment In Vivo and In Vitro. Models for T-Cell Generation and Traffic. Chairmen: ***Eric J. Jenkinson*** *(Birmingham, UK) and* ***Paul Nieuwenhuis*** *(Groningen, The Netherlands).*

Topics discussed in this session of the meeting fell into two broad themes (a) intra-thymic factors regulating the initial stages of T-cell differentiation leading to antigen receptor gene rearrangement and expression and (b) selective interactions within the thymus resulting in shaping of the antigen receptor repertoire.

As an *in vitro* approach to these problems, a number of studies made use of fetal thymus organ cultures which provide an accessible system to study T-cell differentiation *in vitro* (**Eric Jenkinson**, Birmingham, UK). Addition of the lymphokines IL-2 or IL-4 to organ cultures was shown to perturb the normal pattern of development resulting in a reduction of $CD4^+$ $CD8^+$ (double positive) cells and a relative increase in the number of double negative cells **(Magda De**

Smedt, Ghent, Belgium; **Richard Boyd**, Victoria, Australia) whilst IL-1 appeared to have a somewhat similar effect (**Richard Boyd,** Victoria, Australia). The physiological singificance of these results is presently unclear as is the status of the double negative cells generated under these conditions. Further studies are required to determine whether these cells represent an expanded precursor population, or aberrant differentiation or expansion of a normally minor sub- population under the influence of high lymphokine levels.

Attempts to investigate the role of cell surface molecules on stromal cells and T-cell precursors by adding antibodies to organ cultures were also reported. In contrast to the effect seen when antibodies are injected into neonates, addition of antibodies recognising MHC class II molecules on the thymic stroma did not appear to affect T-cell development in organ culture either in terms of cell number or the pattern of differentiation defined by CD4 and CD8 (**Pieter Brekelmans,** Rotterdam, The Netherlands). Similarly, in contrast to previous reports, antibodies to IL-2 receptors did not appear to affect T-cell development in organ culture (**Magda De Smedt, Pieter Brekelmans**). Again, further studies are required using antibodies recognising various epitopes on putative interaction and surface receptor molecules to allow definitive conclusions to be drawn. In this context differences in the mode of action of antibody under different circumstances e.g. receptor blockade versus opsonization of cells expressing a particular receptor need to be borne in mind.

Another strategy for studying stromal cell/stem cell interaction was reported by **Marie Defresne** who showed that nurse cell-like complexes could be reformed by associating isolated epithelial cells with thymocytes in hanging drops. This system may be useful for defining the importance of different stromal cell types in supporting T-cell maturation. Evidence for the importance of stromal cell/stem cell interaction was also presented by **Martine Papiernik** (Paris, France) using isolated populations of double negative T-cell precursor and thymic accessory cell populations. These studies suggest a cascade of events in which intrathymic macrophages mature under the influence of CSF and produce IL-1 which in turn activates dendritic cells enabling the latter to induce IL-2 dependent proliferation within the double negative prothymocyte population. In this context, *in situ* hybridisation studies have indicated that immature double negative cells produce both IL-2 mRNA and IL-4 mRNA in a phase specific pattern during fetal thymic development (**Eric Jenkinson**, Birmingham, UK) indicating that these growth factors are likely to be available within the thymus. In man, an antibody putatively recognising the IL-4 receptor was also described (**Mark Larché**, London, UK) and may also be useful in investigating the growth systems available to mature thymocytes.

Following TCR gene rearrangement and expression, TCR mediated interactions with the thymic environment are thought to influence T-cell development leading to a mature peripheral T-cell population which is self tolerant and MHC restricted, although thymic emigrants may not be fully mature at the time they leave the thymus (**Jaap Kampinga**, Groningen, The Netherlands). Two studies addressed this area using the *in vivo* approach of injecting anti TCR antibodies into neonatal rats or mice. Immunohistological analysis of the thymus of rats undergoing such treatment showed the disappearance of cells expressing the TCR gene product recognised by antibody HIS 42 from the medulla and a reduction in the number of labelled cells in the cortex (**Jaap Kampinga**, Groningen, The Netherlands). Injection of anti-Vβ8 antibody into post-natal mice also prevented the appearance of mature T-cells with receptors using this particular V gene family and produced a parallel inhibition of interaction between Ia^+ cortical epithelial cells and developing thymocytes (**Bruno Kyewski**, Heidelberg, West Germany). One possible interpretation of these findings is that TCR mediated binding to cortical epithelial cells in necessary for further thymocyte maturation (**Bruno Kyewski**, Heidelberg, West Germany). Alternatively, engagement of the TCR by antibody may mimic binding to self antigen resulting in clonal deletion. In this case cells in the cortex as well as in the medulla appear to be susceptible although in normal circumstances deletion may not occur until contact is made with the appropriate stromal cell type. Current evidence suggests that such deletion is more likely to result from interaction with cells of bone marrow origin rather than interaction with thymic epithelial cells.

Evidence on this latter point came from studies using allogeneic grafts of cultured, lymphocyte depleted thymus fragments into nude rats (**Henk-Jan Schuurman**, Utrecht, The Netherlands). The grafts in culture are composed of a unique cell type that after implantation can differentiate to produce all the epithelial types (donor derived) found in the normal thymus and become colonised by lymphoid and other bone marrow derived cells of host origin (**Marion Kendall**, London, UK). Whilst such grafted animals are non-responsive to stimulators of host type they display allo reactivity both to third party and to thymus graft type stimulators in MLC indicating that cells with specificity for thymus graft type MHC have not been deleted (**Henk-Jan Schuurman**, Utrecht, The Netherlands). Despite this, these animals do not reject skin grafts of thymus donor type indicating the existence of some other mechanism preventing rejection *in vivo*. Non rejection of challenge donor type grafts by both mice (**Josselyne Salaün**, Nogent-sur Marne, France) and birds (**Hiroko Ohki**, Nogent-sur Marne, France) grafted with thymic epithelium during fetal life was also reported. The mechanisms for non-rejection of challenge grafts under these circumstances, including the possibility of re-presentation of donor histocompatibility antigen to immature T-cells by host antigen presenting cells, remains to be explored.

SESSION VII

The Thymus in Future: the Molecular Biologic Approach? ***Mary A. Ritter*** *(London) and* ***Jan Rozing*** *(Rijswijk)*

This final session of the Rolduc meeting focussed predominantly on new (molecular) developments in the field of thymus and T cell research. Two major areas were discussed: the expression of various T cell receptor (TCR) genes during T cell differentiation and the impact of transgenic animals in studies on T cell development.

Jacques van Dongen (Rotterdam, The Netherlands) reported as the first topic an elegant set of experiments on the development of γ T cell receptor genes in humans using cell lines derived from a number of acute lymphoblastic leukaemias as a model. In these cell lines he could demonstrate that there was no functional expression of the α-TCR gene at either the molecular or the protein level. There was however production of TCR-γ and β mRNA in these cells. Also the presence of the TCR-γ protein could be shown in association with the CD3 complex. Whereas in one cell line the TCR-γ genes were rearranged to the Cγ1 locus, both these loci were deleted in another line and in this line rearrangement to the Cγ2 locus had occurred. The overall TCR-gene configuration in combination with the observed TCR expression both at the mRNA and the protein level in these cell lines, suggest that during T cell differentiation TCR-γ genes first rearrange to the Cγ1 locus, prior to D-J joining of the TCR-β gene, followed by rearrangement to the γ2 locus and subsequent V-D-J joining of the TCR-β gene. Based on the phenotyping of these cells these events are supposed to occur early during intrathymic processing, although extrathymic TCR-γ rearrangement cannot be excluded. Strikingly one of these lines expressing the TCR-γ protein on its surface and lacking the αβ-TCR proteins was CD8-positive. Although one has to keep in mind that these are leukaemic cells this observation (and others) may well mean that the current belief that cells expressing the γδ TCR genes are only found in the population having a $CD3^+$ $CD4^-$ $CD8^-$ phenotype is not correct.

In the second presentation **Bernard de Geus** (Rijswijk, The Netherlands) reported some recent data on TCR-gene usage of intestinal intra-epithelial lympocytes in mice. Northern blot analysis on total RNA from these cells revealed a high expression of both γ and δ mRNA. This was at least 10-fold higher than the level of TCR-α mRNA-expression in the same material. No δ-mRNA could be detected in cells from the mesenteric lymph nodes and Peyer's patches. In these organs only α-mRNA was found. Preliminary data from FACS analysis experiments on cells from the various tissues using monoclonal antibodies against CD3, CD4 and CD8 markers suggested that a substantial percentage of all CD3-positive cells in the intra-epithelial fraction of the intestine have the $CD4^-$ $CD8^-$ phenotype. Hardly any of such cells were found in mesenteric lymph nodes and Peyer's patches. When extrapolized quantitatively these findings suggest that the total number of γδ receptor bearing cells residing in the various epithelial tissues in the body probably is as high as or even higher than that of αβ-receptor bearing cells. In view of the fact that similar observations were obtained in material from athymic nude mice a significant role for the thymus in the generation of these γδ receptor bearing cells seems to be unlikely. If the thymus is not involved these γδ bearing cells are, by definition, not true "T" cells. The nomenclature of TCR 1 bearing T cells for these cells as recently suggested is therefore perhaps rather premature.

Transgenic mice, produced by the introduction of DNA into the nucleus of fertilised ova, offer the possibility of studying the regulation of gene expression and function of the encoded protein product in an *in vivo* system without the trauma attached to conventional techniques that involve surgery and irradiation. However, even this very elegant technique has its problems in that expression of the trangene product is not always as clear-cut as might be predicted, sometimes making interpretation of the data difficult. **Willem van Ewijk** (Rotterdam, The Netherlands) presented work on the functional role of MHC class II molecules on a variety of cell types using mice transgenic for the Eα gene. The recipient mice possess the Eβ but lack an Eα gene and hence do not express class II E molecules. Normal expression of E molecules can be restored by the presence of a 'wild type' Eα transgene. In order to analyse the function of E molecules on different cell populations an attempt was made to compartmentalize E expression by producing a series of mice that were trangenic for Eα genes with specific promotor/enhancer deletions. Deletion of the X box resulted in mice that had normal class II E expression on thymic medullary dendritic cells, but virtually none on cortical epithelium. In contrast, Y box deletion resulted in normal expression on cortical epithelium, but an almost negative medulla (although a small number of very strongly E positive cells were present in this area of the thymus). The ability of these thymuses to induce tolerance to the E region product was assessed by the mixed lymphocyte reaction assay and by the clonal deletion of T cells bearing the Vβ17a variable region (detected by KJ23 antibody). The results showed that intrathymic tolerance induction to E molecules occurred in all transgenic mice. For the X deleted transgenics, where only medullary dendritic cells were E positive, this very nicely confirms other studies localising tolerance induction to the medullary dendritic cells. The Y deleted mice are harder to interpret. One explanation is that cortical epithelial cells are capable of inducing tolerance to self MHC antigens. Alternatively, the small number of strongly E positive cells that are seen in the medulla of these mice might be sufficient to induce the observed tolerance, since even very small numbers of dendritic cells are known to be highly potent in immune responses. The role, if any, of the cortical epithelium in this process is therefore still unclear.

In the following presentation **Pawel Kisielow** (Wroclaw, Poland) described an elegant study of the mechanism of self-tolerance induction using T cell receptor transgenic mice. These mice were transgenic for both the α and β chain of a T cell receptor that was expressed in many of their T cells and which has specificity for the male (H-Y) antigen in the context of class I H-2D^b MHC molecules. An analysis of T cell phenotypes and the frequency of H-Y antigen-specific precursor cells in these αβ-TCR transgenic mice showed that cells that responded to H-Y antigen were frequent in female but not in male mice. This induction of tolerance to male antigen in male mice involved the deletion of self-reactive thymocytes that expressed high levels of CD8 molecules; such cells were predominantly the CD4, CD8 'double positive' population. Clearly such transgenic mice have already yielded much fascinating and informative data. It will be interesting to see what further insights into the workings of the immune system can be gained using this elegant *in vivo* expression system.

CONCLUDING REMARKS

Thus 'The Thymus' meeting at Rolduc was brought to a final conclusion with this exciting new work. The general consensus was that **Henk-Jan Schuurman**, **Jan Rozing** and their colleagues had worked very hard to produce an excellent meeting, and had laid the foundations for much new collaborative work between thymus research groups throughout Europe. We all look forward to the next Rolduc meeting, to be held 2–6 April, 1989.

'REPORT ON THE 4TH CONFERENCE OF THE INTERNATIONAL SOCIETY OF DEVELOPMENTAL AND COMPARATIVE IMMUNOLOGY, Nottingham University, U.K., 24-29 July, 1988.' - *Marion D. Kendall.*

The delegates to this conference were drawn from all over the world, and although the basic theme was comparative immunology, there were many topics presented that were of direct relevance to the study of the thymus gland, and almost all impinged to some extent on the T cell repertoire of the body. In this report I shall concentrate mainly on the papers dealing directly with the thymus, but a full account of each session of the meeting compiled by each of the co-chairmen, and all of the abstracts will be published in *Developmental and Comparative Immunology*, **13**.

The organisers had arranged a most useful series of plenary lectures throughout the meeting and these combined with introductory chairmen's lectures at the start of each parallel session ensured that all delegates departed with a sound basic knowledge of the areas covered. This, and the comfortable campus hospitality created an exellent environment for the interchange of ideas, and we all look forward to the next meeting in 2 or 3 years time organised through the offices of the new President Laurens N. Ruben (Portland, Oregon, U.S.A.).

John Marchonalis (Charleston, U.S.A.) gave the first plenary lecture on work undertaken with **S.F. Schluter** (Arizona, U.S.A.) on Ig and T cell receptors considered in an evolutionary context and showed us clearly that the J region is the most conserved throughout the vertebrates even to the tunicates thus presenting a useful evolutionary tool.

Parallel session 1 covered recognition, receptors and killing but many of the thymologists attended the parallel session on cytokines and the tissues involved in immune responses. One of the chairmen, **Andrew Gearing** gave a valuable review of cytokines which is published in this volume (Part 2: Current concepts, p. 127-143).

M. Michael Sigel *et al.*, (Columbia, U.S.A.) described work on keratinocytes that produce IL-1-like substances and interferon (fish epidermis) and IL-1 and prostanoids (human cornea). He emphasised the functional role of keratinocytes in immune reactions and made a plea for these cells to be studied more by immunologists. This paper has particular relevance to thymus since thymic epithelium is essentially keratin containing cells derived from the interaction of ectoderm and endoderm.

As co-chairman, **Marion Kendall** (UMDS, London) gave a brief resume of the development, structure, and function of the cells of the thymus concentrating on the six epithelial cells found in man and rodents reminding researchers of the potent erythropoietic function of the avian thymus and finishing with as summary of work undertaken with **Henk-Jan Schuurman** and colleagues in Utrecht, Holland, on the nature of cultured thymic fragments implanted under the kidney capsule of nude rats. Cultures before implantation are largely composed of epithelial cells of unique ultrastructure and phenotype (staining with monoclonal antibodies to mid/deep cortex and medulla) but they have the ability in the implants to develop into the full range of epithelial cells seen in the normal thymus (Kendall *et al.*, *Cell and Tissue Research*, **254**, 283–294 (1988)). Earlier studies have shown that T cell competence in subsequently developed. Subsequent papers in this and other sessions showed that in most vertebrates the epithelial cell types are remarkably similar which must reflect functional attributes of the thymus throughout evolution.

A.G. Zapata (Madrid, Spain) described the development of tissues involved in the immune responses ranging from jawless Agnatha to higher vertebrates. The so called 'thymus' of lampreys seems to be a filtering area, so that the chondrichthyes are the first vertebrates with histologically distinguishable thymus and spleen although the Agnatha exhibit both cell-mediated and humoral immunity. Hassall's corpuscles are less well developed in ectotherms, there is no clear demarcation between cortex and medulla in teleosts, apodans and urodeles, and adult teleosts retain the embryological connection of the thymus with the pharyngeal cavity.

Alberto Villena *et al.*, (Madrid, Spain) has undertaken histochemistry and electron microscopy on fish thymic epithelial cells distinguishing five types that seem to me to correspond to the cells of higher vertebrates.

Studies on thymic development in the frog (**J.G. O'Neill** *et al.*, Leiceter U.K.) showed that the thymic bud develops 2-7 days post hatching and is expanded by lymphocytic infiltration until day 10 at which stage the cortex and medulla differentiated. After 10-30 days the thymocyte population decreased with an increase in reticular cell and vascular endothelial volume, and the infiltration of myoid cells. Thymic involution started at an early tadpole stage, well before the onset of metamorphosis into the froglet/adult stage at 64-90 days. In adult life the myoid cell numbers increased and there was more collagenous connective tissue as the thymus cortex became indistinct and fewer thymocytes were observed.

J.G. O'Neill also studied the effect an evolutionary adaptation to low environmental temperature on the development of lymphoid organs in an antarctic teleost fish. The thymus, spleen and kidney were similar to those of warmer-water species, however the onset and degree of lymphoid organ development were not so advanced. There was no evidence for an advanced stage of thymic regression or involution in the adult fish although the subepithelial connective tissue and mucous cells increased.

Ann G. Clarke (Nottingham, U.K.) described how the mouse thymus changes during pregnancy, with thymus weight and cell number increasing at the time of implantation before falling by 75% in late pregnancy. Both repsonses occur in pregnancy and pseudopregnancy but are greater in allogeneic than syngeneic mating. Using monoclonal antibodies to lineage and functional markers shows that the % of $CD8^+$ cells (Tc/Ts) and $CD4^+$ cells (Th) both decrease by day 3 after pregnancy and pseudopregnancy and continue to fall for the rest of pregnancy but $CD8^+$ levels recover after an infertile mating by day 16-17. In pregnancy there is a general downgrading of fluorescence intensity and the appearance of a unique population of small cells lacking these markers. Single positve cells ($CD4^+CD8^-$), double negative cells and $MEL14^+$ (emigrant) cells remain high throughout pregnancy and pseudopregnancy, while the numbers of double positive cells are depressed. Thus the most mature and the most immature cells of the thymus show the greatest changes in pregnancy.

Functionally related to these observations was the work of **Pirkko Sulila** (Uppsala, Sweden) who reported that the uterus-draining paraaortic nodes in rats and mice enlarge as pregnancy proceeds and have a proportional increase in the numbers of $CD4^+$, $CD8^+$ and B-cells as compared to virgin controls.

The second plenary lecture by **Francoise Dieterlen-Lievre** (Nogent sur Marne, France) outlined much of the group's work using the quail-chick marker to study the development of the immune system, the relationships betweeen vasculogenesis, angiogenesis and haemopoiesis, and thymic development. Of interest here was their work showing the very early formation of diffuse haemopoietic foci in the dorsal mesentary near the aorta resulting in budding of haemopoietic cells into the lumen of the vessels, a process reminiscent of the events producing haemocytoblasts in the yolk sac of mice. These progenitors develop without contact with the ectoderm, unlike the thymic rudiment.

Georg Wick's plenary lecture on work with **J. Penniger** (Innsbruck, Austria) asked the question 'Are thymic nurse cells a school for the acquistion of self-tolerance?'and concluded that the thymocytes in these cells display a dramatic potential for alloreactivity and are able to induce a real graft *versus* host reaction (GvHR) with high efficency. He illustrated this by reference to studies on normal chicken thymic nurse cells (TNC) that generally contain only 4-5 thymocytes within them and to TNC from obese strain chicken which have a deficit of TNC with fewer thymocytes. These obese chicken have a generalized immune hyperreactivity in addition to their characterisitic autoreactivity against thyroid antigens. He obtained single TNC from normal inbred strains and deposited them onto the chorioallantoic membrane (CAM) of 10 day old embryos from the congeneic normal strain. GvHR was assessed by the number of 'pocks' after the 16th embryonic day. Surprisingly TNC lymphocytes gave an efficiency of 1/17 compared with 1/100 for peripheral blood lymphocytes. The pocks resembled classical GvHR at the light and electron microscopical level, and immunohistochemical analysis of frozen, acetone-fixed sections of TNC as well as fluorescence activated cell sorter analysis of suspensions of cells from pocks showed that the foci are composed of cells carrying donor and recipient Ia-like determinants. A majority of donor and some recipient cells displayed the IL2 receptor. Finally some primary pocks were transfered to original donor, primary recipient and unrelated third party recipients and the size of these pocks and the relative spleen index of the secondary recipients were determined on embryonic day 14. These experiments clearly showed preferential pock proliferation and significantly increased relative spleen indices on the congeneic normal strain as compared with the other recipients thus proving specifity and memory.

Istvan Berczi's (Manitoba, Canada) plenary lecture considered immunoregulation by neuroendocrine factors. The pituitary gland secretes both the immunostimulatory (growth hormone and prolactin) and immunosuppressive hormones (ACTH) and thus has the capacity to regulate immune reactions, and growth control in lymphoid tissue is suggested as the major function of these hormones. In general, estrogens stimulate humoral and suppress cell mediated immunity, whereas androgens antagonize estrogens in many systems. Catecholamines suppress allergic and other immune reactions, wheras cholinergic agents are stimulatory. Of the neuropeptides, substance P has been identified as the major mediator of neurogenic inflammation inititated by sensory neurones as it induces mast cell discharge, increased capillary permeability and causes smooth muscle contractions. Somatostatin acts as an antagonist of these pro-inflammatory peptides. This neurogenic pathway of inflammation allows for a rapid host reaction to tissue injury in the absence of antigenic recognition by the immune system. The immune reactions are initiated mainly by IL-1 and other immune derived factors. IL-1 is responsible for the pyrogenic effects of the immune response as well as T cell activation. Thus the ultimate regulator of immunocompetence is the pituitary gland through growth hormone, especially prolactin and ACTH whereas neurotransmitters may be regarded as rapid modulators of immune effector and inflammatory mechanisms.

The plenary lecture by **Norman Ratcliffe** (Swansea, U.K.)discused the biological significance of immunity right across the range of invertebrate and vertebrate animals and considered the evolutionary pressures such as the colonial habit and terrestrial life on its evolution. It was fascinating to learn that mice can sense MHC in some manner, and preferentially mate with 'new' smells. Such is the power of self and non-self!

The session on tolerance, autoimmunity and ageing started with data demonstrating anti-donor alloreactivity in skin graft- tolerant toads presented by **John Horton** *et al.*, (Durham, U.K.). Tolerance can be induced around metamorphosis or in 7-day thymectomised larvae by the transplantation of allogeneic skin. However the potential for anti-donor alloimmune reactivity remains throughout life since host splenocytes can mount a proliferative response when stimulated with irradiated donor strain splenocytes, and *in vitro* mixed leucocyte reactivity is even displayed by host thymocytes developing alongside a persisting donor-derived thymic epithelial environment. The authors believe that tolerance may be brought about by active suppression, not clonal deletion as rejection of tolerated perimetamorphic grafts and newly applied test grafts are in certain donor/host combinations rejected within 3 weeks of a single intra peritoneal injection of cyclophosphamide. This problem of the rejection of skin grafts by early thymectomised toads was discussed by **Michael Balls** *et al* (Nottingham, U.K.) by using the carcinogen nitrosamide *N*-methyl- *N*-nitrosurea (NMU) which temporarily reduces the white cell count, and severely damages lymphoid cell populations in the spleen, thymus, liver and kidneys by removing certain T cell functions, in particular T helper function and IL-2 responsive cells that can be stimulated by PHA. In these experiments allograft rejection capacity is lost whereas that of Xenograft rejection is not and the reactions differ morphologically. However, other workers have shown that xenografts are sometimes retained, so the dependence of the thymus for these reactions is still a debated subject.

The session on regulation by endocrine factors started with an useful overview of immunological cells as a source of factors with neuroendocrine activity by **H. Besedovsky** and **A. del Rey** (Davos, Switzerland). The activity of the central nervous system changes in immune responses, for example nuclei in the ventromedial part of the hypothalamus increase their firing rate. Supernatants from *in vitro* stimulated immune cells can increase ACTH and corticosterone blood levels *in vivo*, and can decrease noradrenaline turnover rates in the hypothalamus thus mimicking two of the neuroendocrine changes seen during immunostimulation. The factors that effect the hypothalamus have not yet been defined. However recent work on IL-1 is important in the immune response as it causes a rapid increase in ACTH and corticosterone blood levels that are integrated at the level of the hypothalamus. *In vivo* injection of IL-1 increases corticosteroid releasing factor (CRF), and if CRF release is blocked then its effects on ACTH and corticosterone is impeded. IL-1 also affects glucose homeostasis as a prolonged hypoglycaemia is induced by IL-1 but not through insulin secretagogue action. CRF seems to be required as an obligatory step in IL-1 release, and IL-1 affects noradrenalin levels in the brain. We await the continuation of this research with great interest.

Catecholamine regulation of and by immunity in the toad *Xenopus* was the subject of the presentation by **Richard Clothier** *et al* (Nottingham, U.K.). By using 6-hydroxydopamine (6-OHDA) to chemically sympathectomise animals, the authors were able to show that all immune responses were elevated suggesting that catecholamines are general inhibitors in the toad. Anti-hapten (TNP-) responses were stimulated using thymus dependent (TD), and thymus independent (TI-1 and TI-2) carriers which activate different T and B cell populations. Dramatic increases of norepinephrine were observed in the spleen after TD and of norepinephrine and epinephrine after TI challenge. The skin glands in *Xenopus* contain high levels of thyrotrophin releasing hormone, 5-hydroxytryptamine and caerulein (a gastrin-like peptide) that can be released by α-adrenergic agonists at high concentrations. Similar glands and mucous cells are found in the thymus after metamorphosis. Their possible involvement in immune reactions awaits clarification.

6th INTERNATIONAL WORKSHOP ON IMMUNE-DEFICIENT ANIMALS IN BIOMEDICAL RESEARCH , Beijing, People's Republic of China, July 3-6, 1988 - *Henk-Jan Schuurman and Hans Petter Hougen*

In 1973, **Dr. Jørgen Rygaard,** Copenhagen, Denmark, took the initiative to organise an international workshop on the nude mice held in Aarhus, Denmark. The success of this meeting formed the basis of subsequent workshops held every three years, under the auspices of the International Council for Laboratory Animal Science (ICLAS). The scope of the meeting was broadened as indicated by the change of its name into "International Workshop on Immune-Deficient Animals in Biomedical Research" (IWIDA). The 6th workshop was held in Beijing, People's Republic of China, July 3-6, 1988. The local organizing committee, with chairman **Prof. Dr. Bing-quan Wu,** Department of Pathology of the Beijing Medical University, did an enormous job in making an excellent organization of this meeting, which was one of the first international conferences in the area of biomedical research held in China after the start of the "open door policy". In the exciting atmosphere of the capital city, about 150 scientists working in breeding immune-deficient animals and their characterization and application in immunologic and oncologic research, gathered for three days to present their experimental data and exchange their ideas. Most participants came from China and Japan, about 30 came from the American continent or Europe. The meeting was particularly fruitful in enabling contacts between scientists from Asian countries and their collegues in USA or Europe. It was very interesting for participants from USA or Europe to recognize that a lot of high-standard research on immune-deficient animals is going on in Asian countries, which unfortunately is not always published in accessible (English-language) scientific journals.

The conference was divided in five sessions, each with Position Papers by invited speakers and with oral presentations, and in addition one poster session. The total number of presentations reached about 100. This report presents a global impression from the workshop and mainly focuses on contributions interesting for readers of Thymus Update. The detailed conference papers will be published in the proceedings by Karger AG, Basel, Switzerland.

Types of genetically-determined immune-deficient animals

The broadening of the scope of the meeting was best illustrated by the first session which comprised a number of presentations on different types of immune-deficient animals. These included both single-gene and double-gene defects generated by specific breeding. **Leonard Shultz** (Bar Harbor, USA) reviewed the presently available genetically determined murine models of Immunodeficiency: a selection is presented in Table I.

TABLE I. Genetically determined murine models of immunodeficiency

Gene name	Gene	Chromosome	Major phenotypic characteristics symbol
Severe Immunologic defects			
Nude	*nu*	11	Congenital thymic aplasia; hairlessness; quantitative and functional T-cell deficiency
Streaker	nu^{str}	""	Remutation of the nu locus in AKR/J; heterozygosity prevents T-cell lymphoma development
Severe combined immunodeficiency	*scid*	16	Absence of functional T- and B-cells; lymphopenia; hypogammaglobulinaemia; defect in rearrangement of genes encoding antigen-specific receptors
Motheaten	*me*	6	Granulocytic skin lesions; impaired humoral and cell-mediated response; autoimmune phenomena (pneumonitis)
Viable	me^{v}	""	Similar to me, but longer lifespan
Some models of less severe immunologic defects			
Dominant hemimelia	*Dh*	1	Congenital asplenism; visceral and skeletal abnormalities; reduced IgM and IgG2
X-linked immunodeficiency	*xid*	X	B-cell abnormalities
Lipopolysaccharide	*lps*	4	Resistance to lethal effects of LPS endotoxin
Beige	*bg*	13	Decreased endogenous NK activity
Viable yellow	A^{vy}	2	Reduced humoral response to tetanus toxoid; reduced GVH reactivity of spleen cells; obesity
Hairless	*hr*	14	Reduced splenic T-cell alloresponse; reduced humoral response to tetanus toxoid; increased incidence of thymic lymphoma

Presented by Leonard Shultz, The Jackson Laboratories, Bar Harbor, USA

A relatively new mutant is the *scid* mouse. **Melvin Bosma** (Philadelphia, USA) reviewed his studies on this interesting mutant, which in future might replace the *nu* mouse in cancer research. The immune defect in this animal concerns primarily lymphoid cells, both of T and B cell lineage. The defect is located at the level of rearrangement of genes encoding the T cell receptor and immunoglobulin. These rearrangements can occur (e.g., induced by Abelson Leukemia Virus), but do not result in productive genes. The observation that non-T/B lymphocytic white blood cell populations are almost normal in the animal (e.g., NK cells, myeloid and macrophage lineage) accords with this basis of the immune defect. The generation of only non-productive gene rearrangement is not absolute: *scid* mice make some antibodies, particularly anti-DNA, which presumably is related to the use of a particular variable gene repertoire. **Sonoko Habu** (Kanagawa, Japan) studied the thymus in *scid* mouse. He showed that thymocytes in this animal were Thy-1 positive and expressed IL-2 receptors. This accords with other studies that the expression of these surface markers is not related with rearranged genes encoding the T-cell receptor. But, the cells did not express Lyt-2, L3T4 and T3 antigens indicating that productive receptor gene rearrangement is required before the cell expresses these markers. In a broader interpretation, these data nicely illustrate that the sole expression of Thy-1 does not make a T cell to be a "true T cell" (see also the report from the workshop "The Thymus. Histophysiology and Dynamics in the Immune System", published elsewhere in this issue). **Norman Reed** (Montana, USA) showed data from a *Nippostrongylus brasiliensis* infection model, that *scid* mouse fail to develop intestinal mast cells despite the demonstration of mast cell precursors. This deficiency could be ascribed to a deficiency in T-cell derived mast cell growth factors.

A severe combined immunodeficiency model is also available in the horse. **Charles Bue** (Pullman, USA) gave his data on bone marrow transplantation in scid horses. Graft failure in non-histocompatible T-cell depleted marrow was ascribed to the presence in the recipients of large granular lymphocytes with natural killer cell activity.

Apart from the introduction of gene defects in different strains - for rat thymologists the extensive list of nude rat strains presented by **Carl Hansen** (Bethesda, USA) is impressive - animal breeding laboratories focus at present on constructing multiple gene defects. This presents interesting features. For instance, motheaten (*me*) mice with the nude (*nu*) mutation manifest the same granulocytic skin lesions and auto-immune pneumonitis as *me* mice without the *nu* gene (**Leonard Shultz**). This indicates that the aberrations in the *me* mice are not T-cell dependent. **Sun Jing** (Beijing, China) reported on a mouse 615/PBI strain with a combined T cell and NK cell deficiency, by mating congenic inbred 615/PBI *nu* mice with C57BL/6 *bg* animals. This construct had a lower NK activity and higher human tumour metastatic potential than C57BL/6 *bg nu* mice. **Mutsushi Matsuyama** (Nagoya, Japan) introduced the *rnu* gene in BUF/Mna rats which strain has a high frequency of spontaneous thymoma development. +/*rnu* constructs showed a smaller thymus than +/+ rats, and had less frequency of spontaneous thymomas. This effect may be ascribed to the *rnu* gene, but this conclusion was not strictly allowed since the congenic constructs differed also at the RT1.A and Es-1 loci. On the other hand, **Akinori Kojima** from the same group presented data on the introduction of the *nu* gene in AKR/Ms mice, a strain with spontaneous thymic lymphoma development. This introduction was associated with an accelerated lymphoma development. It is tempting to speculate on the effect of nude genes on thymic epithelial cells, with these different results in epithelial thymoma and in thymic lymphoma development.

Immunity in nude animals

Nude animals are a continuing study object for immunologists. The absence of a functionally active thymus make the animals excellent tools to study intrathymic T-cell differentiation. **Henk-Jan Schuurman** (Utrecht, The Netherlands) reviewed the Dutch studies on characteristics of the immune defect in nude rats and the induction of immune competence after thymus implantation. In allogeneic combinations, e.g. epithelial/thymocyte chimaeras, the role of the Major Histocompatibility Complex (MHC) in epithelial-lymphocyte interactions was studied. The epithelial MHC haplotype has no role in generation of the allogeneic recognition repertoire of "educated" T cells, but *in vivo* tolerance to the epithelial MHC is introduced. These studied were done using epithelial grafts, thereby showing the biologic activity of the grafted microenvironment on host T-cell precursors. For whole thymus grafts in RT-6 allogeneic combinations it was shown that immune competence induction can be ascribed on one hand to the biologic activity of the epithelial microenvironment, and on the other hand to that of grafted thymocytes. In accord with this finding, **Hidekazu Tamauchi** (Kitasato, Japan) reported that $CD4^+CD8^+$ double-positive thymocytes from Thy-1.1/H-2^k AKR/J mice developed into segregated $CD4^+$ or $CD8^+$ subsets after transfer into irradiated Thy-1.2/H-2^k CBA/N nude animals. Finally, there was a remarkable report by **Osamu Taguchi** (Nagoya, Japan). He implanted embryonic thymus across the species barrier into Balb/C nude mice and found a high incidence of organ-localized auto-immune disease (thyroiditis, gastritis). These data merit further analysis with respect to the role of the thymus in self-tolerance induction which may not exist in these animal models.

Experimental infections in nude animals give a better understanding of the role of specific T-cell immunity in the course of the infection. Hans Petter Hougen (Copenhagen, Denmark) presented data of a *Streptococcus mutans* infection model in (nude) rats, showing that periodontal bone loss develops independently of the thymus-dependent immunity of the animal. In a *Salmonella typhimurium* infection model, the ability of thymic grafts in nude animals to generate antibody production with subsequent protection to infection was shown. The best protection to infection was obtained by passive transfer of spleen cells from immune animals either in euthymic or athymic recipients. This protection exceeded that by active vaccination. Heying Wang (Nanjing, China) developed an experimental leprosy model in nude mice, and showed a higher number of acid-fast bacilli in nude animals than in euthymic animals, one year after infection with *M. leprae*.

Nude animals in cancer research

The main application of nude animals in biomedical research still is in cancer research. A large number of presentations dealt with models of tumour cell inoculation in nude mice. These will not be reported here in detail. We restrict ourselves to some general statements, derived from the lively discussions on the value of these models. These focused on whether tumour growth and its modulation after transfer of human tumour cells in nude animals can directly be compared with the situation in the human cancer patient. Most participants agreed that this direct extrapolation of data of the experimental animal model to the human situation is not allowed. This conclusion is based on a number of facts. First, tumour cell populations are heterogeneous yielding selection/depletion of subsets after inoculation in the nude animal wich differs from the original human situation. For instance, **Robert Clarke** (Bethesda, USA) selected cells from the MCF-7 human breast cancer cell line for estrogen and progesteron receptor expression by passage in nude mice. Dependent on (experimentally modulated) hormone status of the recipient tumour cells obtained after *in vivo* growth differed in hormone receptor status. Second, the metastatic potential of tumour cells in nude animals differs from that in the cancer patient. Third, in studies on chemotherapy large differences exist in drug pharmacokinetics between man and mice. This was nicely illustrated by **Makoto Inaba** (Tokyo, Japan) for methotrexate. Thus, the nude animal tumour cell inoculation model is not appropriate to predict the treatment result in individual patients, or to establish clinically relevant dosages of all chemotherapeutic drugs. This restiction does not diminish its value in development and preclinical evaluation of new chemotherapeutic agents or in the study of biological characteristics of tumours.

Future prospect

It is beyond discussion that the IWIDA has got a well-established place in the present collection of conferences and workshops on biomedical topics. It enables a direct contact between scientists working with immune-deficient animals in different areas ranging from breeding the animals to characterization of the immune defect or to utilization of the immune defect in *in vivo* models of human disease. These research fields are in rapid progress which will influence the scope of the IWIDA. In his concluding remarks, **Jørgen Rygaard** (Copenhagen, Denmark) pointed to this phenomenon. First, he advocated a better definition and comparative analysis of existing immune-deficient animals in different institutes. As an example the nude rat was mentioned. This animal now becomes increasingly popular in biomedical medicine and shows a number of advantages to the nude mice. For a proper interpretation of research data from different groups, a comparative analysis of these animals bred and maintained at different locations is very welcome. Second, new experimental animal models have been introduced. These include new genetically-determined immune deficiencies and multiple gene models, but in addition an immense array of transgenic models. It is worth emphasizing that also models of induced immune deficiency get increasing attention: as an illustation, **Herbert Morse III** (Bethesda, USA) presented his model of murine leukemia virus induced immune deficiency in mice and draw interesting parrallels with the human situation after human immunodeficiency virus infection. So, the field of immune-deficient animals is in rapid movement and may extent beyond its present main attainment in e.g. experimental oncology. For scientists involved in thymus research it is worth following the progess in this field and contribute their expertise to it. The next (7th) IWIDA is planned in 1991, and will be held at The Jackson Laboratories, Bar Harbor, USA, organized by Leonard Shultz.

PART 3: APPENDIX

SELECTED THYMUS PUBLICATIONS, 1987-1988

Aarli-JA; Gilhus-NE; Hofstad-H
CA-antibody: an immunological marker of thymic neoplasia in myasthenia gravis?
Acta-Neurol-Scand. 1987 Jul; 76(1): 55-7

Abdul-Razzak-KK; Garg-L; Wen-L; Reeck-GR
Fetal and newborn calf thymus as a source of chromatin proteins: purification of HMG-1 and HMG-2.
Prep-Biochem. 1987; 17(1): 51-61

Abiko-T
Syntheses and structure-activity relationships of thymopoietin.
Adv-Exp-Med-Biol. 1987; 223: 153-5

Abiko-T; Sekino-H
Synthesis of calf thymosin beta 4 fragment 16-38 and its effect on the impaired blastogenic response of T-lymphocytes of a uremic patient with pneumonia.
Chem-Pharm-Bull-Tokyo. 1987 Sep; 35(9): 3757-65

Abiko-T; Sekino-H
Synthesis of deacetyl-thymosin beta 10 and examination of its immunological effect on T-cell subpopulations of a uremic patient with tuberculosis.
Chem-Pharm-Bull-Tokyo. 1986 Nov; 34(11): 4708-17

Abiko-T; Sekino-H
Synthesis of the revised amino acid sequence of thymopoietin II and examination of its immunological effect on the impaired T-lymphocyte transformation of a uremic patient with pneumonia.
Chem-Pharm-Bull-Tokyo. 1987 May; 35(5): 2016-24

Abiko-T; Sekino-H
The synthesis of peptides with potential thymic hormone activity: the synthesis of endo-Arg38a-deacetylthymosin beta 10.
Biotechnol-Appl-Biochem. 1987 Feb; 9(1): 20-30

Ackland-SP; Bur-ME; Adler-SS; Robertson-M; Baron-JM
White blood cell aplasia associated with thymoma.
Am-J-Clin-Pathol. 1988 Feb; 89(2): 260-3

Acuto-O; Larocca-LM
T-cell antigen receptor expression in the thymus.
Hum-Immunol. 1987 Feb; 18(2): 93-110

Adkins-B; Gandour-D; Strober-S; Weissman-I
Total lymphoid irradiation leads to transient depletion of the mouse thymic medulla and persistent abnormalities among medullary stromal cells.
J-Immunol. 1988 May 15; 140(10): 3373-9

Adkins-B; Tidmarsh-GF; Weissman-IL
Normal thymic cortical epithelial cells developmentally regulate the expression of a B-lineage transformation-associated antigen.
Immunogenetics. 1988; 27(3): 180-6

Adolph-JM; Kimmig-BN; Georgi-P; zum-Winkel-K
Carcinoid tumors: CT and I-131 meta-iodo-benzylguanidine scintigraphy.
Radiology. 1987 Jul; 164(1): 199-203

Aflalo-E; Ofir-R; Apte-RN; Weinstein-Y
Characterization of mouse thymocyte subpopulations by the enzymatic marker 20-alpha-hydroxysteroid dehydrogenase: differential responses to IL-1 and IL-2.
Clin-Exp-Immunol. 1987 Dec; 70(3): 578-84

Aida-Y; Aono-M; Onoue-K
Muramyl dipeptide induced augmentation of the proliferative response of thymocytes to phytohemagglutinin.
Microbiol-Immunol. 1987; 31(2): 169-78

Aldenborg-F
Thymus dependence of compound 48/80-induced mucosal mast cell proliferation.
Int-Arch-Allergy-Appl-Immunol. 1987; 84(3): 298-30

Alguacil-Garcia-A; Halliday-WC
Thymic carcinoma with focal neuroblastoma differentiation.
Am-J-Surg-Pathol. 1987 Jun; 11(6): 474-9

Amagai-T; Kina-T; Hirokawa-K; Nishikawa-S; Imanishi-J; Katsura-Y
Dysfunction of irradiated thymus for the development of helper T cells.
J-Immunol. 1987 Jul 15; 139(2): 358-64

Amari-NM; Meruelo-D
Murine thymomas induced by fractionated-X-irradiation have specific T-cell receptor rearrangements and characteristics associated with day-15 to -16 fetal thymocytes.
Mol-Cell-Biol. 1987 Dec; 7(12): 4159-68

Amiot-M; Dastot-H; Fabbi-M; Degos-L; Bernard-A; Boumsell-L
Intermolecular complexes between three human CD1 molecules on normal thymus cells.
Immunogenetics. 1988; 27(3): 187-95

Amiot-M; Dastot-H; Schmid-M; Bernard-A; Boumsell-L
Analysis of CD1 molecules on thymus cells and leukemic T lymphoblasts identifies discrete phenotypes and reveals that CD1 intermolecular complexes are observed only on normal cells.
Blood. 1987 Sep; 70(3): 676-85

Amor-B; Dougados-M; Mery-C; Dardenne-M; Bach-JF
Nonathymulin in rheumatoid arthritis: two double blind, placebo controlled trials.
Ann-Rheum-Dis. 1987 Jul; 46(7): 549-54

Amoscato-AA; Babcock-GF; Sramkoski-RM; Hynd-BA; Alexander-JW
Synthesis of two biologically active fluorescent probes of thymopentin.
Int-J-Pept-Protein-Res. 1987 Feb; 29(2): 177-86

Andrzejewski-W; Zielinski-KW
Method for vascularized thymus transplantation in rats.
Z-Exp-Chir-Transplant-Kunstliche-Organe. 1987; 20(

Antoni-F; Luat-NN; Csuka-I; Olah-I; Sooki-Toth-A; Banflavi-G
The immunosuppressive effect of acute doses of emetine on murine thymic cells.
Int-J-Immunopharmacol. 1987; 9(3): 333-40

Aoyama-H; Mori-N; Mori-W
Anti-glucocorticoid effects of melatonin on adult rats.
Acta-Pathol-Jpn. 1987 Jul; 37(7): 1143-8

Apostolski-S; Gospavic-J; Isakovic-K; Micic-M; Popeskovic-L; Stojkovic-M
Functional difference between cortical and medullar thymic epithelial cells in patients with myasthenia gravis.
Ann-NY-Acad-Sci. 1987; 496: 707-10

Arliss-J; Scholes-J; Dickson-PR; Messina-JJ
Massive thymic hyperplasia in an adolescent.
Ann-Thorac-Surg. 1988 Feb; 45(2): 220-2

Arnall-JC; Horton-JD
In vivo studies on allotolerance perimetamorphically induced in control and thymectomized Xenopus.
Immunology. 1987 Oct; 62(2): 315-9

Arslan-P; Beltrame-M; Tomasi-A
Intracellular chromium reduction.
Biochim-Biophys-Acta. 1987 Oct 22; 931(1): 10-5

Audhya-T; Schlesinger-DH; Goldstein-G
Isolation and complete amino acid sequence of human thymopoietin and splenin.
Proc-Natl-Acad-Sci-USA. 1987 Jun; 84(11): 3545-9

Auger-G; Blanot-D; Magnin-M; Gastinel-LN; Pleau-JM; Dardenne-M; Bach-JF
Synthesis and biological activity of eight thymulin analogues.
Biol-Chem-Hoppe-Seyler. 1987 May; 368(5): 463-70

Bailey-RO; Dunn-HG; Rubin-AM; Ritaccio-AL
Myasthenia gravis with thymoma and pure red blood cell aplasia.
Am-J-Clin-Pathol. 1988 May; 89(5): 687-93

Ball-JK; Dekaban-GA
Characterization of early molecular biological events associated with thymic lymphoma induction following infection with a thymotropic type-B retrovirus.
Virology. 1987 Dec; 161(2): 357-65

Ballas-ZK; Rasmussen-W
Lymphokine-activated killer (LAK) cells. III. Characterization of LAK precursors and susceptible target cells within the murine thymus.
J-Immunol. 1987 Nov 15; 139(10): 3542-9

Barat-M; Rybak-LP; Dietrich-J
Metastatic thymoma to the head and neck.
Laryngoscope. 1988 Apr; 98(4): 418-21

Barcellini-W; Meroni-PL; Frasca-D; Sguotti-C; Borghi-MO; Uberti-Foppa-C; Buzzetti-P; Lazzarin-A; Dor
Effect of subcutaneous thymopentin treatment in drug addicts with persistent generalized lymphadenopathy.
Clin-Exp-Immunol. 1987 Mar; 67(3): 537-43

Bartik-MM; Takesue-BY; Mokyr-MB
Melphalan-induced enhancement of antitumor immune reactivity in thymocytes of adult BALB/c mice bearing a large MOPC-315 tumor.
Cancer-Res. 1987 Sep 15; 47(18): 4848-55

Bateman-WJ; Jenkinson-EJ; Owen-JJ
The thymus is not a primary site of endogenous Moloney leukaemia virus transcription in Mov 3 and Mov 9 mice.
Br-J-Cancer. 1987 May; 55(5): 521-2

Baxevanis-CN; Reclos-GJ; Panneerselvam-C; Papamichail-M
Enhancement of human T lymphocyte functions by prothymosin alpha. I. Augmentation of mixed lymphocyte culture reactions and soluble protein-induced proliferative responses.
Immunopharmacology. 1988 Mar-Apr; 15(2): 73-84

Baxevanis-CN; Reclos-GJ; Papamichail-M; Tsokos-GC
Prothymosin alpha restores the depressed autologous and allogeneic mixed lymphocyte responses in patients with systemic lupus erythematosus.
Immunopharmacol-Immunotoxicol. 1987; 9(4): 429-40

Baxevanis-CN; Reclos-GJ; Perez-S; Kokkinopoulos-D; Papamichail-M
Immunoregulatory effects of fraction 5 thymus peptides. I. Thymosin alpha 1 enhances while thymosin beta 4 suppresses the human autologous and allogeneic mixed lymphocyte reaction.
Immunopharmacology. 1987 Apr; 13(2): 133-41

Ben-Shahar-M; Rosenblatt-E; Green-J; Cohen-I
Malignant thymoma associated with progressive systemic sclerosis.
Am-J-Med-Sci. 1987 Oct; 294(4): 262-7

Benkovic-B; Burek-B; Jaksic-B; Vitale-B
Modulation of chronic lymphocytic leukemia (CLL) lymphocyte phenotypes by in vitro incubation with alpha 1 thymosin.
Blood-Cells. 1987; 12(2): 441-55

Benson-MT; Buckley-G; Jenkinson-EJ; Owen-JJ
Survival of deoxyguanosine-treated fetal thymus allografts is prevented by priming with dendritic cells.
Immunology. 1987 Apr; 60(4): 593-6

Bernier-L; Alvarez-F; Norgard-EM; Raible-DW; Mentaberry-A; Schembri-JG; Sabatini-DD; Colman-DR
Molecular cloning of a 2',3'-cyclic nucleotide 3'-phosphodiesterase: mRNAs with different 5' ends encode the same set of proteins in nervous and lymphoid tissues.
J-Neurosci. 1987 Sep; 7(9): 2703-10

Berrih-Aknin-S; Morel-E; Raimond-F; Safar-D; Gaud-C; Binet-JP; Levasseur-P; Bach-JF
The role of the thymus in myasthenia gravis: immunohistological and immunological studies in 115 cases.
Ann-NY-Acad-Sci. 1987; 505: 50-70

Beschorner-WE; Di-Gennaro-KA; Hess-AD; Santos-GW
Cyclosporine and the thymus: influence of irradiation and age on thymic immunopathology and recovery.
Cell-Immunol. 1987 Dec; 110(2): 350-64

Beschorner-WE; Hess-AD; Shinn-CA; Santos-GW
Transfer of cyclosporine-associated syngeneic graft-versus-host disease by thymocytes. Resemblance to chronic graft-versus-host disease.
Transplantation. 1988 Jan; 45(1): 209-15

Beschorner-WE; Olson-JL; Hess-AD; Di-Gennaro-KA; Santos-GW
Cyclosporine-induced cell-mediated injury of the thymic medullary epithelium.
Transplantation. 1988 Apr; 45(4): 797-803

Bielory-L; Gascon-P; Lawley-TJ; Young-NS; Frank-MM
Human serum sickness: a prospective analysis of 35 patients treated with equine anti-thymocyte globulin for bone marrow failure.
Medicine-Baltimore. 1988 Jan; 67(1): 40-57

Bigaj-J; Plytycz-B
Interdigitating cells in the thymus of the frog, Rana temporaria.
Folia-Histochem-Cytobiol. 1987; 25(1): 65-8

Binns-RM; Pabst-R; Licence-ST
Subpopulations of T lymphocytes emigrating in venous blood draining pig thymus labelled in vivo with fluorochrome.
Immunology. 1988 Feb; 63(2): 261-7

Bird-TA; Gearing-AJ; Saklatvala-J
Murine interleukin-1 receptor: differences in binding properties between fibroblastic and thymoma cells and evidence for a two-chain receptor model.
FEBS-Lett. 1987 Dec 10; 225(1-2): 21-6

Birr-C; Bohn-B; Jaeger-KH
Biochemical characterization and immunomodulatory action of thymic components as determined by flow cytometry on human lymphocytes.
Thymus. 1987; 10(3-4): 159-68

Blitstein-Willinger-E; Streller-A
In vitro effects of gold sodium thiomalate on IL-1, IL-2 production, IL-2 receptor expression and IL-2 responsiveness in thymocytes and peripheral blood mononuclear cells.
Clin-Exp-Immunol. 1987 May; 68(2): 357-65

Blue-ML; Craig-KA; Daley-JF; Levine-H; Schlossman-SF
Human thymocyte subpopulations: maturational stages defined by the expression of the T3-T cell receptor complex.
Acta-Haematol-Basel. 1987; 78 Suppl 1: 43-9

Blue-ML; Daley-JF; Levine-H; Craig-KA; Schlossman-SF
Activation of immature cortical thymocytes through the T11 sheep erythrocyte binding protein.
J-Immunol. 1987 May 15; 138(10): 3108-13

Blue-ML; Daley-JF; Levine-H; Craig-KA; Schlossman-SF
Identification and isolation of a T4+T8+ cell with high T3 expression in human thymus: a possible late intermediate in thymocyte differentiation.
J-Immunol. 1987 Aug 15; 139(4): 1065-9

Blue-ML; Levine-H; Daley-JF; Craig-KA; Schlossman-SF
Development of natural killer cells in human thymocyte culture: regulation by accessory cells.
Eur-J-Immunol. 1987 May; 17(5): 669-73

Bocchieri-MH
An ELISA for anti-thymocyte autoantibody in NZB mice. Comparison with the cytotoxic assay.
J-Immunol-Methods. 1987 Aug 3; 101(2): 245-9

Bode-U; Scheidt-W
Change of thymic size during and following cytotoxic therapy in young patients.
Pediatr-Radiol. 1988; 18(1): 20-3

Bodey-B; Calvo-W; Prummer-O; Fliedner-TM; Borysenko-M
Development and histogenesis of the thymus in dog. A light and electron microscopical study.
Dev-Comp-Immunol. 1987 Winter; 11(1): 227-38

Boeru-V; Manea-E; Olariu-M
The in vitro effect of a thymic polypeptidic extract on the function of T-cells and macrophages.
Endocrinologie. 1987 Apr-Jun; 25(2): 83-9

Bongertz-V; Chequer-Bou-Habib-D; Galvao-Castro-B; Campos-Neto-A
Production of a monoclonal anti-hamster thymocyte antibody with mitogenic properties.
Hybridoma. 1988 Feb; 7(1): 33-9

Borisova-EA; Chukhlovin-AB; Seiliev-AA; Zherbin-EA; Zhivotovsky-BD; Hanson-KP
Degree of chromatin fragmentation and frequency of nuclear pyknosis in Percoll-fractionated thymocytes of irradiated rats.
Int-J-Radiat-Biol. 1987 Mar; 51(3): 421-8

Born-W; White-J; Kappler-J; Marrack-P
Rearrangement of IgH genes in normal thymocyte development.
J-Immunol. 1988 May 1; 140(9): 3228-32

Botteri-FM; van-der-Putten-H; Wong-DF; Sauvage-CA; Evans-RM
Unexpected thymic hyperplasia in transgenic mice harboring a neuronal promoter fused with simian virus 40 large T antigen.
Mol-Cell-Biol. 1987 Sep; 7(9): 3178-84

Bouchon-B; Iwamori-M; Masago-H; Nagai-Y
Selective expression of cholera toxin-receptor in rat medullary thymocytes.
Cell-Struct-Funct. 1987 Aug; 12(4): 339-44

Boyd-RL; Wilson-TJ; Mitrangas-K; Ward-HA
Characterization of chicken thymic and bursal stromal cells.
Prog-Clin-Biol-Res. 1987; 238: 29-39

Boyer-PD; Rothenberg-EV
IL-2 receptor inducibility is blocked in cortical-type thymocytes.
J-Immunol. 1988 May 1; 140(9): 2886-92

Bramwell-NH
Well-differentiated squamous cell carcinoma of the thymus [letter]
Can-Med-Assoc-J. 1987 Sep 1; 137(5): 376

Brand-K
Role of ornithine decarboxylase on glycolytic enzyme induction during thymocyte proliferation.
J-Biol-Chem. 1987 Nov 5; 262(31): 15232-5

Brand-K; Aichinger-S; Forster-S; Kupper-S; Neumann-B; Nurnberg-W; Ohrisch-G
Cell-cycle-related metabolic and enzymatic events in proliferating rat thymocytes.
Eur-J-Biochem. 1988 Mar 15; 172(3): 695-702

Brand-K; Von-Hintzenstern-J; Langer-K; Fekl-W
Pathways of glutamine and glutamate metabolism in resting and proliferating rat thymocytes: comparison between free and peptide-bound glutamine.
J-Cell-Physiol. 1987 Sep; 132(3): 559-64

Brelinska-R; Seidel-HJ; Warchol-JB; Kreja-L
Murine thymic multicellular complexes during latency period after methylnitrosourea (MNU) injection.
Thymus. 1988; 11(2): 67-75

Brelinska-R; Warchol-JB; Boniver-J; Houben-Defresne-MP
Distribution of Lyt antigens on the surface of thymocytes associated with thymic macrophages and dendritic cells.
Histochemistry. 1988; 88(3-6): 553-6

Brink-JA; DeLuca-SA
Thymoma.
Am-Fam-Physician. 1987 Jul; 36(1): 125-6

Brockbank-KG; Voerman-JS
An in vitro analysis of murine thymic reticular-fibroblasts.
Thymus. 1987; 10(3-4): 169-77

Brocke-S; Takacs-L; Gerdes-J; Osawa-H; Diamantstein-T
The ontogeny of the interleukin 2 receptor expression and of the interleukin 2 responsiveness in the rat thymus.
Immunobiology. 1987 May; 174(3): 266-73

Bron-C; MacDonald-HR
Identification of the plasma membrane receptor for interleukin-1 on mouse thymoma cells.
FEBS-Lett. 1987 Jul 27; 219(2): 365-8

Buckheit-RW Jr; Kurtzberg-J; Bolognesi-DP; Weinhold-KJ
The coenrichment of stem cells, prothymocytes, and stromal elements with ecotropic retrovirus-producing cells from the bone marrow of leukemia-prone AKR mice.
Virology. 1988 Feb; 162(2): 354-61

Buckley-G; Benson-MT; Jenkinson-EJ; Owen-JJ
Factors affecting the acceptance of deoxyguanosine-treated thymus allografts.
Transplantation. 1988 Jan; 45(1): 202-5

Budd-RC; Miescher-GC; Howe-RC; Lees-RK; Bron-C; MacDonald-HR
Developmentally regulated expression of T cell receptor beta chain variable domains in immature thymocytes.
J-Exp-Med. 1987 Aug 1; 166(2): 577-82

Budd-RC; Schreyer-M; Miescher-GC; MacDonald-HR
T cell lineages in the thymus of lpr/lpr mice. Evidence for parallel pathways of normal and abnormal T cell development.
J-Immunol. 1987 Oct 1; 139(7): 2200-10

Buff-DD; Greenberg-SD; Leong-P; Palumbo-FS
Thymoma, Pneumocystis carinii pneumonia, and AIDS.
N-Y-State-J-Med. 1988 May; 88(5): 276-7

Bulloch-K; Bossone-SA
Nerve-related 3S acetylcholinesterase in murine thymus.
Ann-NY-Acad-Sci. 1987; 496: 338-45

Bulloch-K; Cullen-MR; Schwartz-RH; Longo-DL
Development of innervation within syngeneic thymus tissue transplanted under the kidney capsule of the nude mouse: a light and ultrastructural microscope study.
J-Neurosci-Res. 1987; 18(1): 16-27

Bulloch-K; Lucito-R
The effects of cortisone on acetylcholinesterase (AChE) in the neonatal and aged thymus.
Ann-N-Y-Acad-Sci. 1988; 521: 59-71

Buss-WC; Stalter-K; Stepanek-J; Piatt-MK
Studies on the mechanism of glucocorticoid hormone induced alterations in rat thymic transcription--II. Partial purification and characterization of RNA polymerases II from hydrocortisone and cont
J-Steroid-Biochem. 1987 Aug; 28(2): 129-38

Cacciapuoti-F; Arpino-G; D'Avino-M
Reliability of echocardiography in the detection of metastatic malignant pericardial masses.
Int-J-Cardiol. 1988 Jan; 18(1): 109-12

Calviello-G; Bossi-D; Cittadini-A
Further observations on the effect of calcium ionophores on ascites tumor cells.
Arch-Biochem-Biophys. 1987 Nov 15; 259(1): 38-45

Calvo-W; Fliedner-TM; Herbst-EW; Hugl-E; Bodey-B
Degeneration and atrophy of the thymus of lethally irradiated dogs, rescued by transfusion of cryopreserved autologous blood leukocytes.
Exp-Hematol. 1987 Dec; 15(11): 1171-8

Campbell-PA; Collins-JM; Stedman-KE
A spleen-derived maturational factor allows immature thymocytes, prepared as cells bearing low amounts of surface sialic acid, to become cytotoxic T cells.
J-Immunol. 1987 Sep 1; 139(5): 1385-92

Cantelli-Forti-G; Hrelia-P; Scotti-M; Scornavacche-V; Prada-M
Mutagenicity studies on thymomodulin.
Arzneimittelforschung. 1987 Nov; 37(11): 1269-73

Capo-C; Mege-JL; Benoliel-AM; Mishal-Z; Bongrand-P
Quantification of the nonspecific intercellular transfer of fluorescent molecules between labeled and unlabeled rat thymocytes.
Cytometry. 1987 Sep; 8(5): 468-73

Cardarelli-N; Fazely-F
The correspondence between tin and antiproliferative activity in a series of successive thymus extracts.
Thymus. 1988; 11(2): 123-6

Carding-S; Reem-GH
c-myc gene expression and activation of human thymocytes.
Thymus. 1987; 10(3-4): 219-29

Carding-SR; Bottomly-K
IL-4 (B cell stimulatory factor 1) exhibits thymocyte growth factor activity in the presence of IL-2.
J-Immunol. 1988 Mar 1; 140(5): 1519-26

Cassani-F; Fusconi-M; Bianchi-FB; Selleri-L; Baffoni-L; Lenzi-M; Pisi-E
Precipitating antibodies to rabbit thymus extractable antigens in chronic liver disease: relationship with anti-actin antibodies.
Clin-Exp-Immunol. 1987 Jun; 68(3): 588-95

Castillo-MB; Roldan-A
Response ot insulin by thymocytes of adrenalectomized and adrenalectomized-diabetic rats.
Acta-Physiol-Pharmacol-Latinoam. 1986; 36(3): 251-

Cazzola-P; Mazzanti-P; Kouttab-NM
Update and future perspectives of a thymic biological response modifier (Thymomodulin).
Immunopharmacol-Immunotoxicol. 1987; 9(2-3): 195-2

Ceci-A; de-Terlizzi-M; Toma-MG; Calculli-G; Caputo-R; Castello-M; Indolfi-P; Rosati-D
Heterogeneity of immunological patterns in Langerhan's histiocytosis and response to crude calf thymic extract in 11 patients.
Med-Pediatr-Oncol. 1988; 16(2): 111-5

Ceredig-R; Lynch-F; Newman-P
Phenotypic properties, interleukin 2 production, and developmental origin of a "mature" subpopulation of Lyt-2- L3T4- mouse thymocytes.
Proc-Natl-Acad-Sci-USA. 1987 Dec; 84(23): 8578-82

Chan-AS
Effects of hydrocortisone on the ultrastructure of the thymic cysts of chicks.
Exp-Pathol. 1987; 32(1): 23-30

Chen-JL; Weisbrod-GL; Herman-SJ
Computed tomography and pathologic correlations of thymic lesions.
J-Thorac-Imaging. 1988 Jan; 3(1): 61-5

Chen-SJ; Ko-HS
Serum thymosin-alpha 1: lack of association between elevated levels and HIV infection.
Clin-Exp-Immunol. 1987 Nov; 70(2): 263-7

Chervenak-R; Altazan-JD
Influence of MHC on thymus repopulation following intrathymic transfer of mouse T-cell precursors.
Cell-Immunol. 1987 Apr 1; 105(2): 290-300

Chesnokova-VM; Grutenko-EV; Ivanova-LN; Ignatyeva-EV
Evidence for thymus involvement in development of adrenal and gonadal endocrine functions in mice.
Endocrinol-Exp-Bratisl. 1987 Jun; 21(2): 133-42

Chesnokova-VM; Ignat'eva-EV; Ivanova-LN; Tepelina-OM; Livshits-VA; Korotaev-GK
Influence of thymosin on the formation of the functional activity of the adrenals in mice.
Neurosci-Behav-Physiol. 1987 Jul-Aug; 17(4): 353-7

Chien-YH; Iwashima-M; Wettstein-DA; Kaplan-KB; Elliott-JF; Born-W; Davis-MM
T-cell receptor delta gene rearrangements in early thymocytes.
Nature. 1987 Dec 24-31; 330(6150): 722-7

Chilosi-M; Iannucci-A; Menestrina-F; Lestani-M; Scarpa-A; Bonetti-F; Fiore-Donati-L; DiPasquale-B; P
Immunohistochemical evidence of active thymocyte proliferation in thymoma. Its possible role in the pathogenesis of autoimmune diseases.
Am-J-Pathol. 1987 Sep; 128(3): 464-70

Choyke-PL; Zeman-RK; Gootenberg-JE; Greenberg-JN; Hoffer-F; Frank-JA
Thymic atrophy and regrowth in response to chemotherapy: CT evaluation.
AJR. 1987 Aug; 149(2): 269-72

Cikes-N; Momoi-MY; Williams-CL; Howard-FM Jr; Hoagland-HC; Whittingham-S; Lennon-VA
Striational autoantibodies: quantitative detection by enzyme immunoassay in myasthenia gravis, thymoma, and recipients of D-penicillamine or allogeneic bone marrow.
Mayo-Clin-Proc. 1988 May; 63(5): 474-81

Cohen-II; Templeton-A; Philips-AK
Tumors of the thymus.
Med-Pediatr-Oncol. 1988; 16(2): 135-41

Cohen-Kaminsky-S; Berrih-Aknin-S; Savino-W; Dardenne-M
Immunodetection of the thymic epithelial P19 antigen in cultures of normal and pathological human thymic epithelium.
Thymus. 1987; 9(4): 225-38

Cohen-Kaminsky-S; Safar-D; Gaud-C; Berrih-Aknin-S
Role of the thymus in myasthenia gravis: responsiveness to recombinant interleukin 2 of thymic lymphocytes.
Transplant-Proc. 1988 Apr; 20(2): 322-3

Collins-JM; Justement-LB; Stedman-KE; Zlotnik-A; Campbell-PA
B cell stimulatory factor 1 induces lobster agglutinin 1-separated mouse thymocytes to express cytotoxic T lymphocyte activity.
J-Immunol. 1988 Jul 1; 141(1): 145-50

Coltey-M; Jotereau-FV; Le-Douarin-NM
Evidence for a cyclic renewal of lymphocyte precursor cells in the embryonic chick thymus.
Cell-Differ. 1987 Nov; 22(1): 71-82

Compton-MM; Caron-LA; Cidlowski-JA
Glucocorticoid action on the immune system.
J-Steroid-Biochem. 1987; 27(1-3): 201-8

Compton-MM; Cidlowski-JA
Identification of a glucocorticoid-induced nuclease in thymocytes. A potential "lysis gene" product.
J-Biol-Chem. 1987 Jun 15; 262(17): 8288-92

Compton-MM; Haskill-JS; Cidlowski-JA
Analysis of glucocorticoid actions on rat thymocyte deoxyribonucleic acid by fluorescence-activated flow cytometry.
Endocrinology. 1988 May; 122(5): 2158-64

Consolini-R; Cini-P; Cei-B; Bottone-E
Thymic dysfunction in histiocytosis-X.
Am-J-Pediatr-Hematol-Oncol. 1987 Summer; 9(2): 146

Conti-JA; Colicchio-AR; Howard-LM
Thymoma, myasthenia gravis, and relapsing polychondritis.
Ann-Intern-Med. 1988 Jul 15; 109(2): 163-4

Cook-JC; Dold-KM; Greenlee-WF
An in vitro model for studying the toxicity of 2,3,7,8-tetrachlorodibenzo-p-dioxin to human thymus.
Toxicol-Appl-Pharmacol. 1987 Jun 30; 89(2): 256-68

Cotton-DW; Cooper-C; Barrett-DF; Leppard-BJ
Severe atypical molluscum contagiosum infection in an immunocompromised host.
Br-J-Dermatol. 1987 Jun; 116(6): 871-6

Crispe-IN; Bevan-MJ
Expression and functional significance of the J11d marker on mouse thymocytes.
J-Immunol. 1987 Apr 1; 138(7): 2013-8

Crispe-IN; Moore-MW; Husmann-LA; Smith-L; Bevan-MJ; Shimonkevitz-RP
Differentiation potential of subsets of CD4-8- thymocytes.
Nature. 1987 Sep 24-30; 329(6137): 336-9

Crispe-IN; Shimonkevitz-RP; Husmann-LA; Kimura-J; Allison-JP
Expression of T cell antigen receptor beta-chains on subsets of mouse thymocytes. Analysis by three-color flow cytometry.
J-Immunol. 1987 Dec 1; 139(11): 3585-9

Cruz-JR; Chew-F; Fernandez-RA; Torun-B; Goldstein-AL; Keusch-GT
Effects of nutritional recuperation on E-rosetting lymphocytes and in vitro response to thymosin in malnourished children.
J-Pediatr-Gastroenterol-Nutr. 1987 May-Jun; 6(3):

Cruz-Martinez-A; Jimenez-MD
Thymoma without myasthenia gravis. Electrophysiological study after thymectomy.
J-Neurol-Neurosurg-Psychiatry. 1987 Apr; 50(4): 50

Csaba-G; Fulop-AK
Localisation of 3H-corticosterone inside thymocytes at different stages of ontogenetic development.
J-Submicrosc-Cytol. 1987 Oct; 19(4): 567-71

Csermely-P; Fodor-P; Somogyi-J
The tumor promoter tetradecanoylphorbol-13-acetate elicits the redistribution of heavy metals in subcellular fractions of rabbit thymocytes as measured by plasma emission spectroscopy.
Carcinogenesis. 1987 Nov; 8(11): 1663-6

Csermely-P; Gueth-S; Somogyi-J
The tumor promoter tetradecanoyl-phorbol-acetate (TPA) elicits the redistribution of zinc in subcellular fractions of rabbit thymocytes measured by X-ray fluorescence.
Biochem-Biophys-Res-Commun. 1987 Apr 29; 144(2): 8

Cung-MT; Marraud-M; Lefrancier-P; Dardenne-M; Bach-JF; Laussac-JP
NMR study of a lymphocyte differentiating thymic factor. An investigation of the Zn(II)-nonapeptide complexes (thymulin).
J-Biol-Chem. 1988 Apr 25; 263(12): 5574-80

D'Andrea-E; Saggioro-D; Fleissner-E; Chieco-Bianchi-L
Abelson murine leukemia virus-induced thymic lymphomas: transformation of a primitive lymphoid precursor.
JNCI. 1987 Jul; 79(1): 189-95

Dabrowski-MP; Dabrowska-Bernstein-BK; Stasiak-A; Gajkowski-K; Korniluk-S
Immunologic and clinical evaluation of multiple sclerosis patients treated with corticosteroids and/or calf thymic hormones.
Ann-NY-Acad-Sci. 1987; 496: 697-706

Dardenne-M; Savino-W; Feutren-G; Bach-JF
Stimulatory effects of cyclosporin A on human and mouse thymic epithelial cells.
Eur-J-Immunol. 1987 Feb; 17(2): 275-9

De-Angelo-L; Michael-SD
The effect of antisera to thymosin alpha 1 on the course of autoimmune ovarian dysgenesis in neonatally thymectomized mice.
J-Reprod-Immunol. 1987 May; 11(1): 41-53

De-la-Hera-A; Toribio-ML; Marcos-MA; Marquez-C; Martinez-C
Interleukin 2 pathway is autonomously activated in human T11+3-4-6-8- thymocytes.
Eur-J-Immunol. 1987 May; 17(5): 683-7

Delebassee-S; Van-Weydevelt-FC; Gualde-N
Effect of arachidonic acid metabolites on thymocyte tolerance.
Ann-N-Y-Acad-Sci. 1988; 524: 227-39

Denes-L; Hajos-G; Szporny-L; Szende-B; Lapis-K; Sipka-S; Szegedi-G
The effect of thymopoietin 32-34 (TP3) on suramin-induced inhibition on delayed type hypersensitivity in guinea pigs.
Immunol-Invest. 1987 Jun; 16(4): 275-9

Denes-L; Szende-B; Ember-J; Major-J; Szporny-L; Hajos-G; Nyeki-O; Schon-I; Lapis-K; Kisfaludy-L
Immunoregulating peptides II. In vitro effects of TP5 analogs on E-rosette formation and cell division.
Immunopharmacol-Immunotoxicol. 1987; 9(1): 1-18

Denes-L; Szende-B; Hajos-G; Szporny-L; Lapis-K
Therapeutic possibilities of thymopoietin fragments (TP3 and TP4) based on experimental animal models.
Drugs-Exp-Clin-Res. 1987; 13(5): 279-87

Denning-SM; Haynes-BF
Differentiation of human T cells.
Clin-Lab-Med. 1988 Mar; 8(1): 1-14

Denning-SM; Kurtzberg-J; Le-PT; Tuck-DT; Singer-KH; Haynes-BF
Human thymic epithelial cells directly induce activation of autologous immature thymocytes.
Proc-Natl-Acad-Sci-U-S-A. 1988 May; 85(9): 3125-9

Denning-SM; Tuck-DT; Vollger-LW; Springer-TA; Singer-KH; Haynes-BF
Monoclonal antibodies to CD2 and lymphocyte function-associated antigen 3 inhibit human thymic epithelial cell-dependent mature thymocyte activation.
J-Immunol. 1987 Oct 15; 139(8): 2573-8

Deschaux-P; Rouabhia-M
The thymus. Key organ between endocrinologic and immunologic systems.
Ann-NY-Acad-Sci. 1987; 496: 49-55

Dewes-W; Chandler-WF; Gormanns-R; Ebhardt-G
Brain metastasis of an invasive thymoma.
Neurosurgery. 1987 Mar; 20(3): 484-6

Di-Marino-V; Argeme-M; Brunet-C; Coppens-R; Bonnoit-J
Macroscopic study of the adult thymus.
Surg-Radiol-Anat. 1987; 9(1): 51-62

DiMario-FJ Jr; Lisak-RP; Kornstein-MJ; Brooks-JJ
Myasthenia gravis and primary squamous cell carcinoma of the thymus: a case report.
Neurology. 1988 Apr; 38(4): 580-2

Diezel-W; Eckert-R; Forner-K; Gruner-S
Functional effects of splenin32-34 on antibody formation in immunosuppressed mice.
Biomed-Biochim-Acta. 1986; 45(10): 1349-52

Dillman-RO; Beauregard-J; Royston-I; Zavanelli-MI
Phase II trial of thymosin fraction 5 and thymosin alpha 1.
J-Biol-Response-Mod. 1987 Jun; 6(3): 263-7

Dimery-IW; Lee-JS; Blick-M; Pearson-G; Spitzer-G; Hong-WK
Association of the Epstein-Barr virus with lymphoepithelioma of the thymus.
Cancer. 1988 Jun 15; 61(12): 2475-80

Dumontet-C; Dodon-MD; Gazzolo-L; Gerlier-D
Human T-cell leukemia virus type I-induced proliferation of human thymocytes requires the presence of a comitogen.
Cell-Immunol. 1988 Apr 1; 112(2): 391-401

Dy-C; Calvo-FA; Mindan-JP; Aparicio-LA; Algarra-SM; Gil-A; Gonzalez-F; Harguindey-S
Undifferentiated epithelial-rich invasive malignant thymoma: complete response to cisplatin, vinblastine, and bleomycin therapy.
J-Clin-Oncol. 1988 Mar; 6(3): 536-42

Eckert-R; Diezel-W; Forner-K; Pfuller-U; Sonnichsen-N; von-Baehr-R
Histamine receptor-bearing lymphocytes (HRL). VIII. Splenopentin-accelerated reconstitution of the immunoresponsiveness after elimination of HRL.
Allerg-Immunol-Leipz. 1986; 32(4): 267-70

el-Rouby-S; Ezine-S; Papiernik-M
Repopulation potential of thymocytes forming rosettes with phagocytic cells of the thymic reticulum.
Scand-J-Immunol. 1988 Apr; 27(4): 433-8

el-Rouby-S; Papiernik-M
Con A response and interleukin 2 receptor expression of thymocytes rosetting with phagocytic cells of the thymic reticulum.
Thymus. 1987; 9(4): 239-51

Ellingsworth-LR; Nakayama-D; Segarini-P; Dasch-J; Carrillo-P; Waegell-W
Transforming growth factor-beta s are equipotent growth inhibitors of interleukin-1-induced thymocyte proliferation.
Cell-Immunol. 1988 Jun; 114(1): 41-54

Elliott-JF; Rock-EP; Patten-PA; Davis-MM; Chien-YH
The adult T-cell receptor delta-chain is diverse and distinct from that of fetal thymocytes.
Nature. 1988 Feb 18; 331(6157): 627-31

Eren-R; Zharhary-D; Abel-L; Globerson-A
Age-related changes in the capacity of bone marrow cells to differentiate in thymic organ cultures.
Cell-Immunol. 1988 Apr 1; 112(2): 449-55

Eren-R; Zharhary-D; Abel-L; Globerson-A
Ontogeny of T cells: development of pre-T cells from fetal liver and yolk sac in the thymus microenvironment.
Cell-Immunol. 1987 Aug; 108(1): 76-84

Ernstrom-U; Karlsson-P; Soder-O
Isolation of a thymocyte growth peptide from human thymus.
Int-Arch-Allergy-Appl-Immunol. 1988; 85(4): 434-40

Ernstrom-U; Soder-O; Schmidt-W
Autocrine growth stimulation of thymocyte precursors in the thymus.
Thymus. 1987; 9(3): 141-8

Estopinan-V; Varela-C; Riobo-P; Dominguez-JR; Sancho-J
Ectopic Cushing's syndrome with periodic hormonogenesis--a case suggesting a pathogenetic mechanism.
Postgrad-Med-J. 1987 Oct; 63(744): 887-9

Ewald-SJ; Frost-WW
Effect of prenatal exposure to ethanol on development of the thymus.
Thymus. 1987; 9(4): 211-5

Ewert-DL; Weber-WT
Effect of phorbol myristate acetate on in vitro growth of lymphocytes from the chicken thymus and bursa of Fabricius.
Prog-Clin-Biol-Res. 1987; 238: 67-78

Ewing-T; Egerton-M; Wilson-A; Scollay-R; Shortman-K
Subpopulations of CD4- CD8- murine thymocytes: differences in proliferation rate in vivo and proliferative responses in vitro.
Eur-J-Immunol. 1988 Feb; 18(2): 261-8

Ezine-S; Jerabek-L; Weissman-I
The phenotype of thymocytes derived from a single clonogenic precursor.
J-Immunol. 1987 Oct 1; 139(7): 2195-9

Fabien-N; Auger-C; Monier-JC
Immunolocalization of thymosin alpha 1, thymopoietin and thymulin in mouse thymic epithelial cells at different stages of culture: a light and electron microscopic study.
Immunology. 1988 Apr; 63(4): 721-7

Fabris-N; Mocchegiani-E; Galli-M; Irato-L; Lazzarin-A; Moroni-M
AIDS, zinc deficiency, and thymic hormone failure [letter]
JAMA. 1988 Feb 12; 259(6): 839-40

Fabris-N; Mocchegiani-E; Mariotti-S; Caramia-G; Braccili-T; Pacini-F; Pinchera-A
Thymulin deficiency and low 3,5,3'-triiodothyronine syndrome in infants with low birth weight syndromes.
J-Clin-Endocrinol-Metab. 1987 Aug; 65(2): 247-52

Fabris-N; Mocchegiani-E; Muzzioli-M; Provinciali-M
Neuroendocrine-thymus interactions: perspectives for intervention in aging.
Ann-N-Y-Acad-Sci. 1988; 521: 72-87

Faist-E; Riedel-A; Bolla-K
Influence of thymopentin on postsurgical immune deficiency: a clinical pilot study.
Int-J-Clin-Pharmacol-Res. 1987; 7(1): 83-7

Farah-JM Jr; Hall-NR; Bishop-JF; Goldstein-AL; O'Donohue-TL
Thymosin fraction 5 stimulates secretion of immunoreactive beta-endorphin in mouse corticotropic tumor cells.
J-Neurosci-Res. 1987; 18(1): 140-6

Farr-AG; Anderson-SK; Braddy-SC; Mejino-JL Jr
Selective binding of Dolichos biflorus agglutinin to L3T4-, Lyt-2- thymocytes. Expression of terminal alpha-linked N-acetyl-D-galactosamine residues defines a subpopulation of fetal and adult muri
J-Immunol. 1988 Feb 15; 140(4): 1014-21

Farwell-AP; Devlin-JT; Stewart-JA
Total suppression of cortisol excretion by ketoconazole in the therapy of the ectopic adrenocorticotropic hormone syndrome.
Am-J-Med. 1988 Jun; 84(6): 1063-6

Fasanelli-S; Barbuti-D; Graziani-M; Donfrancesco-A
Thymic involvement in acute disseminated histiocytosis X.
Rays. 1987 Jan-Apr; 12(1): 39-45

Fatani-JA; Qayyum-MA; Mehta-L; Singh-U
Parasympathetic innervation of the thymus: a histochemical and immunocytochemical study.
J-Anat. 1986 Aug; 147: 115-9

Faure-GC; Bene-MC; Thomas-P; Tamisier-JN
An open trial of thymulin (FTS-Zn) in rheumatoid arthritis patients: sequential clinical and immunological follow-up [letter]
Clin-Exp-Rheumatol. 1987 Jul-Sep; 5(3): 291-3

Fava-RA; Piltch-AS
Histological distribution of the 35-KD protein substrate of the epidermal growth factor receptor/kinase in thymus.
J-Histochem-Cytochem. 1987 Nov; 35(11): 1309-15

Fiorica-E; Spector-S
Opioid binding site in EL-4 thymoma cell line.
Life-Sci. 1988; 42(2): 199-206

Fitzpatrick-FT; Greenstein-BD
Effects of various steroids on the thymus, spleen, ventral prostate and seminal vesicles in old orchidectomized rats.
J-Endocrinol. 1987 Apr; 113(1): 51-5

Ford-EG; Lockhart-SK; Sullivan-MP; Andrassy-RJ
Mediastinal mass following chemotherapeutic treatment of Hodgkin's disease: recurrent tumor or thymic hyperplasia?
J-Pediatr-Surg. 1987 Dec; 22(12): 1155-9

Forrest-D; Onions-D; Lees-G; Neil-JC
Altered structure and expression of c-myc in feline T-cell tumours.
Virology. 1987 May; 158(1): 194-205

Fowlkes-BJ; Kruisbeek-AM; Ton-That-H; Weston-MA; Coligan-JE; Schwartz-RH; Pardoll-DM
A novel population of T-cell receptor alpha beta-bearing thymocytes which predominantly expresses a single V beta gene family.
Nature. 1987 Sep 17-23; 329(6136): 251-4

Frangou-Lazaridis-M; Clinton-M; Goodall-GJ; Horecker-BL
Prothymosin alpha and parathymosin: amino acid sequences deduced from the cloned rat spleen cDNAs.
Arch-Biochem-Biophys. 1988 Jun; 263(2): 305-10

Frasca-D; Adorini-L; Doria-G
Enhanced frequency of mitogen-responsive T cell precursors in old mice injected with thymosin alpha 1.
Eur-J-Immunol. 1987 May; 17(5): 727-30

Fujimura-S; Kondo-T; Handa-M; Shiraishi-Y; Tamahashi-N; Nakada-T
Results of surgical treatment for thymoma based on 66 patients.
J-Thorac-Cardiovasc-Surg. 1987 May; 93(5): 708-14

Fukayama-M; Maeda-Y; Funata-N; Koike-M; Saito-K; Sakai-T; Ikeda-T
Pulmonary and pleural thymoma. Diagnostic application of lymphocyte markers to the thymoma of unusual site.
Am-J-Clin-Pathol. 1988 May; 89(5): 617-21

Furuoka-H; Taniyama-H; Matsui-T; Takahashi-T; Ichijo-S; Ono-T
Malignant thymoma with multiple metastases in a mare.
Nippon-Juigaku-Zasshi. 1987 Jun; 49(3): 577-9

Gallagher-PF; Miller-JF
Immunoglobulin gene expression is a normal differentiation event in embryonic thymocytes.
Eur-J-Immunol. 1988 Jan; 18(1): 183-6

Gambon-F; Kreisler-M; Diaz-Espada-F
Correlated expression of surface antigens in human thymocytes. Evidence of class I HLA modulation in thymic maturation.
Eur-J-Immunol. 1988 Jan; 18(1): 153-9

Gause-WC; Mountz-JD; Steinberg-AD
Characterization and differentiation of CD4-, CD8- thymocytes sorted with the Ly-24 marker.
J-Immunol. 1988 Jan 1; 140(1): 1-7

Gause-WC; Mountz-JD; Steinberg-AD
Subpopulations of CD4-, CD8- thymocytes.
Eur-J-Immunol. 1987 Sep; 17(9): 1387-90

Gaz-RD; Doubler-PB; Wang-CA
The management of 50 unusual hyperfunctioning parathyroid glands.
Surgery. 1987 Dec; 102(6): 949-57

Geenen-V; Defresne-MP; Robert-F; Legros-JJ; Franchimont-P; Boniver-J
The neurohormonal thymic microenvironment: immunocytochemical evidence that thymic nurse cells are neuroendocrine cells.
Neuroendocrinology. 1988 Apr; 47(4): 365-8

Geenen-V; Legros-JJ; Franchimont-P; Defresne-MP; Boniver-J; Ivell-R; Richter-D
The thymus as a neuroendocrine organ. Synthesis of vasopressin and oxytocin in human thymic epithelium.
Ann-NY-Acad-Sci. 1987; 496: 56-66

Gengozian-N; Good-RA; Day-NK
Guinea pig and gerbil erythrocytes rosette with different cells in the blood, bone marrow, and thymus of the cat.
Cell-Immunol. 1988 Mar; 112(1): 1-13

Georgiou-HM; Lagarde-AC; Bellgrau-D
T cell dysfunction in the diabetes-prone BB rat. A role for thymic migrants that are not T cell precursors.
J-Exp-Med. 1988 Jan 1; 167(1): 132-48

Geppetti-P; Maggi-CA; Zecchi-Orlandini-S; Santicioli-P; Meli-A; Frilli-S; Spillantini-MG; Amenta-F
Substance P-like immunoreactivity in capsaicin-sensitive structures of the rat thymus.
Regul-Pept. 1987 Sep; 18(5-6): 321-9

Gerli-R; Bertotto-A; Spinozzi-F; Cernetti-C; Battaglia-A; Falchetti-R; Grignani-F; Rambotti-P
Thymic hormone modulation of CD38 (T10) antigen on human cord blood lymphocytes.
Clin-Immunol-Immunopathol. 1987 Dec; 45(3): 323-32

Gibson-LE; Muller-SA
Dermatologic disorders in patients with thymoma.
Acta-Derm-Venereol-Stockh. 1987; 67(4): 351-6

Girdlestone-J; Milstein-C
Differential expression and interferon response of HLA class I genes in thymocyte lines and response variants.
Eur-J-Immunol. 1988 Jan; 18(1): 139-43

Goff-BL; Roth-JA; Arp-LH; Incefy-GS
Growth hormone treatment stimulates thymulin production in aged dogs.
Clin-Exp-Immunol. 1987 Jun; 68(3): 580-7

Goldberg-RE; Haaga-JR; Yulish-BS
Serial CT scans in thymic hyperplasia.
J-Comput-Assist-Tomogr. 1987 May-Jun; 11(3): 539-4

Golsteyn-EJ; Fritzler-MJ
The role of the thymus-hypothalamus-pituitary-gonadal axis in normal immune processes and autoimmunity.
J-Rheumatol. 1987 Oct; 14(5): 982-90

Gomes-RM; Shinzato-TO; Grynberg-NF
Growth control of experimental tumor by syngeneic thymocyte extracts.
Braz-J-Med-Biol-Res. 1987; 20(3-4): 449-51

Gomez-Marquez-J; Segade-F
Prothymosin alpha is a nuclear protein.
FEBS-Lett. 1988 Jan 4; 226(2): 217-9

Gondo-H; Kudo-J; White-JW; Barr-C; Selvanayagam-P; Saunders-GF
Differential expression of the human thymosin-beta 4 gene in lymphocytes, macrophages, and granulocytes.
J-Immunol. 1987 Dec 1; 139(11): 3840-8

Gonzalez-MD; Rondinone-CM; Burton-G; Lantos-CP; Roldan-A
Mechanism of the inhibitory effect of 11 beta-hydroxypregna-1, 4-diene-3, 20-dione (delta HOP) at high concentrations in RNA synthesis.
Acta-Physiol-Pharmacol-Latinoam. 1986; 36(3): 303-

Goodall-GJ; Horecker-BL
Molecular cloning of the cDNA for rat spleen thymosin beta 10 and the deduced amino acid sequence.
Arch-Biochem-Biophys. 1987 Jul; 256(1): 402-5

Gossrau-R; Merker-HJ; Gunther-T; Graf-R; Vormann-J
Enzymatic and morphological response of the thymus to drugs in normal and zinc-deficient pregnant rats and their fetuses.
Histochemistry. 1987; 86(3): 321-9

Goya-RG; Takahashi-S; Quigley-KL; Sosa-YE; Goldstein-AL; Meites-J
Immune-neuroendocrine interactions during aging: age-dependent thyrotropin-inhibiting activity of thymosin peptides.
Mech-Ageing-Dev. 1987 Dec; 41(3): 219-27

Gradzka-I; Lukaszkiewicz-J; Lorenc-R
Application of the rat thymus receptor in a specific, sensitive and simplified assay of 1,25-dihydroxyvitamin D in blood serum.
Biochim-Biophys-Acta. 1988 Apr 14; 965(1): 52-9

Granata-AL; Vecchi-C; De-Toni-M; Corsi-A
Life span of lymphocytes in thymectomized rats.
Pathologica. 1986 Sep-Oct; 78(1057): 573-7

Green-WR; Pressoir-R; Gumbs-RV; Warner-O; Naab-T; Qayumi-M
Intrapulmonary thymoma.
Arch-Pathol-Lab-Med. 1987 Nov; 111(11): 1074-6

Greenstein-BD; Fitzpatrick-FT; Kendall-MD; Wheeler-MJ
Regeneration of the thymus in old male rats treated with a stable analogue of LHRH.
J-Endocrinol. 1987 Mar; 112(3): 345-50

Grooten-J; Leroux-Roels-G; Fiers-W
Specific suppression elicited by EL4 lymphoma cells in syngeneic mice. Specificity includes self-antigens on EL4.
Eur-J-Immunol. 1987 May; 17(5): 605-11

Grossman-Z; Mietz-JA; Kuff-EL
Nearly identical members of the heterogeneous IAP gene family are expressed in thymus of different mouse strains.
Nucleic-Acids-Res. 1987 May 11; 15(9): 3823-34

Haar-JL; Grogan-WM; Popp-JD; Taubenberger-JK
Chemoattraction enrichment of thymus-homing bone marrow cells.
Cell-Tissue-Kinet. 1987 Mar; 20(2): 227-31

Habbal-O; Leeming-G; Morgan-ET; McLean-JM
A dioestrous increase in thymocyte proliferation during the oestrous cycle.
Experientia. 1987 Sep 15; 43(9): 1010-2

Hadden-JW; Hadden-EM; Coffey-RG
Interleukin II increases cyclic GMP levels in immature thymocytes and mitogen-primed T-lymphocytes.
Int-J-Immunopharmacol. 1987; 9(7): 851-7

Hadzija-M; Slijepcevic-M; Sverko-V; Poljak-Blazi-M
Influence of the thymus extract on the immunological function of animals with experimental diabetes.
Horm-Metab-Res. 1987 Jan; 19(1): 6-10

Hagiwara-H; Yokota-T; Luh-J; Lee-F; Arai-K; Arai-N; Zlotnik-A
The AKR thymoma BW5147 is able to produce lymphokines when stimulated with calcium ionophore and phorbol ester.
J-Immunol. 1988 Mar 1; 140(5): 1561-5

Halbach-M; Homberg-V; Freund-HJ
Neuromuscular, autonomic and central cholinergic hyperactivity associated with thymoma and acetylcholine receptor-binding antibody.
J-Neurol. 1987 Aug; 234(6): 433-6

Hale-ML; Greiner-DL; McCarthy-KF
Characterization of rat prothymocyte with monoclonal antibodies recognizing rat lymphocyte membrane antigenic determinants.
Cell-Immunol. 1987 Jun; 107(1): 188-200

Hanazawa-S; Hirose-K; Ohmori-Y; Amano-S; Kitano-S
Bacteroides gingivalis fimbriae stimulate production of thymocyte-activating factor by human gingival fibroblasts.
Infect-Immun. 1988 Jan; 56(1): 272-4

Handzel-ZT; Berner-Y; Segal-O; Burstein-Y; Buchner-V; Pecht-M; Levin-S; Burstein-R; Milchan-R; Bentw
Immunoreconstitution of T-cell impairments in asymptomatic male homosexuals by thymic humoral factor (THF).
Int-J-Immunopharmacol. 1987; 9(2): 165-73

Hannappel-E; Kalbacher-H; Voelter-W
Thymosin beta 4Xen: a new thymosin beta 4-like peptide in oocytes of Xenopus laevis.
Arch-Biochem-Biophys. 1988 Feb 1; 260(2): 546-51

Hannappel-E; van-Kampen-M
Determination of thymosin beta 4 in human blood cells and serum.
J-Chromatogr. 1987 Jun 26; 397: 279-85

Hannum-C; Marrack-P; Kubo-R; Kappler-J
Thymocytes with the predicted properties of pre-T cells.
J-Exp-Med. 1987 Oct 1; 166(4): 874-89

Haritos-AA
Alpha-thymosins: relationships in structure, distribution, and function.
Isozymes-Curr-Top-Biol-Med-Res. 1987; 14: 123-52

Haritos-AA; Yialouris-PP; Heimer-EP; Felix-AM; Rosemeyer-MA
On the molecular size of thymosins.
FEBS-Lett. 1987 Jun 22; 218(1): 107-12

Harms-G; van-Goor-H; Koudstaal-J; de-Ley-L; Hardonk-MJ
Simultaneous immunohistochemical demonstration of antigen expression and 5-bromodeoxyuridine incorporation in plastic embedded sections.
Histochemistry. 1987; 86(4): 393-5

Harris-GJ; Harman-PK; Trinkle-JK; Grover-FL
Standard biplane roentgenography is highly sensitive in documenting mediastinal masses.
Ann-Thorac-Surg. 1987 Sep; 44(3): 238-41

Hashimoto-S; Itoh-Y; Kawaguchi-A; Mizuno-Y; Muraoka-M; Tsuchiya-M
In vitro effect of a thymic epithelial culture supernate or thymosin fraction 5 on rabbit platelet aggregation and intracellular cyclic AMP levels.
J-Cell-Physiol. 1987 Dec; 133(3): 499-506

Hashimura-H; Toyokawa-T; Hato-F; Oshitani-N; Kimura-S; Kinoshita-Y
Separation of biologically active polypeptides from rat thymus-epithelial-cell-culture-supernatant by high-performance liquid chromatography.
Cell-Mol-Biol. 1987; 33(3): 375-86

Hassoun-EA
In vivo and in vitro interactions of TCDD and other ligands of the Ah-receptor: effect on embryonic and fetal tissues.
Arch-Toxicol. 1987 Dec; 61(2): 145-9

Hattori-A; Kunz-HW; Gill-TJ 3d; Pan-SF; Shinozuka-H
Diversity of the promoting action of cyclosporine on the induction of murine lymphoid tumors.
Carcinogenesis. 1988 Jun; 9(6): 1091-4

Hattori-A; Kunz-HW; Gill-TJ 3d; Shinozuka-H
Thymic and lymphoid changes and serum immunoglobulin abnormalities in mice receiving cyclosporine.
Am-J-Pathol. 1987 Jul; 128(1): 111-20

Havran-WL; Poenie-M; Kimura-J; Tsien-R; Weiss-A; Allison-JP
Expression and function of the CD3-antigen receptor on murine CD4+8+ thymocytes.
Nature. 1987 Nov 12-18; 330(6144): 170-3

Heavner-GA; Kroon-DJ; Audhya-T; Goldstein-G
Biologically active analogs of thymopentin with enhanced enzymatic stability.
Peptides. 1986 Nov-Dec; 7(6): 1015-9

Heimann-A; Sneige-N; Shirkhoda-A; DeCaro-LF
Fine needle aspiration cytology of thymolipoma. A case report.
Acta-Cytol-Baltimore. 1987 May-Jun; 31(3): 335-9

Helland-DE; Male-R; Kleppe-K
Separation of damage specific DNA endonuclease activities present in calf thymus.
FEBS-Lett. 1987 Mar 9; 213(1): 215-20

Hemmick-LM; Bidlack-JM
Beta-endorphin modulation of mitogen-stimulated calcium uptake by rat thymocytes.
Life-Sci. 1987 Oct 19; 41(16): 1971-8

Henner-WD; Kiker-NP; Jorgensen-TJ; Munck-JN
Purification and amino-terminal amino acid sequence of an apurinic/apyrimidinic endonuclease from calf thymus.
Nucleic-Acids-Res. 1987 Jul 24; 15(14): 5529-44

Henry-L; Anderson-G
Epithelial-cell architecture during involution of the human thymus.
J-Pathol. 1987 Jul; 152(3): 149-55

Herbst-WM; Kummer-W; Hofmann-W; Otto-H; Heym-C
Carcinoid tumors of the thymus. An immunohistochemical study.
Cancer. 1987 Nov 15; 60(10): 2465-70

Heron-CW; Husband-JE; Williams-MP
Hodgkin disease: CT of the thymus.
Radiology. 1988 Jun; 167(3): 647-51

Hiai-S; Sasayama-Y; Oguro-C
Chronic effects of ginseng saponin, glycyrrhizin and flavin adenine dinucleotide on adrenal and thymus weight in normal and dexamethasone-treated rats.
Chem-Pharm-Bull-Tokyo. 1987 Jan; 35(1): 241-8

Hildreth-NG; Schneider-AB; Cave-WT Jr
A comparative study between individuals receiving thymic irradiation in infancy and their nontreated siblings: clinical and laboratory thyroid abnormalities.
Radiat-Res. 1987 Jun; 110(3): 458-67

Hirokawa-K; Utsuyama-M; Katsura-Y; Sado-T
Influence of age on the proliferation and peripheralization of thymic T cells.
Arch-Pathol-Lab-Med. 1988 Jan; 112(1): 13-21

Hirose-W; Kawagoe-M; Hara-M; Kitani-A; Hirose-T; Norioka-K; Harigai-M; Nakamura-H
Production of thymocyte-stimulating activity by cultured human thyroid epithelial cells.
Clin-Exp-Immunol. 1987 Oct; 70(1): 102-9

Hitt-ME; Shaw-DP; Hogan-PM; Lennon-VA; Amann-JF
Radiation treatment for thymoma in a dog.
J-Am-Vet-Med-Assoc. 1987 May 1; 190(9): 1187-90

Ho-AD; Stehle-B; Birr-C; Heinzel-W; Nebe-TC
Differentiation changes in cord blood T lymphocytes induced by synthetic C-terminal peptides of thymosin alpha 1.
Thymus. 1987; 9(2): 77-84

Hodgkin-PD; Bond-MW; O'Garra-A; Frank-G; Lee-F; Coffman-RL; Zlotnik-A; Howard-M
Identification of IL-6 as a T cell-derived factor that enhances the proliferative response of thymocytes to IL-4 and phorbol myristate acetate.
J-Immunol. 1988 Jul 1; 141(1): 151-7

Hofert-JF; Mahowald-TA; Mahowald-NA; Heidrick-ML
The binding of FITC-insulin to ANAE-positive cells in rat thymus.
Experientia. 1988 Jan 15; 44(1): 37-8

Hofmann-WJ; Moller-P; Otto-HF
Thymic hyperplasia. I. True thymic hyperplasia. Review of the literature.
Klin-Wochenschr. 1987 Jan 15; 65(2): 49-52

Hofmann-WJ; Moller-P; Otto-HF
Thymic hyperplasia. II. Lymphofollicular hyperplasia of the thymus. An immunohistologic study.
Klin-Wochenschr. 1987 Jan 15; 65(2): 53-60

Hofmann-WJ; Momburg-F; Moller-P; Otto-HF
Intra- and extrathymic B cells in physiologic and pathologic conditions. Immunohistochemical study on normal thymus and lymphofollicular hyperplasia of the thymus.
Virchows-Arch-[A]. 1988; 412(5): 431-42

Hohn-KT; Grosse-F
Processivity of the DNA polymerase alpha-primase complex from calf thymus.
Biochemistry. 1987 May 19; 26(10): 2870-8

Holtfreter-HB; Cohen-N
In vitro behavior of thymic nurse cell-like complexes from mechanically and enzymatically dissociated frog tadpole thymuses.
Am-J-Anat. 1987 Aug; 179(4): 342-55

Horton-JD; Russ-JH; Aitchison-P; Horton-TL
Thymocyte/stromal cell chimaerism in allothymus-grafted Xenopus: developmental studies using the X. borealis fluorescence marker.
Development. 1987 May; 100(1): 107-17

Hosono-M; Hosokawa-T; Kina-T; Katsura-Y
Neonatal tolerance induction in the thymus to MHC-class II-associated antigens. III. Significance of hemopoietic stem cells for induction and maintenance of Mls tolerance by continuous supply of t
Cell-Immunol. 1987 Aug; 108(1): 162-74

Hougen-HP; Klausen-B
The thymus reconstituted nude rat: lymphocyte subpopulations and immunological characteristics.
Thymus. 1987; 10(3-4): 207-17

Hougen-HP; Klausen-B; Stenvang-JP; Kraemmer-J; Rygaard-J
Effects of xenogeneic, allogeneic and isogeneic thymus grafts on lymphocyte populations in peripheral lymphoid organs of the nude rat.
Lab-Anim. 1987 Apr; 21(2): 103-11

Houssaint-E; Diez-E; Pink-JR
Ontogeny and tissue distribution of the chicken Bu-1a antigen.
Immunology. 1987 Nov; 62(3): 463-70

Howe-RC; MacDonald-HR
Heterogeneity of immature (Lyt-2-/L3T4-) thymocytes. Identification of four major phenotypically distinct subsets differing in cell cycle status and in vitro activation requirements.
J-Immunol. 1988 Feb 15; 140(4): 1047-55

Hubscher-U; Gassmann-M; Spadari-S; Brown-NC; Ferrari-E; Buhk-HJ
Mammalian DNA polymerase alpha: a replication-competent holoenzyme form from calf thymus.
Philos-Trans-R-Soc-Lond-[Biol]. 1987 Dec 15; 317(1

Hui-YZ; Guo-QX
Thymolipoma simulating pericardial cyst. A clinicopathologic rarity.
Chin-Med-J-[Engl]. 1987 Oct; 100(10): 809-12

Huiskamp-R; Davids-JA; van-Ewijk-W
The effect of graded doses of fission neutrons or X rays on the stromal compartment of the thymus in mice.
Radiat-Res. 1988 Jan; 113(1): 25-39

Huiskamp-R; van-Ewijk-W
Unilateral T cell maturation arrest in the thymus of CBA/H mice as a long-term effect after neutron irradiation.
Radiat-Res. 1987 Apr; 110(1): 95-107

Ikehara-S; Shimizu-J; Yasumizu-R; Nakamura-T; Inaba-M; Inoue-S; Oyaizu-N; Sugiura-K; Oo-MM; Hamashim
Thymic rudiments are responsible for induction of functional T cells in nu/nu mice.
Thymus. 1987; 10(3-4): 193-205

Imaizumi-A; Torisu-M; Yoshida-T
A chemotactic factor for rat thymocytes may regulate T-lymphocyte migration toward the thymic microenvironment.
Cell-Immunol. 1987 Aug; 108(1): 53-63

Inaba-K; Witmer-Pack-MD; Inaba-M; Muramatsu-S; Steinman-RM
The function of Ia+ dendritic cells and Ia- dendritic cell precursors in thymocyte mitogenesis to lectin and lectin plus interleukin 1.
J-Exp-Med. 1988 Jan 1; 167(1): 149-62

Insler-MS; Shelin-RG
Sjogren's syndrome and thymoma.
Am-J-Ophthalmol. 1987 Jul 15; 104(1): 90-1

Isaacson-PG; Norton-AJ; Addis-BJ
The human thymus contains a novel population of B lymphocytes.
Lancet. 1987 Dec 26; 2(8574): 1488-91

Ishikura-H; Ishikawa-N; Aizawa-M
Differential expression of HLA-class II antigens in the human thymus. Relative paucity of HLA-DQ antigens in the thymic medulla.
Transplantation. 1987 Aug; 44(2): 314-7

Ishizuka-T; Ito-Y; Mori-H; Miura-K; Nagao-S; Nagata-K; Nozawa-Y
Concanavalin A- or phorbol ester-induced translocation of protein kinase C in thymoma cells from a patient with myasthenia gravis.
Clin-Chim-Acta. 1988 Feb 29; 172(1): 109-15

Ito-M; Taki-T; Miyake-M; Mitsuoka-A
Lymphocyte subsets in human thymoma studied with monoclonal antibodies.
Cancer. 1988 Jan 15; 61(2): 284-7

Itoh-T
Development of Thy 1 positive mouse thymocytes. Analysis of early fetal thymocytes by flow microfluorometry.
Thymus. 1987; 9(4): 217-23

Itoh-T; Doi-H; Chin-S; Nishimura-T; Kasahara-S
Establishment of mouse thymic nurse cell clones from a spontaneous BALB/c thymic tumor.
Eur-J-Immunol. 1988 May; 18(5): 821-4

Itoh-T; Kasahara-S
Development of Thy 1 positive rat thymocytes: analysis of fetal and neonatal thymocytes by flow microfluorometry.
Thymus. 1987; 9(1): 13-23

Jankovic-BD; Korolija-P; Isakovic-K; Popeskovic-LJ; Pesic-MC; Horvat-J; Jeremic-D; Vajs-V
Immunorestorative effects in elderly humans of lipid and protein fractions from the calf thymus: a double-blind study.
Ann-N-Y-Acad-Sci. 1988; 521: 247-59

Jankovic-BD; Maric-D
Restoration of in vivo humoral and cell-mediated immune responses in neonatally thymectomized and aged rats by means of lipid and protein fractions from the calf thymus.
Ann-N-Y-Acad-Sci. 1988; 521: 228-46

Jaretzki-A 3d; Penn-AS; Younger-DS; Wolff-M; Olarte-MR; Lovelace-RE; Rowland-LP
"Maximal" thymectomy for myasthenia gravis. Results.
J-Thorac-Cardiovasc-Surg. 1988 May; 95(5): 747-57

Jenkinson-EJ; Kingston-R; Owen-JJ
Importance of IL-2 receptors in intra-thymic generation of cells expressing T-cell receptors.
Nature. 1987 Sep 10-16; 329(6135): 160-2

Jensen-DE; Sando-JJ
Absence of protein kinase C in nuclei of EL4 mouse thymoma cells.
Cancer-Res. 1987 Jul 15; 47(14): 3868-72

Jenski-LJ; Belloni-ML; Miller-BA
Thymic influence on the T-lymphocyte self MHC repertoire. I. Helper T-lymphocyte precursors.
Thymus. 1988; 11(3): 135-50

Jenski-LJ; Miller-BA
Thymic influence on the T-lymphocyte self MHC repertoire. II. Cytotoxic T-lymphocyte precursors.
Thymus. 1988; 11(3): 151-71

Johnson-MA; Johns-DG; Fridland-A
2',3'-Dideoxynucleoside phosphorylation by deoxycytidine kinase from normal human thymus extracts: activation of potential drugs for AIDS therapy.
Biochem-Biophys-Res-Commun. 1987 Nov 13; 148(3): 1

Jones-GW; Laukkanen-E; Miller-RR
Well-differentiated squamous cell carcinoma of the thymus.
Can-Med-Assoc-J. 1987 Jul 1; 137(1): 43-4

Joshi-NN; Sainis-KB; Phondke-GP
Concanavalin A-induced electrokinetic changes and redistribution of receptors in mouse thymocytes.
Thymus. 1987; 9(2): 65-75

Judd-RL
Massive thymic hyperplasia with myoid cell differentiation.
Hum-Pathol. 1987 Nov; 18(11): 1180-3

Just-G; Holler-E
The binding of adenosine(5')tetraphospho(5')adenosine to calf thymus histones measured by non-equilibrium dialysis.
Biochem-J. 1987 Sep 15; 246(3): 681-6

Kaffenberger-W; Gruber-DF; MacVittie-TJ
Flow cytometric characterization of rat thymus cells in a radiation-dominated model of combined injury.
J-Trauma. 1988 May; 28(5): 593-601

Kamiyama-T; Cortes-GT; Rubio-Z
The role of thymocytes and IgG antibody in protection against malaria in nude rats.
Zentralbl-Bakteriol-Mikrobiol-Hyg-[A]. 1987 May; 2

Kanazawa-K; Higuchi-M; Okamoto-M; Fukuyama-K; Epstein-WL
Induction of granuloma-dependent angiotensin-converting enzyme and eosinophil chemotactic factor in the skin of athymic nude mice.
J-Cell-Biochem. 1987 May; 34(1): 61-9

Kanellos-J; Pietersz-GA; Cunningham-Z; McKenzie-IF
Anti-tumour activity of aminopterin-monoclonal antibody conjugates; in vitro and in vivo comparison with methotrexate-monoclonal antibody conjugates.
Immunol-Cell-Biol. 1987 Dec; 65 (Pt 6): 483-93

Kaneshima-H; Ito-M; Asai-J; Taguchi-O; Hiai-H
Thymic epithelial reticular cell subpopulations in mice defined by monoclonal antibodies.
Lab-Invest. 1987 Apr; 56(4): 372-80

Kaneshima-H; Itoh-M; Asai-J; Taguchi-O; Hiai-H
Reorganization of thymic microenvironments during development and lymphomagenesis.
Jpn-J-Cancer-Res. 1987 Aug; 78(8): 799-806

Kappler-JW; Roehm-N; Marrack-P
T cell tolerance by clonal elimination in the thymus.
Cell. 1987 Apr 24; 49(2): 273-80

Karttunen-T
Basement membrane proteins and reticulin in a normal thymus and the thymus in myasthenia gravis.
Virchows-Arch-[A]. 1987; 411(3): 245-52

Kast-WM; Voordouw-AC; Leupers-T; Visser-JW; Melief-CJ
Thymic immune response gene function in radiation chimeras reconstituted with purified hemopoietic stem cells.
Eur-J-Immunol. 1987 Apr; 17(4): 471-5

Kato-S; Nakamura-H
Suppression of acute experimental allergic encephalomyelitis by synthetic serum thymic factor. Clinical, histopathological, and immunohistochemical studies.
Acta-Neuropathol-Berl. 1988; 75(4): 337-44

Kato-S; Schoefl-GI
The vasculature of the guinea-pig thymus: topographic studies by light and electron microscopy.
Arch-Histol-Jpn. 1987 Jul; 50(3): 299-314

Katorza-E; Pecht-M; Apte-RN; Benharroch-D; Burstein-Y; Trainin-N; Rager-Zisman-B
Restoration of immunological responses by THF, a thymic hormone, in mice infected with murine cytomegalovirus (MCMV).
Clin-Exp-Immunol. 1987 Nov; 70(2): 268-75

Katzin-WE; Fishleder-AJ; Linden-MD; Tubbs-RR
Immunoglobulin and T-cell receptor genes in thymomas: genotypic evidence supporting the nonneoplastic nature of the lymphocytic component.
Hum-Pathol. 1988 Mar; 19(3): 323-8

Kawanami-S; Kakuno-T; Horio-T
Purification of acetylcholine receptor-like protein from fetal calf thymus.
Jpn-J-Psychiatry-Neurol. 1987 Mar; 41(1): 97-104

Kawanami-S; Kamei-H; Oita-J; Kurokawa-M; Uchida-Y; Hayashi-K
Acetylcholine-receptor-like protein from human thymoma associated with myasthenia gravis.
J-Neurol. 1987 May; 234(4): 207-10

Kelley-KW; Brief-S; Westly-HJ; Novakofski-J; Bechtel-PJ; Simon-J; Walker-ER
Hormonal regulation of the age-associated decline in immune function.
Ann-NY-Acad-Sci. 1987; 496: 91-7

Kelley-KW; Davila-DR; Brief-S; Simon-J; Arkins-S
A pituitary-thymus connection during aging.
Ann-N-Y-Acad-Sci. 1988; 521: 88-98

Kendall-MD
The syncytial nature of epithelial cells in the thymic cortex.
J-Anat. 1986 Aug; 147: 95-106

Kerekgyarto-C; Erdei-J; Mandi-B; Olveti-E; Fachet-J
Genetic factors and efficiency of thymosin.
Thymus. 1987; 9(1): 61-3

Kerekgyarto-C; Erdei-J; Mandi-B; Olveti-E; Fachet-J
The role of genetic factors in the immunomodulating effect of thymosin.
Folia-Biol-Praha. 1987; 33(4): 237-45

Kerr-RS; Kaye-AH; Bartlett-PF
A quantitative assessment by flow cytometry of the preservation of the Thy-1 antigen by various fixation procedures.
Neurosci-Lett. 1988 May 3; 87(3): 259-65

Kikuchi-A; Kamo-I; Fujisawa-K; Nonaka-I
Semi-quantitative immunohistochemical studies on Thy-1 antigen expressed by thymic myoid cells.
J-Neuroimmunol. 1988 Feb; 17(3): 217-28

Kimura-N; Toyonaga-B; Yoshikai-Y; Du-RP; Mak-TW
Sequences and repertoire of the human T cell receptor alpha and beta chain variable region genes in thymocytes.
Eur-J-Immunol. 1987 Mar; 17(3): 375-83

Kina-T
Differentiation and maturation of functional T lymphocyte subsets in the thymus. II. Generation of T cell specificities and functions from a single stem cell.
Bull-Chest-Dis-Res-Inst-Kyoto-Univ. 1987 Mar; 20(1

Kina-T
Differentiation and maturation of functional T lymphocyte subsets in the thymus. I. Effect of irradiation of the thymus.
Bull-Chest-Dis-Res-Inst-Kyoto-Univ. 1986 Mar; 19(1

Kina-T; Nishikawa-S; Amagai-T; Katsura-Y
Concanavalin A-mediated polyclonal helper assay of normal thymocytes and its use for the analysis of ontogeny.
Thymus. 1987; 9(3): 159-72

Kincaid-RL; Takayama-H; Billingsley-ML; Sitkovsky-MV
Differential expression of calmodulin-binding proteins in B, T lymphocytes and thymocytes.
Nature. 1987 Nov 12-18; 330(6144): 176-8

Kinnon-C; McGuire-KL; Rothenberg-EV
Differential regulation of T cell receptor gamma genes in immature thymocyte populations.
Eur-J-Immunol. 1987 Sep; 17(9): 1265-9

Kinohara-N; Makinodan-T
Thymic stromal cells produce inhibitors of proliferation of thymic, but not splenic T cells.
Thymus. 1987; 10(3-4): 179-92

Kirchner-T; Hoppe-F; Schalke-B; Muller-Hermelink-HK
Microenvironment of thymic myoid cells in myasthenia gravis.
Virchows-Arch-[B]. 1988; 54(5): 295-302

Kirchner-T; Tzartos-S; Hoppe-F; Schalke-B; Wekerle-H; Muller-Hermelink-HK
Pathogenesis of myasthenia gravis. Acetylcholine receptor-related antigenic determinants in tumor-free thymuses and thymic epithelial tumors.
Am-J-Pathol. 1988 Feb; 130(2): 268-80

Klein-AS; Lang-R; Eshel-I; Sharabi-Y; Shoham-J
Modulation of immune response and tumor development in tumor-bearing mice treated by the thymic factor thymostimulin.
Cancer-Res. 1987 Jul 1; 47(13): 3351-6

Knight-JA; Rezuke-WN; Wong-SH; Hopfer-SM; Zaharia-O; Sunderman-FW Jr
Acute thymic involution and increased lipoperoxides in thymus of nickel chloride-treated rats.
Res-Commun-Chem-Pathol-Pharmacol. 1987 Mar; 55(3):

Knobler-RL; Oldstone-MB
Infection and involution of mouse thymus by MHV-4.
Adv-Exp-Med-Biol. 1987; 218: 451-3

Knopp-J; Strakova-Z; Brtko-J
Effects of forskolin and triiodothyronine on cyclic AMP production in rat thymocytes.
Endocrinol-Exp-Bratisl. 1988 Mar; 22(1): 29-34

Koga-Y; Lindstrom-E; Fenyo-EM; Wigzell-H; Mak-TW
High levels of heterodisperse RNAs accumulate in T cells infected with human immunodeficiency virus and in normal thymocytes.
Proc-Natl-Acad-Sci-U-S-A. 1988 Jun; 85(12): 4521-5

Kohler-C; Jeanvoine-G; Pierrez-J; Olive-D; Gerard-H
Modifications of the thymus and splenic thymic dependent zones after in utero exposure to phenytoin: qualitative and quantitative analysis in C3H mice.
Dev-Pharmacol-Ther. 1987; 10(6): 405-12

Kokkinopoulos-D; Perez-S; Baxevanis-C; Papamichail-M
Effect of two thymosin fraction 5 polypeptides on human peripheral blood lymphocytes.
Immunopharmacol-Immunotoxicol. 1987; 9(4): 409-19

Komschlies-KL; Greiner-DL; Shultz-L; Goldschneider-I
Defective lymphopoiesis in the bone marrow of motheaten (me/me) and viable motheaten (mev/mev) mutant mice. III. Normal mouse bone marrow cells enable mev/mev prothymocytes to generate thymocytes
J-Exp-Med. 1987 Oct 1; 166(4): 1162-7

Kosmidis-PA; Iliopoulos-E; Pentea-S
Combination chemotherapy with cyclophosphamide, adriamycin, and vincristine in malignant thymoma and myasthenia gravis.
Cancer. 1988 May 1; 61(9): 1736-40

Kouassi-E; Li-YS; Flacher-M; Revillard-JP
Cyclic adenosine 3',5'-monophosphate response to dopamine in mouse lymphoid cells and cell lines.
Life-Sci. 1987 Jun 22; 40(25): 2385-92

Kovacs-WJ; Olsen-NJ
Androgen receptors in human thymocytes.
J-Immunol. 1987 Jul 15; 139(2): 490-3

Koyama-H; Kunita-S; Uehara-N; Saito-H
Preparation of monoclonal antibodies reacting with bovine cortical thymocytes.
Nippon-Juigaku-Zasshi. 1987 Jun; 49(3): 433-7

Kraft-R; Fankhauser-G; Gerber-H; Hess-MW; Cottier-H
Age-related involution and terminal disorganization of the human thymus.
Int-J-Radiat-Biol. 1988 Jan; 53(1): 169-76

Kringsholm-B; Christoffersen-P
Lymph-node and thymus pathology in fatal drug addiction.
Forensic-Sci-Int. 1987 Aug; 34(4): 245-54

Kurihara-M; Castren-E; Gutkind-JS; Saavedra-JM
Lower number of atrial natriuretic peptide receptors in thymocytes and spleen cells of spontaneously hypertensive rats.
Biochem-Biophys-Res-Commun. 1987 Dec 31; 149(3): 1

Kurihara-M; Katamine-S; Saavedra-JM
Atrial natriuretic peptide, ANP(99-126), receptors in rat thymocytes and spleen cells.
Biochem-Biophys-Res-Commun. 1987 Jun 15; 145(2): 7

Kyewski-BA
Seeding of thymic microenvironments defined by distinct thymocyte-stromal cell interactions is developmentally controlled.
J-Exp-Med. 1987 Aug 1; 166(2): 520-38

Kyewski-BA; Momburg-F; Schirrmacher-V
Phenotype of stromal cell-associated thymocytes in situ is compatible with selection of the T cell repertoire at an "immature" stage of thymic T cell differentiation.
Eur-J-Immunol. 1987 Jul; 17(7): 961-7

Lacaze-Masmonteil-T; de-Keyzer-Y; Luton-JP; Kahn-A; Bertagna-X
Characterization of proopiomelanocortin transcripts in human nonpituitary tissues.
Proc-Natl-Acad-Sci-USA. 1987 Oct; 84(20): 7261-5

Ladisch-S; Kitada-S; Hays-EF
Gangliosides shed by tumor cells enhance tumor formation in mice.
J-Clin-Invest. 1987 Jun; 79(6): 1879-82

Lahiri-TK; Misra-RC; Agarwal-SP; Singh-A; Khanna-MN
Myasthenia gravis due to thymoma in an infant.
Indian-J-Chest-Dis-Allied-Sci. 1985 Apr-Jun; 27(2)

Laigret-F; Repaske-R; Boulukos-K; Rabson-AB; Khan-AS
Potential progenitor sequences of mink cell focus-forming (MCF) murine leukemia viruses: ecotropic, xenotropic, and MCF-related viral RNAs are detected concurrently in thymus tissues of AKR mice.
J-Virol. 1988 Feb; 62(2): 376-86

Lakin-Thomas-PL; Brand-MD
Mitogenic stimulation transiently increases the exchangeable mitochondrial calcium pool in rat thymocytes.
Biochem-J. 1987 Aug 15; 246(1): 173-7

Lamballe-F; Maniey-D; Boscher-MY; Fauchet-R; le-Prise-PY; David-JC
Effects of clinical combinations of antileukemic drugs on DNA ligase from human thymocytes and normal, stimulated, or leukemic lymphocytes.
Leukemia. 1988 Jun; 2(6): 363-70

Lamesch-AJ
Massive thymic hyperplasia in infants.
Bull-Soc-Sci-Med-Grand-Duche-Luxemb. 1988; 125 Spe

Larocca-LM; Piantelli-M; Maggiano-N; Musiani-P
T cell surface markers expression by immature human thymocytes in in vitro culture: role of Ia+ accessory cells.
J-Immunol. 1987 Apr 15; 138(8): 2410-6

Larsson-Sciard-EL; Coutinho-A
A compartment of effector helper and suppressor T cells in the normal mouse thymus.
Ann-Inst-Pasteur-Immunol. 1988 Jan-Feb; 139(1): 55

Lautz-J; Kessler-H; Boelens-R; Kaptein-R; van-Gunsteren-WF
Conformational analysis of a cyclic thymopoietin-analogue by 1H n.m.r. spectroscopy and restrained molecular dynamics simulations.
Int-J-Pept-Protein-Res. 1987 Sep; 30(3): 404-14

Lazarus-GM; Henrich-JP; Kelly-WG; Schmitz-SA; Castora-FJ
Purification and characterization of a type I DNA topoisomerase from calf thymus mitochondria.
Biochemistry. 1987 Sep 22; 26(19): 6195-203

Lazo-PA; Tsichlis-PN
Recombination between two integrated proviruses, one of which was inserted near c-myc in a retrovirus-induced rat thymoma: implications for tumor progression.
J-Virol. 1988 Mar; 62(3): 788-94

Lazzarin-A; Barcellini-W; Uberti-Foppa-C; Borghi-MO; Franzetti-R; Cinque-P; Moroni-M
Experiences with immunomodulant agents in HIV infections.
Acta-Haematol-Basel. 1987; 78 Suppl 1: 84-90

Le-PT; Tuck-DT; Dinarello-CA; Haynes-BF; Singer-KH
Human thymic epithelial cells produce interleukin 1.
J-Immunol. 1987 Apr 15; 138(8): 2520-6

Lee-HH; Mutchnick-MG
Quantitative analysis of mitogen-stimulated cellular proliferation: effect of concanavalin A and thymosin interaction.
Int-J-Biomed-Comput. 1987 May; 20(4): 265-73

Lee-NE; Davis-MM
T cell receptor beta-chain genes in BW5147 and other AKR tumors. Deletion order of murine V beta gene segments and possible 5' regulatory regions.
J-Immunol. 1988 Mar 1; 140(5): 1665-75

Lefebvre-P; Formstecher-P; Richard-C; Dautrevaux-M
RU 486 stabilizes a high molecular weight form of the glucocorticoid receptor containing the 90K non-steroid binding protein in intact thymus cells.
Biochem-Biophys-Res-Commun. 1988 Feb 15; 150(3): 1

Lefrancois-L
Expression of carbohydrate differentiation antigens during ontogeny of the murine thymus.
J-Immunol. 1987 Oct 1; 139(7): 2220-9

Lelievre-V; Bertin-B; Chanoine-F; Bure-J; Junien-JL
Correlation between binding activity, inhibition of lymphoblastic transformation and metabolism of tixocortol 21 pivalate in mouse thymocytes.
Agents-Actions. 1987 Aug; 21(3-4): 262-5

Lemaitre-L; Leclerc-F; Marconi-V; Taboureau-O; Avni-FE; Remy-J
Ultrasonographic findings in thymic lymphoma in children.
Eur-J-Radiol. 1987 May; 7(2): 125-9

Lemaitre-L; Marconi-V; Avni-F; Remy-J
The sonographic evaluation of normal thymus in infants and children.
Eur-J-Radiol. 1987 May; 7(2): 130-6

Lenkei-R; Norder-H; Biberfeld-G; Magnius-LO
Autoantibodies to thymic epithelial cells in HDV infection.
Prog-Clin-Biol-Res. 1987; 234: 257-65

Leo-O; Foo-M; Forman-J; Shivakumar-S; Rabinowitz-R; Bluestone-JA
Selective expression of the Ly-6C.2 epitope on Lyt-2+ T cells that can develop in the absence of thymic influence.
J-Immunol. 1988 Jul 1; 141(1): 37-44

Lesley-J; Schulte-R; Trotter-J; Hyman-R
Qualitative and quantitative heterogeneity in Pgp-1 expression among murine thymocytes.
Cell-Immunol. 1988 Mar; 112(1): 40-54

Levin-N; Akana-SF; Jacobson-L; Kuhn-RW; Siiteri-PK; Dallman-MF
Plasma adrenocorticotropin is more sensitive than transcortin production or thymus weight to inhibition by corticosterone in rats.
Endocrinology. 1987 Sep; 121(3): 1104-10

Levine-SR; Diaczok-IM; Deegan-MJ; Kieran-SN; Feit-H; Elias-SB; Welch-KM
Recurrent stroke associated with thymoma and anticardiolipin antibodies.
Arch-Neurol. 1987 Jun; 44(6): 678-9

Levy-RB; Cotterell-AH; Jones-M; Malek-TR
Graft-versus-host reaction-induced immune modulation. I. Donor-recipient genetic disparity and the differential expression of Lyt-2, L3T4, and Ly-6 during acute reactions in the host thymus.
J-Immunol. 1988 Mar 15; 140(6): 1717-25

Lewis-JE; Wick-MR; Scheithauer-BW; Bernatz-PE; Taylor-WF
Thymoma. A clinicopathologic review.
Cancer. 1987 Dec 1; 60(11): 2727-43

Liberati-AM; Ballatori-E; Fizzotti-M; Schippa-M; Cini-L; Cinieri-S; Proietti-MG; Di-Marzio-R; Senato
A randomized trial to evaluate the immunorestorative properties of thymostimulin in patients with Hodgkin's disease in complete remission.
Cancer-Immunol-Immunother. 1988; 26(1): 87-93

Lie-TS; Nakajima-Y; Hofer-M; Lee-KS; Otani-Y; Nakano-H
Alteration of thymus and of immune regulatory-antifibroblast factors in acute and chronic hepatic damage.
Res-Exp-Med-Berl. 1987; 187(3): 195-201

Lie-TS; Yasuda-K; Hofer-M; Otani-Y
Thymus alteration in hepatic regeneration.
Res-Exp-Med-Berl. 1987; 187(5): 379-84

Lillehoj-HS
Production and flow cytometric analysis of monoclonal antibodies reactive with subpopulations of chicken lymphocytes.
Prog-Clin-Biol-Res. 1987; 238: 87-97

Lin-CY; Hsu-HC; Chen-CL; Shen-EY
Treatment of combined immunodeficiency with thymic extract (Thymostimulin).
Ann-Allergy. 1987 May; 58(5): 379-84

Lin-CY; Kuo-YC; Lin-CC; Ou-BR
Enhancement of interleukin-2 and gamma-interferon production in vitro on cord blood lymphocytes and in vivo on primary cellular immunodeficiency patients with thymic extract (thymostimulin).
J-Clin-Immunol. 1988 Mar; 8(2): 103-7

Linder-J
The thymus gland in secondary immunodeficiency.
Arch-Pathol-Lab-Med. 1987 Dec; 111(12): 1118-22

Lindsten-T; Fowlkes-BJ; Samelson-LE; Davis-MM; Chien-YH
Transient rearrangements of the T cell antigen receptor alpha locus in early thymocytes.
J-Exp-Med. 1987 Sep 1; 166(3): 761-75

Lisak-RP; Levinson-AI; Zweiman-B; Kornstein-MJ
In vitro synthesis of IgG and antibodies to AChR by peripheral and thymic lymphocytes.
Ann-NY-Acad-Sci. 1987; 505: 39-49

Lo-D; Ron-Y; Sprent-J
Induction of MHC-restricted specificity and tolerance in the thymus.
Immunol-Res. 1986; 5(3): 221-32

Lobach-DF; Haynes-BF
Ontogeny of the human thymus during fetal development.
J-Clin-Immunol. 1987 Mar; 7(2): 81-97

Lobach-DF; Itoh-T; Singer-KH; Haynes-BF
The thymic microenvironment. Characterization of in vitro differentiation of the IT26R21 rat thymic epithelial cell line.
Differentiation. 1987; 34(1): 50-9

Lockhart-F; Burton-AF
An effect of corticosteroids on thymocytes not mediated by macromolecule synthesis.
Mol-Cell-Biochem. 1987 Jun; 75(2): 133-6

Lotz-M; Jirik-F; Kabouridis-P; Tsoukas-C; Hirano-T; Kishimoto-T; Carson-DA
B cell stimulating factor 2/interleukin 6 is a costimulant for human thymocytes and T lymphocytes.
J-Exp-Med. 1988 Mar 1; 167(3): 1253-8

Loudon-MM; Thompson-EN
Severe combined immunodeficiency syndrome, tissue transplant, leukaemia, and Q fever.
Arch-Dis-Child. 1988 Feb; 63(2): 207-9

Lowenthal-JW; Ransom-J; Howard-M; Zlotnik-A
Up-regulation of interleukin 4 receptor expression on immature (Lyt-2-/L3T4-) thymocytes.
J-Immunol. 1988 Jan 15; 140(2): 474-8

Lussier-G; Guenette-D; Shek-WR; Descoteaux-JP
Detection of antibodies to mouse thymic virus by enzyme-linked immunosorbent assay.
Can-J-Vet-Res. 1988 Apr; 52(2): 236-8

Lycke-N; Eriksen-L; Holmgren-J
Protection against cholera toxin after oral immunization is thymus-dependent and associated with intestinal production of neutralizing IgA antitoxin.
Scand-J-Immunol. 1987 Apr; 25(4): 413-9

MacDonald-HR; Budd-RC; Howe-RC
A CD3- subset of CD4-8+ thymocytes: a rapidly cycling intermediate in the generation of CD4+8+ cells.
Eur-J-Immunol. 1988 Apr; 18(4): 519-23

MacDonald-HR; Miescher-GC; Howe-RC
Anti-Thy-1-induced proliferation of immature thymocytes expressing the CD3-associated gamma/delta heterodimer.
Eur-J-Immunol. 1987 Nov; 17(11): 1677-80

Machi-M; Itoyama-Y; Goto-I; Kuroiwa-Y
Surface phenotypes of lymphoid cells altered in the human myasthenic thymus.
Neurology. 1988 Apr; 38(4): 592-6

Mackova-N
Thymus repair compared with hemopoiesis repair in spleen after protracted irradiation.
Folia-Haematol-Leipz. 1987; 114(6): 817-23

Mackova-N; Suliova-J
Repair processes of hemopoiesis after applying cyclophosphamide. I. Morphological changes in the bone marrow, spleen and thymus.
Folia-Haematol-Leipz. 1986; 113(5): 596-604

Maddox-JF; Mackay-CR; Brandon-MR
Ontogeny of ovine lymphocytes. I. An immunohistological study on the development of T lymphocytes in the sheep embryo and fetal thymus.
Immunology. 1987 Sep; 62(1): 97-105

Maekawa-R; Kitagawa-T; Hojo-K; Sato-K
Differential efficacies of recombinant murine interferon-gamma and recombinant human interleukin 2 against EL4-bearing mice.
J-Interferon-Res. 1988 Apr; 8(2): 241-9

Maeno-M; Nakamura-T; Tochinai-S; Katagiri-C
Analysis of allotolerance in thymectomized Xenopus restored with semiallogeneic thymus grafts.
Transplantation. 1987 Aug; 44(2): 308-14

Maggiano-N; Larocca-LM; Piantelli-M; Acuto-O; Musiani-P
Expression of T cell receptor-alpha and -beta subunits in human thymocytes. An immunocytologic study.
J-Immunol. 1987 Sep 1; 139(5): 1438-45

Magni-F; Bruschi-F; Kasti-M
The afferent innervation of the thymus gland in the rat.
Brain-Res. 1987 Oct 27; 424(2): 379-85

Magno-J; Kaneshiro-M; Doney-K; Nakamura-J
Anti-thymocyte globulin therapy of hematologic disorders.
Hawaii-Med-J. 1987 Mar; 46(3): 85-91

Malaise-MG; Bastings-M; Reuter-AM; Vrindts-Gevaert-Y; Bernier-J; Sondag-D; Frere-MC; Goldstein-G; Fr
In vitro effects of thymopentin on the gamma-interferon production by peripheral blood mononuclear cells from normal subjects and from patients with rheumatoid arthritis.
Immunol-Lett. 1987 Aug; 15(4): 323-7

Malaise-MG; Hazee-Hagelstein-MT; Reuter-AM; Vrinds-Gevaert-Y; Goldstein-G; Franchimont-P
Thymopoietin and thymopentin enhance the levels of ACTH, beta-endorphin and beta-lipotropin from rat pituitary cells in vitro.
Acta-Endocrinol-Copenh. 1987 Aug; 115(4): 455-60

Male-R; Haukanes-BI; Helland-DE; Kleppe-K
Substrate specificity of 3-methyladenine-DNA glycosylase from calf thymus.
Eur-J-Biochem. 1987 May 15; 165(1): 13-9

Manickasundari-M; Pitchappan-RM
Characteristics of heterologous anti lizard (Calotes versicolor) thymocyte serum which identify the functions of thymus-derived cell lineage.
Dev-Comp-Immunol. 1987 Summer; 11(3): 605-12

Marrack-P; Kushnir-E; Born-W; McDuffie-M; Kappler-J
The development of helper T cell precursors in mouse thymus.
J-Immunol. 1988 Apr 15; 140(8): 2508-14

Marrack-P; Lo-D; Brinster-R; Palmiter-R; Burkly-L; Flavell-RH; Kappler-J
The effect of thymus environment on T cell development and tolerance.
Cell. 1988 May 20; 53(4): 627-34

Marrack-P; McDuffie-M; Born-W; Blackman-M; Hannum-C; Kappler-J
The T cell receptor: its repertoire and role in thymocyte development.
Adv-Exp-Med-Biol. 1987; 213: 1-12

Martin-GM; Ogburn-CE; Au-K; Disteche-CM
Altered differentiation, indefinite growth potential, diminished tumorigenicity, and suppressed chimerization potential of hybrids between mouse teratocarcinoma cells and thymocytes.
J-Exp-Pathol. 1984; 1(2): 103-33

Martin-KW; McAlister-WH
Intratracheal thymus: a rare cause of airway obstruction.
AJR. 1987 Dec; 149(6): 1217-8

Martin-LH; Calabi-F; Lefebvre-FA; Bilsland-CA; Milstein-C
Structure and expression of the human thymocyte antigens CD1a, CD1b, and CD1c.
Proc-Natl-Acad-Sci-USA. 1987 Dec; 84(24): 9189-93

Martinez-Valdez-H; Doherty-PJ; Thompson-E; Benedict-SH; Gelfand-EW; Cohen-A
Antagonistic effects of calcium ionophores and phorbol esters on T cell receptor mRNA levels in human thymocytes.
J-Immunol. 1988 Jan 15; 140(2): 361-6

Martinez-Valdez-H; Thompson-E; Cohen-A
Phorbol esters and calcium ionophores differentially regulate the transcription of gamma-T-cell antigen receptor gene in human thymocytes.
J-Biol-Chem. 1988 Mar 25; 263(9): 4043-6

Maslinski-W; Grabczewska-E; Laskowska-Bozek-H; Ryzewski-J
Expression of muscarinic cholinergic receptors during T cell maturation in the thymus.
Eur-J-Immunol. 1987 Jul; 17(7): 1059-63

Maslinski-W; Kullberg-M; Nordstrom-O; Bartfai-T
Muscarinic receptors and receptor-mediated actions on rat thymocytes.
J-Neuroimmunol. 1988 Mar; 17(4): 265-74

Maslinski-W; Ryzewski-J; Bartfai-T
Rat thymocytes release a factor which inhibits muscarinic ligand binding.
J-Neuroimmunol. 1988 Mar; 17(4): 275-85

Masuda-H; Ogata-T; Kikuchi-K; Takagi-K; Tanaka-S
Total replacement of superior vena cava because of invasive thymoma: seven years' survival [letter]
J-Thorac-Cardiovasc-Surg. 1988 Jun; 95(6): 1083-4

Matsuzaki-G; Yoshikai-Y; Kishihara-K; Nomoto-K
Expression of T cell antigen receptor genes in the thymus of irradiated mice after bone marrow transplantation.
J-Immunol. 1988 Jan 15; 140(2): 384-7

Matthiessen-HP; Echarti-C; Gerber-J; Maurer-HR
T-lymphocyte colony formation of murine thymocytes in agar contained in glass capillaries.
Blut. 1987 Dec; 55(6): 517-22

Matthiessen-HP; Maurer-HR
Low molecular mass inhibitors from calf thymus selective for T-lymphocyte proliferation.
Z-Naturforsch-[C]. 1986 Nov-Dec; 41(11-12): 1131-6

McCaughey-ES; Walker-V; Rolles-CJ; Scheurmier-NI; Hale-AC; Rees-LH
Ectopic ACTH production by a thymic carcinoid tumour.
Eur-J-Pediatr. 1987 Nov; 146(6): 590-1

McCreary-V; Kartha-S; Bell-GI; Toback-FG
Sequence of a human kidney cDNA clone encoding thymosin beta 10.
Biochem-Biophys-Res-Commun. 1988 Apr 29; 152(2): 8

McDuffie-M; Roehm-N; Born-W; Marrack-P; Kappler-JW
T cell receptor/MHC interactions in the thymus and the shaping of the T cell repertoire.
Transplant-Proc. 1987 Dec; 19(6 Suppl 7): 111-6

McGillis-JP; Hall-NR; Goldstein-AL
Thymosin fraction 5 (TF5) stimulates secretion of adrenocorticotropic hormone (ACTH) from cultured rat pituitaries.
Life-Sci. 1988; 42(22): 2259-68

McGuire-LJ; Huang-DP; Teoh-R; Arnold-M; Wong-K; Lee-JC
Epstein-Barr virus genome in thymoma and thymic lymphoid hyperplasia.
Am-J-Pathol. 1988 Jun; 131(3): 385-90

Melez-KA; Deleargyros-N; Bellanti-JA; Goldstein-AL; Smathers-P; Steinberg-AD
Effect of partial testosterone replacement or thymosin on anti-DNA in castrated (NZB X NZW)F1 males.
Clin-Immunol-Immunopathol. 1987 Mar; 42(3): 319-27

Mellors-JW; Bartiss-AH; Coleman-DL
Stimulation of macrophage H2O2 release by resident thymocytes: effect of a soluble factor distinct from interferon-gamma.
Cell-Immunol. 1987 Dec; 110(2): 391-9

Melms-A; Schalke-BC; Kirchner-T; Muller-Hermelink-HK; Albert-E; Wekerle-H
Thymus in myasthenia gravis. Isolation of T-lymphocyte lines specific for the nicotinic acetylcholine receptor from thymuses of myasthenic patients.
J-Clin-Invest. 1988 Mar; 81(3): 902-8

Menard-L; Poirier-GG
Rapid assay of poly(ADP-ribose) glycohydrolase.
Biochem-Cell-Biol. 1987 Jul; 65(7): 668-73

Merendino-A; Vecchi-A; Spreafico-F; Sironi-M; Borrelli-F; Antonetti-F; Cantelmo-A; Falchetti-R
Effect of thymostimulin in models of cell-mediated and humoral autoreactivity and on T-dependent suppression.
Int-J-Immunopharmacol. 1987; 9(8): 937-45

Merrill-BM; Lopresti-MB; Stone-KL; Williams-KR
Amino acid sequence of UP1, an hnRNP-derived single-stranded nucleic acid binding protein from calf thymus.
Int-J-Pept-Protein-Res. 1987 Jan; 29(1): 21-39

Metcalfe-JC; Smith-GA; Moore-JP; Hesketh-R
The early mitogenic pathway in mouse thymocytes: an analysis of the dual signal hypothesis.
Adv-Exp-Med-Biol. 1987; 213: 29-44

Metreau-E; Pleau-JM; Dardenne-M; Bach-JF; Pradelles-P
An enzyme immunoassay for synthetic thymulin.
J-Immunol-Methods. 1987 Sep 24; 102(2): 233-42

Michon-JM; Caligiuri-MA; Hazanow-SM; Levine-H; Schlossman-SF; Ritz-J
Induction of natural killer effectors from human thymus with recombinant IL-2.
J-Immunol. 1988 May 15; 140(10): 3660-7

Micklem-LR; McCann-MC; James-K
The use of rat mixed-thymocyte culture-conditioned medium for hybridoma production, cloning and revival.
J-Immunol-Methods. 1987 Nov 23; 104(1-2): 81-6

Miescher-GC; Howe-RC; Lees-RK; MacDonald-HR
CD3-associated alpha/beta and gamma/delta heterodimeric receptors are expressed by distinct populations of CD4- CD8- thymocytes.
J-Immunol. 1988 Mar 15; 140(6): 1779-82

Mihara-M; Ohsugi-Y; Saito-K; Miyai-T; Togashi-M; Ono-S; Murakami-S; Dobashi-K; Hirayama-F; Hamaoka-T
Immunologic abnormality in NZB/NZW F1 mice. Thymus-independent occurrence of B cell abnormality and requirement for T cells in the development of autoimmune disease, as evidenced by an analysis of
J-Immunol. 1988 Jul 1; 141(1): 85-90

Milicevic-NM; Milicevic-Z; Colic-M; Mujovic-S
Ultrastructural study of macrophages in the rat thymus, with special reference to the cortico-medullary zone.
J-Anat. 1987 Feb; 150: 89-98

Millar-J; Allen-R; Wakefield-JS; Buchanan-AJ; Gupta-RK
Diagnosis of thymoma by fine-needle aspiration cytology: light and electron microscopic study of a case.
Diagn-Cytopathol. 1987 Jun; 3(2): 166-9

Miller-H; Shenstone-BD; Joffe-R; Kannangara-S
A case of invasive thymoma associated with myasthenia gravis, myositis and demyelinating neuropathy.
Clin-Exp-Neurol. 1986; 22: 13-8

Mingari-MC; Varese-P; Bottino-C; Merli-A; Moretta-L
Interleukin-2-induced proliferation of CD4-CD8- human thymocytes. In vitro expression of CD3 and CD8 antigens and cytolytic activity.
Ric-Clin-Lab. 1988 Jan-Mar; 18(1): 67-73

Minoda-M; Horiuchi-A
The effects of macrophages on interleukin 2 production in thymocytes of New Zealand black mice.
J-Clin-Lab-Immunol. 1987 Jan; 22(1): 29-34

Minoda-M; Senda-S; Horiuchi-A
The relationship between the defect in the syngeneic mixed lymphocyte reaction and thymic abnormality in New Zealand mice.
J-Clin-Lab-Immunol. 1987 Jun; 23(2): 101-8

Mittal-A; Woodward-B; Chandra-RK
Involution of thymic epithelium and low serum thymulin bioactivity in weanling mice subjected to severe food intake restriction or severe protein deficiency.
Exp-Mol-Pathol. 1988 Apr; 48(2): 226-35

Mizuochi-T; Singer-A
Intrathymic differentiation and tolerance induction of lymphokine-secreting Lyt-2+ T helper cells.
J-Immunol. 1988 Mar 1; 140(5): 1421-4

Mizutani-S; Watt-SM; Greaves-MF
Cloning of human thymic subcapsular cortex epithelial cells by SV40 ori-transfection.
Hamatol-Bluttransfus. 1987; 31: 252-6

Mizutani-S; Watt-SM; Robertson-D; Hussein-S; Healy-LE; Furley-AJ; Greaves-MF
Cloning of human thymic subcapsular cortex epithelial cells with T-lymphocyte binding sites and hemopoietic growth factor activity.
Proc-Natl-Acad-Sci-USA. 1987 Jul; 84(14): 4999-500

Mjaaland-S; Fossum-S
The localization of antigen in lymph node follicles of congenitally athymic nude rats.
Scand-J-Immunol. 1987 Aug; 26(2): 141-7

Moerman-P; Dumoulin-M; Lauweryns-J; Van-der-Hauwaert-LG
Interrupted right aortic arch in DiGeorge syndrome.
Br-Heart-J. 1987 Sep; 58(3): 274-8

Mohri-S; Handa-S; Tateishi-J
Lack of effect of thymus and spleen on the incubation period of Creutzfeldt-Jakob disease in mice.
J-Gen-Virol. 1987 Apr; 68 (Pt 4): 1187-9

Mondola-P; Santangelo-F; Falconi-C; Belfiore-A
The serum apo B and apo E in rats following cholesterol diet and thymus treatment.
Horm-Metab-Res. 1987 Sep; 19(9): 407-10

Monier-JC; Auger-C; Corvee-N; Stahli-C; Fabien-N
Immunoreactivity of aqueous extracts of rat and mouse tissue with anti-thymosin alpha 1, anti-bovine thymopoietin and anti-thymulin antibodies. Studies using immunoblotting.
Thymus. 1988; 11(3): 173-83

Monroe-WE; Roth-JA; Grier-RL; Arp-LH; Naylor-PH
Effects of growth hormone on the adult canine thymus.
Thymus. 1987; 9(3): 173-87

Montagna-D; Maccario-R; Nespoli-L; Mazzanti-P; Cazzola-P
Thymomodulin enhances in vitro natural killer (NK) activity of human cord blood lymphocytes (CBL).
Thymus. 1988; 11(3): 201-5

Moody-Corbett-F; Brehm-P
Acetylcholine reduces inward rectification on thymus-derived macrophage cells in culture.
Can-J-Physiol-Pharmacol. 1987 Mar; 65(3): 348-51

Morel-E; Vernet-der-Garabedian-B; Raimond-F; Audhya-TK; Goldstein-G; Bach-JF
Myasthenic sera recognize the human acetylcholine receptor bound to thymopoietin.
Eur-J-Immunol. 1987 Aug; 17(8): 1109-13

Mori-K; Eguchi-K; Moriyama-H; Miyazawa-N; Kodama-T
Computed tomography of anterior mediastinal tumors. Differentiation between thymoma and germ cell tumor.
Acta-Radiol. 1987 Jul-Aug; 28(4): 395-8

Morimoto-K; Tanaka-A; Yamaha-T; Takahashi-A
Alkylation of DNA in F344 rat thymus following administration of 1-n-propyl-1-nitrosourea in vivo and comparison of O6-alkylguanine DNA alkyltransferase activity in thymus from F344 and Long-Evans
Carcinogenesis. 1988 May; 9(5): 853-6

Morinaga-S; Sato-Y; Shimosato-Y; Sinkai-T; Tsuchiya-R
Multiple thymic squamous cell carcinomas associated with mixed type thymoma.
Am-J-Surg-Pathol. 1987 Dec; 11(12): 982-8

Morishima-Y; Morishita-Y; Adachi-K; Tanimoto-M; Ohno-R; Saito-H
Phorbol ester induces interleukin-2 receptor on the cell surface of precursor thymocyte leukemia with no rearrangement of T cell receptor beta and gamma genes.
Blood. 1987 Nov; 70(5): 1291-6

Morris-M; Barclay-AN; Williams-AF
Analysis of T cell receptor beta chains in rat thymus, and rat C alpha and C beta sequences.
Immunogenetics. 1988; 27(3): 174-9

Morse-SS
Mouse thymic necrosis virus: a novel murine lymphotropic agent.
Lab-Anim-Sci. 1987 Dec; 37(6): 717-25

Morselt-AF; Leene-W; De-Groot-C; Kipp-JB; Evers-M; Roelofsen-AM; Bosch-KS
Differences in immunological susceptibility to cadmium toxicity between two rat strains as demonstrated with cell biological methods. Effect of cadmium on DNA synthesis of thymus lymphocytes.
Toxicology. 1988 Feb; 48(2): 127-39

Mroczek-EC; Seemayer-TA; Grierson-HL; Markin-RS; Linder-J; Brichacek-B; Purtilo-DT
Thymic lesions in fatal infectious mononucleosis.
Clin-Immunol-Immunopathol. 1987 May; 43(2): 243-55

Mueller-C; Hess-MW; Cottier-H
Macrophages containing amylase-resistant, PAS-positive material in the thymic cortex of conventional and germ-free mice.
Thymus. 1987; 9(4): 253-6

Mueller-C; Tschumper-A; Tschumper-JC; Hess-MW; Cottier-H
Parathymic lymph node: oriented proliferative response of the murine thymic cortex to intraperitoneal stimulation.
Thymus. 1987; 9(1): 3-12

Mulder-AH; Visser-JW
The entry of the prothymocyte into the thymus after lethal irradiation and bone marrow transplantation. I. Seeding of bone marrow cells into the thymus.
Thymus. 1988; 11(1): 15-27

Mulder-AH; Visser-JW; Zoetelief-J; van-Bekkum-DW
The entry of the prothymocyte into the thymus after lethal irradiation and bone marrow transplantation. II. Time of entry.
Thymus. 1988; 11(1): 29-41

Muller-Hermelink-HK; Sale-GE; Borisch-B; Storb-R
Pathology of the thymus after allogeneic bone marrow transplantation in man. A histologic immunohistochemical study of 36 patients.
Am-J-Pathol. 1987 Nov; 129(2): 242-56

Munno-I; Pellegrino-NM; Fumo-G; Barbieri-G; Polimeno-G; de-Filippis-V; Jirillo-E
Effects of a synthetic extract (thymopentin) on the immune system of lepromatous leprosy patients.
Cytobios. 1987; 52(210-211): 167-73

Munoz-Blay-T; Benedict-CV; Picciano-PT; Cohen-S
Substrate requirements for the isolation and purification of thymic epithelial cells.
J-Exp-Pathol. 1987; 3(3): 251-8

Munoz-Blay-T; Nieburgs-AC; Cohen-S
Thymic epithelium in vitro. V. Binding of thymocytes to cultured thymic epithelial cells.
Cell-Immunol. 1987 Oct 15; 109(2): 371-83

Mutchnick-MG; Lee-HH; Hollander-DI; Haynes-GD; Chua-DC
Defective in vitro gamma interferon production and elevated serum immunoreactive thymosin beta 4 levels in patients with inflammatory bowel disease.
Clin-Immunol-Immunopathol. 1988 Apr; 47(1): 84-92

Muto-M; Kubo-E; Sado-T
Development of prelymphoma cells committed to thymic lymphomas during radiation-induced thymic lymphomagenesis in B10 mice.
Cancer-Res. 1987 Jul 1; 47(13): 3469-72

Nabarra-B; Andrianarison-I
Pattern of secretion in thymic epithelial cells: ultrastructural studies of the effect of blockage at various levels.
Cell-Tissue-Res. 1987 Jul; 249(1): 171-8

Nabarra-B; Andrianarison-I
Ultrastructural studies of thymic reticulum: I. Epithelial component.
Thymus. 1987; 9(2): 95-121

Nabarra-B; Papiernik-M
Phenotype of thymic stromal cells. An immunoelectron microscopic study with anti-IA, anti-MAC-1, and anti-MAC-2 antibodies.
Lab-Invest. 1988 May; 58(5): 524-31

Nakahara-K; Ohno-K; Hashimoto-J; Maeda-H; Miyoshi-S; Sakurai-M; Monden-Y; Kawashima-Y
Thymoma: results with complete resection and adjuvant postoperative irradiation in 141 consecutive patients.
J-Thorac-Cardiovasc-Surg. 1988 Jun; 95(6): 1041-7

Nakamura-K; Hashimoto-Y; Yamakawa-T; Suzuki-A
Genetic polymorphism of ganglioside expression in mouse organs.
J-Biochem-Tokyo. 1988 Jan; 103(1): 201-8

Nakamura-T; Matsuyama-M; Kojima-A; Ogiu-T; Kubota-A; Suzuki-Y; Arakawa-M; Shimizu-F
The effect of thymectomy on the development of nephropathy in spontaneous thymoma rats of the BUF/Mna strain.
Clin-Exp-Immunol. 1988 Feb; 71(2): 350-2

Nakano-N; Hardy-RR; Kishimoto-T
Identification of intrathymic T progenitor cells by expression of Thy-1, IL2 receptor and CD3.
Eur-J-Immunol. 1987 Nov; 17(11): 1567-71

Nasheuer-HP; Grosse-F
DNA polymerase alpha-primase from calf thymus. Determination of the polypeptide responsible for primase activity.
J-Biol-Chem. 1988 Jun 25; 263(18): 8981-8

Naylor-PH; Goldstein-AL
Thymosin modulates immune and endocrine events.
Prog-Clin-Biol-Res. 1988; 256: 489-502

Naylor-PH; Naylor-CW; Badamchian-M; Wada-S; Goldstein-AL; Wang-SS; Sun-DK; Thornton-AH; Sarin-PS
Human immunodeficiency virus contains an epitope immunoreactive with thymosin alpha 1 and the 30-amino acid synthetic p17 group-specific antigen peptide HGP-30.
Proc-Natl-Acad-Sci-USA. 1987 May; 84(9): 2951-5

Naz-RK; Naylor-PH; Goldstein-AL
Thymosin alpha 1 levels in human seminal plasma and follicular fluid: implication in germ cell function.
Int-J-Fertil. 1987 Sep-Oct; 32(5): 375-9

Nelipovich-PA; Nikonova-LV; Umansky-SR
Inhibition of poly(ADP-ribose) polymerase as a possible reason for activation of Ca2+/Mg2+-dependent endonuclease in thymocytes of irradiated rats.
Int-J-Radiat-Biol. 1988 May; 53(5): 749-65

Nemoto-K; Ishikawa-H; Ohnishi-Y; Nakamura-T; Ohsaki-N
Hodgkin's disease accompanied with thymoma.
Acta-Pathol-Jpn. 1987 Sep; 37(9): 1505-12

Newton-WA Jr; Hamoudi-AB; Shannon-BT
Role of the thymus in histiocytosis-X.
Hematol-Oncol-Clin-North-Am. 1987 Mar; 1(1): 63-74

Nghiem-LN; Banfalvi-G; Csuka-I; Olah-I; Antoni-F
In vitro effect of emetine and chloroquine on the macromolecular biosynthesis of murine thymus cells.
Acta-Biochim-Biophys-Hung. 1987; 22(1): 75-83

Nguyen-TD; Scheving-LA
Thymosin alpha 1: amino acid homology with peptide T from the human immunodeficiency virus envelope.
Biochem-Biophys-Res-Commun. 1987 Jun 15; 145(2): 8

Nieburgs-AC; Korn-JH; Picciano-PT; Cohen-S
Thymic epithelium in vitro. IV. Regulation of growth and mediator production by epidermal growth factor.
Cell-Immunol. 1987 Sep; 108(2): 396-404

Nielsen-J
Ontogeny of T cells in the pig fetus.
Acta-Vet-Scand. 1986; 27(3): 343-50

Nogues-A; Tovar-JA; Sunol-M; Palacio-M; Albisu-J
Hodgkin's disease of the thymus: a rare mediastinal cystic mass.
J-Pediatr-Surg. 1987 Nov; 22(11): 996-7

Nomori-H; Horinouchi-H; Kaseda-S; Ishihara-T; Torikata-C
Evaluation of the malignant grade of thymoma by morphometric analysis.
Cancer. 1988 Mar 1; 61(5): 982-8

Nordlind-K; Ernstrom-U; Henze-A
Enhanced mitogenic reactivity of human T cells after passage through the thymus. Response to phytohemagglutinin and mercuric chloride.
Int-Arch-Allergy-Appl-Immunol. 1987; 83(2): 190-2

Novogrodsky-A; Patya-M; Avissar-S; Stenzel-KH
Osmotic enhancement of mitogenic stimulation in PNA-negative murine thymocytes.
Cell-Immunol. 1987 Apr 1; 105(2): 262-9

Oates-KK; Naylor-PH; Goldstein-AL
Localization of thymosin alpha 1 production to thymus medullary epithelial cells by use of monoclonal antibodies.
Hybridoma. 1987 Feb; 6(1): 47-59

Ober-MN; Prahlad-KV
Ultrastructure of thymus and liver following adrenalectomy and replacement adrenal hormone therapy in rats.
Cytobios. 1987; 52(209): 71-82

Oda-K; Miyasaki-K; Ohta-M; Shibasaki-H
A case of myasthenia gravis associated with thymoma, multiple schwannomas and monoclonal IgA gammopathy.
Jpn-J-Med. 1987 May; 26(2): 223-5

Oehme-P; Hecht-K; Jumatov-J; Repke-H; Hilse-H; Cordova-A
Prevention of stress-induced involution of the thymus in rats by substance P (SP1-11) and its N-terminal fragment SP1-4.
Pharmazie. 1987 Jan; 42(1): 34-6

Ogata-M; Sato-S; Sano-H; Hamaoka-T; Doi-H; Nakanishi-K; Asano-Y; Itoh-T; Fujiwara-H
Thymic stroma-derived T cell growth factor (TSTGF). I. Functional distinction of TSTGF from interleukins 2 and 4 and its preferential growth-promoting effect on helper T cell clones.
J-Immunol. 1987 Oct 15; 139(8): 2675-82

Ogiu-T; Fukami-H
Existence of N-nitroso-N-propylurea target cells in the thymus of F344 rats in thymic lymphomagenesis.
JNCI. 1987 Jul; 79(1): 179-83

Oguchi-Y; Morita-I; Fujii-T; Matsunaga-K; Yoshikumi-C; Kawai-Y; Tsuru-S; Nomoto-K
Involution of the thymus in tumor-bearing mice and its restoration by PSK. II. Mechanism of the involution and its restoration.
J-Clin-Lab-Immunol. 1987 Oct; 24(2): 93-9

Ohki-H; Martin-C; Corbel-C; Coltey-M; Le-Douarin-NM
Tolerance induced by thymic epithelial grafts in birds.
Science. 1987 Aug 28; 237(4818): 1032-5

Ohmori-Y; Hanazawa-S; Amano-S; Miyoshi-T; Hirose-K; Kitano-S
Spontaneous production of thymocyte-activating factor by human gingival fibroblasts and its autoregulatory effect on their proliferation.
Infect-Immun. 1987 Apr; 55(4): 947-54

Ohta-Y; Tezuka-E; Tamura-S; Sugawara-M; Nihira-S; Imai-S; Yagi-Y
Thymosin-alpha 1 increases the capability to produce interleukin-3 but not interleukin-2 in nu/nu mice.
J-Biol-Response-Mod. 1987 Apr; 6(2): 181-93

Ohtaki-S
Quantitative interactions in weight of lymphoid organs and steroid hormonal organs in hamsters under several experimental conditions.
Br-J-Exp-Pathol. 1988 Feb; 69(1): 1-16

Okazaki-K; Davis-DD; Sakano-H
T cell receptor beta gene sequences in the circular DNA of thymocyte nuclei: direct evidence for intramolecular DNA deletion in V-D-J joining.
Cell. 1987 May 22; 49(4): 477-85

Or-R; Raz-I; Raveh-D; Lichovitzki-G; Kleinman-Y
Thymoma presenting as a superior vena cava syndrome remission following therapy.
Klin-Wochenschr. 1987 Jul 1; 65(13): 617-9

Osband-ME
Immunotherapy of histiocytosis-X.
Hematol-Oncol-Clin-North-Am. 1987 Mar; 1(1): 131-4

Ovadia-H; Lubetzki-Korn-I; Abramsky-O
Dopamine receptors on isolated membranes of rat thymocytes.
Ann-NY-Acad-Sci. 1987; 496: 211-6

Owen-JJ; Owen-MJ; Williams-GT; Kingston-R; Jenkinson-EJ
The effects of anti-CD3 antibodies on the development of T-cell receptor alpha beta + lymphocytes in embryonic thymus organ cultures.
Immunology. 1988 Apr; 63(4): 639-42

Owen-MJ; Jenkinson-EJ; Brown-MH; Sewell-WA; Krissansen-GW; Crumpton-MJ; Owen-JJ
Murine CD2 gene expression during fetal thymus ontogeny.
Eur-J-Immunol. 1988 Jan; 18(1): 187-9

Palacios-R; Sideras-P; von-Boehmer-H
Recombinant interleukin 4/BSF-1 promotes growth and differentiation of intrathymic T cell precursors from fetal mice in vitro.
EMBO-J. 1987 Jan; 6(1): 91-5

Paladini-G; Tosato-F
Effects of thymopentin on the anaemia of chronic lymphoproliferative disorders [letter]
Haematologica-Pavia. 1986 Nov-Dec; 71(6): 519-20

Palestro-G; Valente-G; Novero-D; Stramignoni-D; Geuna-M; Rosai-J
Relationship between structure and T-lymphocyte maturation in human thymomas. Enzyme histochemical and immunohistological studies.
Virchows-Arch-[B]. 1987; 52(5): 389-402

Palmisano-L; Chisesi-T; Galli-M; Gritti-FM; Ielasi-G; Lazzarin-A; Mezzaroma-I; Moroni-M; Raise-E; Va
Thymostimulin treatment in AIDS-related complex.
Clin-Immunol-Immunopathol. 1988 Jun; 47(3): 253-61

Pandey-V; Modak-MJ
Purification of high molecular mass species of calf thymus terminal deoxynucleotidyltransferase.
Prep-Biochem. 1987; 17(4): 359-77

Panneerselvam-C; Caldarella-J; Horecker-BL
A radioimmunoassay for parathymosin.
J-Immunol-Methods. 1987 Nov 23; 104(1-2): 131-6

Panneerselvam-C; Haritos-AA; Caldarella-J; Horecker-BL
Prothymosin alpha in human blood.
Proc-Natl-Acad-Sci-USA. 1987 Jul; 84(13): 4465-9

Papatestas-AE; Genkins-G; Kornfeld-P; Eisenkraft-JB; Fagerstrom-RP; Pozner-J; Aufses-AH Jr
Effects of thymectomy in myasthenia gravis.
Ann-Surg. 1987 Jul; 206(1): 79-88

Papatestas-AE; Pozner-J; Genkins-G; Kornfeld-P; Matta-RJ
Prognosis in occult thymomas in myasthenia gravis following transcervical thymectomy.
Arch-Surg. 1987 Nov; 122(11): 1352-6

Papiernik-M; Lehuen-A; Savino-W
Definition of a differentiation antigen on the surface of phagocytic cells of thymic reticulum which is down-regulated by interferon gamma.
Cell-Immunol. 1987 Apr 1; 105(2): 280-9

Papiernik-M; Penit-C; el-Rouby-S
Control of prothymocyte proliferation by thymic accessory cells.
Eur-J-Immunol. 1987 Sep; 17(9): 1303-10

Papiernik-M; el-Rouby-S
Control of helper-T-cell proliferation by recognition of Ia and Mac-1 antigens on phagocytic cells of the thymic reticulum.
Cell-Immunol. 1988 Apr 15; 113(1): 95-106

Pardoll-DM; Fowlkes-BJ; Lew-AM; Maloy-WL; Weston-MA; Bluestone-JA; Schwartz-RH; Coligan-JE; Kruisbee
Thymus-dependent and thymus-independent developmental pathways for peripheral T cell receptor-gamma delta-bearing lymphocytes.
J-Immunol. 1988 Jun 15; 140(12): 4091-6

Parekh-RB; Tse-AG; Dwek-RA; Williams-AF; Rademacher-TW
Tissue-specific N-glycosylation, site-specific oligosaccharide patterns and lentil lectin recognition of rat Thy-1.
EMBO-J. 1987 May; 6(5): 1233-44

Parmentier-HK; Teppema-JS; van-Loveren-H; Tas-J; Ruitenberg-EJ
Effect of a Trichinella spiralis infection on the distribution of mast cell precursors in tissues of thymus-bearing and non-thymus-bearing (nude) mice determined by an in vitro assay.
Immunology. 1987 Apr; 60(4): 565-71

Paterson-DJ; Williams-AF
An intermediate cell in thymocyte differentiation that expresses CD8 but not CD4 antigen.
J-Exp-Med. 1987 Nov 1; 166(5): 1603-8

Pavia-CS
Thymocyte-dependent immunity to toxoplasmosis in the normal and immunocompromised guinea-pig host.
Parasite-Immunol. 1987 Mar; 9(2): 205-18

Pavoni-D; Crosti-F; Ciboddo-GF; Sabbadini-MG; Baldini-V; Navone-P; Bellone-M; Rugarli-C
Thymopentine treatment improves clinical and immunological functions in aged subjects with chronic bronchitis.
Boll-Ist-Sieroter-Milan. 1987; 66(4): 308-15

Pelkonen-J; Sideras-P; Rammensee-HG; Karjalainen-K; Palacios-R
Thymocyte clones from 14-day mouse embryos. I. State of T cell receptor genes, surface markers, and growth requirements.
J-Exp-Med. 1987 Nov 1; 166(5): 1245-58

Penit-C
In vivo proliferation and differentiation of prothymocytes in the thymus.
Immunol-Res. 1987; 6(4): 271-8

Penit-C
Localization and phenotype of cycling and post-cycling murine thymocytes studied by simultaneous detection of bromodeoxyuridine and surface antigens.
J-Histochem-Cytochem. 1988 May; 36(5): 473-8

Penit-C; Vasseur-F
Sequential events in thymocyte differentiation and thymus regeneration revealed by a combination of bromodeoxyuridine DNA labeling and antimitotic drug treatment.
J-Immunol. 1988 May 15; 140(10): 3315-23

Phan-DT; Phuong-DT; An-VT; Khue-P; Petranyi-GG; Hollan-SR
Effect of thymus transplantation on the number of granulocytes, antibody forming cells and on the life-span of aged mice [letter]
Thymus. 1987; 9(2): 123-5

Philips-AV; Robbins-DJ; Coleman-MS; Barkely-MD
Immunoaffinity purification and fluorescence studies of human adenosine deaminase.
Biochemistry. 1987 May 19; 26(10): 2893-903

Phillips-JH; Lanier-LL
Acquisition of non-MHC restricted cytotoxic function by IL 2 activated thymocytes with an "immature" antigenic phenotype.
J-Immunol. 1987 Aug 1; 139(3): 683-7

Pierart-ME; Najdovski-T; Appelboom-TE; Deschodt-Lanckman-MM
Effect of human endopeptidase 24.11 ("enkephalinase") on IL-1-induced thymocyte proliferation activity.
J-Immunol. 1988 Jun 1; 140(11): 3808-11

Pierres-M; Barbet-J
Two novel phospholipid-linked mouse thymocyte surface molecules released by phosphatidylinositol-specific phospholipase C.
Mol-Immunol. 1987 Dec; 24(12): 1273-80

Pietersz-GA; Smyth-MJ; McKenzie-IF
Immunochemotherapy of a murine thymoma with the use of idarubicin monoclonal antibody conjugates.
Cancer-Res. 1988 Feb 15; 48(4): 926-31

Pillai-PS; Reynolds-SD; Scott-DW; Gauldie-J; Sauder-DN
Role of epidermal cell thymocyte-activating factor in the proliferation and differentiation of murine B cells.
J-Leukocyte-Biol. 1987 Sep; 42(3): 222-9

Piltch-A; Naylor-P; Hayashi-J
A cloned rat thymic epithelial cell line established from serum-free selective culture.
In-Vitro-Cell-Dev-Biol. 1988 Apr; 24(4): 289-93

Pipino-F; Vittore-D
Experience with thymopentin in the treatment of periarthritis humeroscapularis, tendinitis, and gonarthrosis by local infiltration.
Arzneimittelforschung. 1988 Jan; 38(1): 116-9

Platanias-L; Gascon-P; Bielory-L; Griffith-P; Nienhuis-A; Young-N
Lymphocyte phenotype and lymphokines following anti-thymocyte globulin therapy in patients with aplastic anaemia.
Br-J-Haematol. 1987 Aug; 66(4): 437-43

Plum-J; De-Smedt-M
Differentiation of thymocytes in fetal organ culture: lack of evidence for the functional role of the interleukin 2 receptor expressed by prothymocytes.
Eur-J-Immunol. 1988 May; 18(5): 795-9

Plum-J; de-Smedt-M
Analysis of human prenatal and postnatal thymocytes with the anti-Tac monoclonal antibody and recombinant interleukin-2.
Thymus. 1987; 9(3): 149-58

Pozzilli-P; Visalli-N; Cavallo-MG; Beales-P; Negri-M; Ghirlanda-G; Papa-V; Andreani-D
Normalization of the CD4/CD8 lymphocyte ratio and increased B lymphocytes in long standing diabetic patients following therapy with thymopentin.
Diabetes-Res. 1987 Oct; 6(2): 51-6

Pritchard-GC
'Goitre' in goat kids [letter]
Vet-Rec. 1987 Oct 31; 121(18): 430

Pross-S; Newton-C; Klein-TW; Friedman-H
Age-associated differences in cannabinoid-induced suppression of murine spleen, lymph node and thymus cell blastogenic responses.
Immunopharmacology. 1987 Dec; 14(3): 159-68

Provinciali-L; Giovagnoli-AR; Di-Bella-P; Baroni-M; Dellantonio-R
A therapeutic trial of thymic factor in amyotrophic lateral sclerosis (ALS).
Adv-Exp-Med-Biol. 1987; 209: 293-6

Pugliese-A; Biglino-A; Mascolo-M; Schioppacassi-G; Falchetti-R; Tovo-PA
Effect of thymostimulin on K-562 cells proliferation [letter]
Boll-Ist-Sieroter-Milan. 1987; 66(5): 418-20

Ramarli-D; Fox-DA; Reinherz-EL
Selective inhibition of interleukin 2 gene function following thymocyte antigen/major histocompatibility complex receptor crosslinking: possible thymic selection mechanism.
Proc-Natl-Acad-Sci-USA. 1987 Dec; 84(23): 8598-602

Ramsdell-FJ; Golub-SH
Generation of lymphokine-activated killer cell activity from human thymocytes.
J-Immunol. 1987 Sep 1; 139(5): 1446-53

Ramsdell-FJ; Gray-JD; Golub-SH
Similarities between LAK cells derived from human thymocytes and peripheral blood lymphocytes: expression of the NKH-1 and CD3 antigens.
Cell-Immunol. 1988 Jun; 114(1): 209-21

Ranges-GE; Zlotnik-A; Espevik-T; Dinarello-CA; Cerami-A; Palladino-MA Jr
Tumor necrosis factor alpha/cachectin is a growth factor for thymocytes. Synergistic interactions with other cytokines.
J-Exp-Med. 1988 Apr 1; 167(4): 1472-8

Ransom-J; Fischer-M; Mercer-L; Zlotnik-A
Lymphokine-mediated induction of antigen-presenting ability in thymic stromal cells.
J-Immunol. 1987 Oct 15; 139(8): 2620-8

Ransom-J; Fischer-M; Mosmann-T; Yokota-T; DeLuca-D; Schumacher-J; Zlotnik-A
Interferon-gamma is produced by activated immature mouse thymocytes and inhibits the interleukin 4-induced proliferation of immature thymocytes.
J-Immunol. 1987 Dec 15; 139(12): 4102-8

Rappaport-J; Reinberg-D; Zandomeni-R; Weinmann-R
Purification and functional characterization of transcription factor SII from calf thymus. Role in RNA polymerase II elongation.
J-Biol-Chem. 1987 Apr 15; 262(11): 5227-32

Ready-AR; Jenkinson-EJ
Deoxyguanosine-induced survival of class II-expressing fetal mouse thymic allografts is associated with the depletion of donor lymphoid and dendritic cells.
Transplant-Proc. 1987 Oct; 19(5): 4225-6

Reardon-CL; Lucas-DO
Heavy-metal mitogenesis: thymocyte activation by Zn++ requires 2-mercaptoethanol and lipopolysaccharide as cofactors.
Immunobiology. 1987 Mar; 174(2): 233-43

Reclos-GJ; Baxevanis-CN; Sfagos-C; Papageorgiou-C; Tsokos-GC; Papamichail-M
Multiple sclerosis: II. Effects of prothymosin alpha on the autologous and allogeneic MLR in patients with multiple sclerosis.
Clin-Exp-Immunol. 1987 Nov; 70(2): 336-44

Reddy-MM; Grieco-MH
In vitro immunomodulatory effects of interleukin-2 and thymosin fraction V in acquired immune deficiency syndrome.
Int-J-Immunopharmacol. 1987; 9(4): 483-8

Reem-GH; Carding-S; Reinherz-EL
Lymphokine synthesis is induced in human thymocytes by activation of the CD 2 (T11) pathway.
J-Immunol. 1987 Jul 1; 139(1): 130-4

Rendina-EA; Pescarmona-EO; Venuta-F; Nardi-S; De-Rosa-G; Martelli-M; Ricci-C
Thymoma: a clinico-pathologic study based on newly developed morphologic criteria.
Tumori. 1988 Feb 29; 74(1): 79-84

Rendina-EA; Venuta-F; Martelli-M; Pescarmona-E; Ricci-C
Successful resection of a thymoma in an elderly patient.
Ital-J-Surg-Sci. 1988; 18(1): 79-82

Revah-F; Mulle-C; Pinset-C; Audhya-T; Goldstein-G; Changeux-JP
Calcium-dependent effect of the thymic polypeptide thymopoietin on the desensitization of the nicotinic acetylcholine receptor.
Proc-Natl-Acad-Sci-USA. 1987 May; 84(10): 3477-81

Richie-ER
N-methyl-N-nitrosourea-induced and spontaneous AKR/J thymic lymphomas express distinct differentiation antigen phenotypes.
Leuk-Res. 1988; 12(3): 233-42

Richie-ER; Angel-JM; McEntire-B
N-methyl-N-nitrosourea-induced T-lymphomas of AKR/J mice contain somatically acquired ecotropic-like murine leukemia proviruses.
Virology. 1988 May; 164(1): 211-9

Richie-ER; McEntire-B; Crispe-N; Kimura-J; Lanier-LL; Allison-JP
Alpha/beta T-cell antigen receptor gene and protein expression occurs at early stages of thymocyte differentiation.
Proc-Natl-Acad-Sci-USA. 1988 Feb; 85(4): 1174-8

Ritter-J; Sepetjan-M; Monier-JC
Lack of reactivity of anti-human immunodeficiency virus (HIV) P17/18 antibodies against alpha 1 thymosin and of anti-alpha 1 thymosin monoclonal antibody against P17/18 protein.
Immunol-Lett. 1987 Nov; 16(2): 97-100

Robbins-DJ; Barkley-MD; Coleman-MS
Interaction of terminal transferase with single-stranded DNA.
J-Biol-Chem. 1987 Jul 15; 262(20): 9494-502

Rocha-B; Lehuen-A; Papiernik-M
IL-2-dependent proliferation of thymic accessory cells.
J-Immunol. 1988 Feb 15; 140(4): 1076-80

Rollins-NK; Currarino-G
MR imaging of posterior mediastinal thymus.
J-Comput-Assist-Tomogr. 1988 May-Jun; 12(3): 518-2

Romeo-F; Palmisano-L; Arcoria-D
Thymostimulin in the treatment of hepatitis B surface antigen-positive chronic active hepatitis. Controlled clinical trial--two years follow-up.
Arzneimittelforschung. 1987 Apr; 37(4): 450-6

Rosai-J
The pathology of thymic neoplasia.
Monogr-Pathol. 1987(29): 161-83

Rosen-P
Aging of the immune system.
Med-Hypotheses. 1985 Oct; 18(2): 157-61

Rossi-A
Developmental expression and cellular localization of the beta-adrenergic receptor in the murine thymus.
Ann-NY-Acad-Sci. 1987; 496: 307-9

Roth-JA
Possible association of thymus dysfunction with fading syndromes in puppies and kittens.
Vet-Clin-North-Am-[Small-Anim-Pract]. 1987 May; 17

Rouse-RV; Weiss-LM
Human thymomas: evidence of immunohistologically defined normal and abnormal microenvironmental differentiation.
Cell-Immunol. 1988 Jan; 111(1): 94-106

Rusquet-R; Maniey-D; Logeais-Y; Merdrignac-G; David-JC
Differentiation of thymocytes during human ontogeny: stage-specific DNA ligase in relation to terminal deoxynucleotidyl transferase, cell size and surface antigen.
Immunology. 1987 Aug; 61(4): 509-14

Russ-JH; Horton-JD
Cytoarchitecture of the Xenopus thymus following gamma-irradiation.
Development. 1987 May; 100(1): 95-105

Sacchi-N; de-Klein-A; Showalter-SD; Bigi-G; Papas-TS
High expression of ets-1 gene in human thymocytes and immature T leukemic cells.
Leukemia. 1988 Jan; 2(1): 12-8

Sakaguchi-S; Sakaguchi-N
Thymus and autoimmunity. Transplantation of the thymus from cyclosporin A-treated mice causes organ-specific autoimmune disease in athymic nude mice.
J-Exp-Med. 1988 Apr 1; 167(4): 1479-85

Sakiyama-Y; Ishizaka-A; Watanabe-T; Ariga-T; Matsumoto-S
Induction of the CD4-8+ suppressor phenotype in CD4+8+ human thymocytes by phorbol myristate acetate.
Tohoku-J-Exp-Med. 1988 Feb; 154(2): 195-203

Salvin-SB; Horecker-BL; Pan-LX; Rabin-BS
The effect of dietary zinc and prothymosin alpha on cellular immune responses of RF/J mice.
Clin-Immunol-Immunopathol. 1987 Jun; 43(3): 281-8

Sandberg-G; Ernstrom-U; Soder-O; Kolare-S
Effect of growth factors on the proliferation of thymocytes in different functional and anatomical compartments.
Int-Arch-Allergy-Appl-Immunol. 1987; 82(3-4): 490-

Sandberg-G; Rydlander-L; Stenvinkel-C
Purification, identification and elimination of a natural human serum antibody with cytotoxic effect on guinea pig thymocytes.
Int-Arch-Allergy-Appl-Immunol. 1987; 83(1): 57-63

Sanders-DB; Howard-JF Jr
Thymoma without myasthenia gravis: electrophysiological study after thymectomy [letter]
J-Neurol-Neurosurg-Psychiatry. 1988 Jan; 51(1): 16

Saphier-D; Kidron-D; Ovadia-H; Trainin-N; Pecht-M; Burstein-Y; Abramsky-O
Neurophysiological changes in the brain following central administration of immunomodulatory factors.
Isr-J-Med-Sci. 1988 Apr-May; 24(4-5): 261-3

Savino-W; Emonard-H; Grimaud-JA
Thymic extracellular matrix in myasthenia gravis. II. Immunohistochemical evidence for increased type I collagen degradation.
J-Neuroimmunol. 1988 Jun; 18(3): 223-30

Schalke-BC; Mertens-HG; Kirchner-T; Wegener-S; Muller-Hermelink-HK
Long-term treatment with azathioprine abolishes thymic lymphoid follicular hyperplasia in myasthenia gravis [letter]
Lancet. 1987 Sep 19; 2(8560): 682

Scher-HI; Shank-B; Chapman-R; Geller-N; Pinsky-C; Gralla-R; Kelsen-D; Bosl-G; Golbey-R; Petroni-G; e
Randomized trial of combined modality therapy with and without thymosin fraction V in the treatment of small cell lung cancer.
Cancer-Res. 1988 Mar 15; 48(6): 1663-70

Schiavon-M; Chiarelli-A; Lohr-G; Mazzoleni-F
Thymostimulin in the antiinfectious treatment in patients with burns.
Arzneimittelforschung. 1987 May; 37(5): 557-60

Schleuning-M; Duggan-A; Reem-GH
Cyclosporine does not inhibit the early transducing signals generated by the activation of human thymocytes.
Transplant-Proc. 1988 Apr; 20(2 Suppl 2): 63-8

Schluep-M; Willcox-N; Vincent-A; Dhoot-GK; Newsom-Davis-J
Acetylcholine receptors in human thymic myoid cells in situ: an immunohistological study.
Ann-Neurol. 1987 Aug; 22(2): 212-22

Schmitt-D; Zambruno-G; Staquet-MJ; Dezutter-Dambuyant-C; Ohrt-C; Brochier-J; Thivolet-J
Antigenic thymus-epidermis relationships. Reactivity of a panel of anti-thymic cell monoclonal antibodies on human keratinocytes and Langerhans cells.
Dermatologica. 1987; 175(3): 109-20

Schnyder-P; Candardjis-G
Computed tomography of thymic abnormalities.
Eur-J-Radiol. 1987 May; 7(2): 107-13

Schulof-RS; Naylor-PH; Sztein-MB; Goldstein-AL
Thymic physiology and biochemistry.
Adv-Clin-Chem. 1987; 26: 203-92

Schultz-RM; Altom-MG
Modulation of interleukin-1 activity on murine thymocytes by various inhibitors of arachidonic acid oxygenation.
Immunopharmacol-Immunotoxicol. 1988; 10(1): 21-34

Schuurman-HJ; Vaessen-LM; Broekhuizen-R; Brandt-CJ; Holewijn-MC; Vos-JG; Rozing-J
Implantation of cultured thymic fragments in congenitally athymic (nude) rats. Influx of lymphocytes and dendritic cells.
Scand-J-Immunol. 1987 Aug; 26(2): 129-39

Scollay-R
Differences in Ly 2- L3T4- thymocytes from foetal and adult murine thymus.
Immunol-Lett. 1987 Jun; 15(2): 171-7

Segal-A; Seidman-I; Melchionne-S
Induction of thymic lymphomas and squamous cell carcinomas following topic application of isopropyl methanesulfonate to female Hsd:(ICR)BR mice.
Cancer-Res. 1987 Jul 1; 47(13): 3402-5

Segal-J
Adrenergic inhibition of the stimulatory effect of 3,5,3'-triiodothyronine on calcium accumulation and cytoplasmic free calcium concentration in rat thymocytes. Further evidence in support of the
Endocrinology. 1988 May; 122(5): 2240-6

Segal-J
Effect of hypothyroidism on the in-vivo and in-vitro uptake of sugar by rat thymocytes and on the stimulatory response to 3,5,3'-tri-L-iodothyronine.
J-Endocrinol. 1988 Jan; 116(1): 107-14

Segal-J
Opposite regulatory effects of cAMP and cGMP on sugar uptake in rat thymocytes.
Am-J-Physiol. 1987 May; 252(5 Pt 1): E588-94

Segal-J
The effect of trypsin on sugar uptake in rat thymocytes. Modulation of cellular cyclic AMP concentration and the sugar-transport system.
Biochem-J. 1987 Sep 15; 246(3): 561-6

Serra-HM; Krowka-JF; Ledbetter-JA; Pilarski-LM
Loss of CD45R (Lp220) represents a post-thymic T cell differentiation event.
J-Immunol. 1988 Mar 1; 140(5): 1435-41

Serrate-SA; Schulof-RS; Leondaridis-L; Goldstein-AL; Sztein-MB
Modulation of human natural killer cell cytotoxic activity, lymphokine production, and interleukin 2 receptor expression by thymic hormones.
J-Immunol. 1987 Oct 1; 139(7): 2338-43

Seto-F
PFC formation in immature avian hosts grafted with thymic cells from juvenile chicken donors.
Prog-Clin-Biol-Res. 1987; 238: 171-85

Sgonc-R; Hala-K; Wick-G
Relationship between the expression of class I antigen and reactivity of chicken thymocytes.
Immunogenetics. 1987; 26(3): 150-4

Shachor-Y; Radnay-J; Bernheim-J; Rozenszajn-A; Bruderman-I; Klajman-A; Steiner-ZP
Malignant thymoma with peripheral blood lymphocytosis.
Cancer. 1988 Mar 15; 61(6): 1222-7

Sheng-PG; Wang-HL; Shen-ZM
A spin labeling study of calf thymus DNA.
Sci-Sin-[B]. 1987 Sep; 30(9): 946-59

Sheu-MM; Tsai-RK; Kaplan-HJ
The immunohistologic identification of mononuclear cells and Ia antigen in rabbit spleen and thymus by monoclonal antibodies.
Kao-Hsiung-I-Hsueh-Ko-Hsueh-Tsa-Chih. 1988 Mar; 4(

Shiau-AL; Wu-CL; Huang-KY
The effect of thymosin on experimental herpes simplex virus infections.
Taiwan-I-Hsueh-Hui-Tsa-Chih. 1988 Jan; 87(1): 34-4

Shimada-T
Lack of correlation between formation of reactive metabolites and thymic atrophy caused by 3, 4, 3', 4'-tetrachlorobiphenyl in C57BL/6N mice.
Arch-Toxicol. 1987 Feb; 59(5): 301-6

Shimidzu-M; Shindo-H; Takahashi-K; Taniguchi-S; Matsumoto-U
In vivo phosphorylation of histones H1 and H5 in calf thymus and chicken erythrocyte as studied by 31P nuclear magnetic resonance spectroscopy.
J-Biochem-Tokyo. 1987 Aug; 102(2): 351-8

Shimonkevitz-RP; Husmann-LA; Bevan-MJ; Crispe-IN
Transient expression of IL-2 receptor precedes the differentiation of immature thymocytes.
Nature. 1987 Sep 10-16; 329(6135): 157-9

Shipman-PM; Schmidt-RR; Chepenik-KP
Relation between arachidonic acid metabolism and development of thymocytes in fetal thymic organ cultures.
J-Immunol. 1988 Apr 15; 140(8): 2714-20

Shivatcheva-TM; Hadjioloff-AI
Adaptive seasonal involution of the ground squirrel thymus.
Thymus. 1987; 10(3-4): 251-5

Shortman-K; Wilson-A; Egerton-M; Pearse-M; Scollay-R
Immature CD4- CD8+ murine thymocytes.
Cell-Immunol. 1988 May; 113(2): 462-79

Shuangshoti-S; Seksarn-P
Histiocytosis X: overlapping form between Letterer-Siwe disease and Hand-Schuller-Christian disease with advanced thymic involvement and obstructive jaundice.
J-Med-Assoc-Thai. 1987 Jun; 70(6): 344-50

Shultz-LD
Pleiotropic mutations causing abnormalities in the murine immune system and the skin.
Curr-Probl-Dermatol. 1987; 17: 236-50

Sideras-P; Funa-K; Zalcberg-Quintana-I; Xanthopoulos-KG; Kisielow-P; Palacios-R
Analysis by in situ hybridization of cells expressing mRNA for interleukin 4 in the developing thymus and in peripheral lymphocytes from mice.
Proc-Natl-Acad-Sci-USA. 1988 Jan; 85(1): 218-21

Simmons-VP
Thymic "atrophy" at puberty is a myth [letter]
Med-Hypotheses. 1987 Mar; 22(3): 299-301

Singer-A
Experimentation and thymic selection.
J-Immunol. 1988 Apr 15; 140(8): 2481-3

Singer-A; Munitz-TI; Gress-RE
Specificity of thymic selection and the role of self antigens.
Transplant-Proc. 1987 Dec; 19(6 Suppl 7): 107-10

Singer-KH; Haynes-BF
Epithelial-thymocyte interactions in human thymus.
Hum-Immunol. 1987 Oct; 20(2): 127-44

Singh-U; Fatani-JA; Mohajir-AM
Ontogeny of cholinergic innervation of thymus in mouse.
Dev-Comp-Immunol. 1987 Summer; 11(3): 627-35

Singla-R; Talwar-D; Bhattacharya-D; Malhotra-V; Malik-R; Beohar-PC; Narayanan-PS; Menon-MP
Silent locally invasive thymoma in a young patient.
Indian-J-Chest-Dis-Allied-Sci. 1986 Oct-Dec; 28(4)

Sinopoli-NT; Pesce-A; Baldassarre-AM; Trivers-GE; Frati-L; Rinaldi-Garaci-C
MicroELISA detection of thymosin alpha 1 released in thymic organ cultures.
J-Immunol-Methods. 1988 Jun 13; 110(2): 261-5

Skinner-M; Le-Gros-G; Marbrook-J; Watson-JD
Development of fetal thymocytes in organ cultures. Effect of interleukin 2.
J-Exp-Med. 1987 Jun 1; 165(6): 1481-93

Skinner-M; Marbrook-J
Effect of interleukin 2 on fetal thymocytes in organ cultures: generation of lymphokine-activated killer cells.
Cell-Immunol. 1988 Mar; 112(1): 104-11

Smyth-MJ; Pietersz-GA; McKenzie-IF
Increased antitumor effect of immunoconjugates and tumor necrosis factor in vivo.
Cancer-Res. 1988 Jul 1; 48(13): 3607-12

Smyth-MJ; Pietersz-GA; McKenzie-IF
The mode of action of methotrexate-monoclonal antibody conjugates.
Immunol-Cell-Biol. 1987 Apr; 65 (Pt 2): 189-200

Smyth-MJ; Pietersz-GA; McKenzie-IF
Use of vasoactive agents to increase tumor perfusion and the antitumor efficacy of drug-monoclonal antibody conjugates.
JNCI. 1987 Dec; 79(6): 1367-73

Snoeij-NJ; Bol-Schoenmakers-M; Penninks-AH; Seinen-W
Differential effects of tri-n-butyltin chloride on macromolecular synthesis and ATP levels of rat thymocyte subpopulations obtained by centrifugal elutriation.
Int-J-Immunopharmacol. 1988; 10(1): 29-37

Soder-O; Hellstrom-PM
Neuropeptide regulation of human thymocyte, guinea pig T lymphocyte and rat B lymphocyte mitogenesis.
Int-Arch-Allergy-Appl-Immunol. 1987; 84(2): 205-11

Sooki-Toth-A; Csuzi-S; Altmann-H; Antoni-F; Banfalvi-G
Poly(ADP-ribose) and replicative DNA synthesis studied in permeable mouse thymocytes.
Acta-Biochim-Biophys-Hung. 1986; 21(4): 313-25

Spangelo-BL; Hall-NR; Goldstein-AL
Biology and chemistry of thymosin peptides. Modulators of immunity and neuroendocrine circuits.
Ann-NY-Acad-Sci. 1987; 496: 196-204

Spangelo-BL; Judd-AM; Ross-PC; Login-IS; Jarvis-WD; Badamchian-M; Goldstein-AL; MacLeod-RM
Thymosin fraction 5 stimulates prolactin and growth hormone release from anterior pituitary cells in vitro.
Endocrinology. 1987 Dec; 121(6): 2035-43

Spangrude-GJ; Muller-Sieburg-CE; Heimfeld-S; Weissman-IL
Two rare populations of mouse Thy-1lo bone marrow cells repopulate the thymus.
J-Exp-Med. 1988 May 1; 167(5): 1671-83

Spath-A; Tempel-K
Diethyldithiocarbamate inhibits scheduled and unscheduled DNA synthesis of rat thymocytes in vitro and in vivo--dose-effect relationships and mechanisms of action.
Chem-Biol-Interact. 1987; 64(1-2): 151-66

Spiazzi-A; Corradin-G; Nagasawa-R; Tridente-G; MacDonald-HR; Bron-C
Biochemical characterization of a T-lymphoma-specific 90,000 molecular weight disulfide linked dimeric glycoprotein.
Mol-Immunol. 1987 Jul; 24(7): 719-27

Sprent-J; Lo-D; Gao-EK; Ron-Y
T cell selection in the thymus.
Immunol-Rev. 1988 Jan; 101: 173-90

Springer-O; Grgic-Z
Effect of thymic humoral factor (THF) on hemopoiesis in neonatally thymectomized mice [letter]
Thymus. 1987; 9(4): 257-9

Srivastava-US; Thakur-ML; Majumdar-PK; Bhatnagar-GM; Supakar-PC
Lymphoid organ mRNA translatability in rats: effect of protein energy undernutrition in early life.
J-Nutr. 1987 Feb; 117(2): 242-6

St.-Amour-TE; Siegel-MJ; Glazer-HS; Nadel-SN
CT appearances of the normal and abnormal thymus in childhood.
J-Comput-Assist-Tomogr. 1987 Jul-Aug; 11(4): 645-5

Staal-SP; Hartley-JW
Thymic lymphoma induction by the AKT8 murine retrovirus.
J-Exp-Med. 1988 Mar 1; 167(3): 1259-64

Stanulis-Praeger-BM; Yaar-M; Redziniak-G; Meybeck-A; Gilchrest-BA
An extract of bovine thymus stimulates human keratinocyte growth in vitro.
J-Invest-Dermatol. 1988 May; 90(5): 749-54

Stephens-M; Khalil-J; Gibbs-AR
Primary clear cell carcinoma of the thymus gland.
Histopathology. 1987 Jul; 11(7): 763-5

Stet-RJ; Wagenaar-Hilbers-JP; Nieuwenhuis-P
Thymus localization of monoclonal antibodies circumventing the blood-thymus barrier.
Scand-J-Immunol. 1987 May; 25(5): 441-6

Stout-RD; Schwarting-GA; Suttles-J
Evidence that expression of asialo-GM1 may be associated with cell activation. Correlation of asialo-GM1 expression with increased total cellular RNA and protein content in normal thymocyte and sp
J-Immunol. 1987 Oct 1; 139(7): 2123-9

Stumpf-WE; Downs-TW
Nuclear receptors for 1,25(OH)2 vitamin D3 in thymus reticular cells studied by autoradiography.
Histochemistry. 1987; 87(4): 367-9

Suliova-J; Mackova-N; Hvizdosova-M; Gabor-J
Repair processes of hemopoiesis after applying cyclophosphamide. II. Quantitative changes in basic proteins and DNA in the bone marrow, spleen, and thymus.
Folia-Haematol-Leipz. 1986; 113(5): 605-14

Sun-L; Serrero-G; Piltch-A; Hayashi-J
EGF receptors on TEA3A1 endocrine thymic epithelial cells.
Biochem-Biophys-Res-Commun. 1987 Oct 29; 148(2): 6

Sungurov-AYu; Sharlaeva-TM
Thymocyte membrane changes and modifications of interphase death.
Int-J-Radiat-Biol. 1988 Mar; 53(3): 501-6

Svenson-M; Bendtzen-K
Inhibitor of interleukin 1 in normal human urine. Different effects on mouse thymocytes and on a murine T-cell line.
Scand-J-Immunol. 1988 May; 27(5): 593-9

Swerdlow-SH; Angermeier-PA; Hartman-AL
Intrathymic ontogeny of the T cell receptor associated CD3 (T3) antigen.
Lab-Invest. 1988 Apr; 58(4): 421-7

Szamel-M; Goppelt-Strube-M; Bessler-W; Wiesmuller-KH; Resch-K
Separation of plasma membrane domains of calf thymocytes by affinity chromatography on ouabain-Sepharose.
Biochim-Biophys-Acta. 1987 May 29; 899(2): 247-57

Szende-B; Lapis-K; Pal-K; Sipos-L; Denes-L; Kisfaludy-L
The effect of TP-3 (Arg-Lys-Asp), TP-4 (Arg-Lys-Asp-Val), and TP-5 on the metastatic capacity of intravenously injected Lewis lung tumor cells.
Immunopharmacol-Immunotoxicol. 1987; 9(1): 19-24

Szkaradkiewicz-A
The influence of thymic preparations on natural cytotoxicity of human monocytes.
Allerg-Immunol-Leipz. 1987; 33(3): 169-73

Szobor-A; Fornet-B
Myasthenia gravis: 75seleno-methionine scanning of thymus gland.
Acta-Med-Hung. 1986; 43(3): 243-8

Szobor-A; Molnar-J; Szepeshazy-K
Myasthenia gravis: influence of thymic changes on clinical parameters. A study on the basis of 364 thymectomies.
Acta-Med-Hung. 1986; 43(3): 229-41

Sztein-MB; Simon-GL; Parenti-DM; Scheib-R; Goldstein-AL; Goodman-R; DiGioia-R; Paxton-H; Skotnicki-A
In vitro effects of thymosin and lithium on lymphoproliferative responses of normal donors and HIV seropositive male homosexuals with AIDS-related complex.
Clin-Immunol-Immunopathol. 1987 Jul; 44(1): 51-62

Takagi-N
Requirement of mitoses for the reversal of X-inactivation in cell hybrids between murine embryonal carcinoma cells and normal female thymocytes.
Exp-Cell-Res. 1988 Apr; 175(2): 363-75

Takahashi-T; Ueda-R; Nishida-K; Namikawa-R; Fukami-H; Matsuyama-M; Masaoka-A; Imaizumi-M; Takahashi-
Immunohistological analysis of thymic tumors with PE-35 monoclonal antibody reactive with medullary thymic epithelium.
Cancer-Res. 1988 Apr 1; 48(7): 1896-903

Tamir-M; Rozenszajn-LA; Malik-Z; Zipori-D
Thymus-derived stromal cell lines.
Int-J-Cell-Cloning. 1987 Jul; 5(4): 289-301

Tan-CK; So-MJ; Downey-KM; So-AG
Apparent stimulation of calf thymus DNA polymerase alpha by ATP.
Nucleic-Acids-Res. 1987 Mar 11; 15(5): 2269-78

Tanaka-K; Koga-Y; Taniguchi-K; Nomoto-K
T-cell recruitment from the thymus to the spleen in tumor bearing mice: phenotypical alteration and recruitment of thymocytes raised in a tumor bearing state.
Cancer-Res. 1987 Apr 15; 47(8): 2136-41

Tanaka-Y; Koide-SS; Yoshihara-K; Kamiya-T
Poly (ADP-ribose) synthetase is phosphorylated by protein kinase C in vitro.
Biochem-Biophys-Res-Commun. 1987 Oct 29; 148(2): 7

Taubenberger-JK; Haar-JL
In vitro bone marrow cell migration to supernatants prepared from thymic epithelial cell cultures.
Thymus. 1987; 9(1): 45-60

Taubenberger-JK; Haar-JL
The initial characterization of a thymus factor chemotactic to bone marrow cells.
Thymus. 1987; 9(2): 85-93

Taylor-A; Allende-CC; Weinmann-R; Allende-JE
The phosphorylation of nucleoplasmin by casein kinase-2 is resistant to heparin inhibition.
FEBS-Lett. 1987 Dec 21; 226(1): 109-14

Taylor-HG; Butler-WM; Karcher-DS; Zaloznik-AJ
Thymic carcinoma: clinical findings in two patients with extrathoracic metastases.
South-Med-J. 1988 May; 81(5): 664-6

Taylor-MV; Hesketh-TR; Metcalfe-JC
Phosphoinositide metabolism and the calcium response to concanavalin A in S49 T-lymphoma cells. A comparison with thymocytes.
Biochem-J. 1988 Feb 1; 249(3): 847-55

Tempel-K; Spath-A
Stimulation of DNA repair synthesis of rat thymocytes by novobiocin and nalidixic acid in vitro without detectable DNA damage.
Arch-Toxicol. 1987 Jun; 60(4): 287-92

Tempelis-LD
Thymic epithelium controls thymocyte expression of preleukemic phenotype and leukemogenic retroviruses.

J-Immunol. 1987 May 15; 138(10): 3555-65

Templeton-PA; Vainright-JR; Rodriguez-A; Diaconis-JN
Mediastinal tumors presenting as spontaneous hemothorax, simulating aortic dissection.
Chest. 1988 Apr; 93(4): 828-30

Tentori-L; Pardoll-DM; Zuniga-JC; Hu-Li-J; Paul-WE; Bluestone-JA; Kruisbeek-AM
Proliferation and production of IL-2 and B cell stimulatory factor 1/IL-4 in early fetal thymocytes by activation through Thy-1 and CD3.
J-Immunol. 1988 Feb 15; 140(4): 1089-94

Teraoka-H; Tsukada-K
Influence of polyethylene glycol on the ligation reaction with calf thymus DNA ligases I and II.
J-Biochem-Tokyo. 1987 Jan; 101(1): 225-31

Thomas-CV; Manivel-JC
Thymic carcinoma and aplastic anemia: report of a previously undocumented association.
Am-J-Hematol. 1987 Jul; 25(3): 333-5

Thomas-NB; Gupta-SC
Unilobar enlargement of normal thymus gland causing mass effect.
Br-J-Radiol. 1988 Mar; 61(723): 244-6

Thorne-AW; Sautiere-P; Briand-G; Crane-Robinson-C
The structure of ubiquitinated histone H2B.
EMBO-J. 1987 Apr; 6(4): 1005-10

Thorvinger-B; Lyttkens-K; Samuelsson-L
Computed tomography of the thymus gland in myasthenia gravis.
Acta-Radiol. 1987 Jul-Aug; 28(4): 399-401

Tomazic-VJ; Sacasa-CR; Loftus-A; Suter-CS; Elias-GE
Thymic factor-induced reduction of pulmonary metastases in mice with FSA-1 fibrosarcoma.
Clin-Exp-Metastasis. 1988 Jan-Feb; 6(1): 17-25

Tomooka-S; Matsuzaki-G; Kishihara-K; Tanaka-K; Yoshikai-Y; Taniguchi-K; Himeno-K; Nomoto-K
Sequential appearance of thymocyte subpopulations and T cell antigen receptor gene messages in the mouse thymus after sublethal irradiation.
J-Immunol. 1987 Dec 15; 139(12): 3986-90

Toniolo-C; Bonora-GM; Heimer-EP; Felix-AM
Structure, solubility and reactivity of peptides. A conformational study of two protected key intermediates from a large-scale synthesis of thymosin alpha 1.
Int-J-Pept-Protein-Res. 1987 Aug; 30(2): 232-9

Touraine-JL; Roncarolo-MG; Royo-C; Touraine-F
Fetal tissue transplantation, bone marrow transplantation and prospective gene therapy in severe immunodeficiencies and enzyme deficiencies.
Thymus. 1987; 10(1-2): 75-87

Toyokawa-T; Hato-F; Mizoguchi-S; Kimura-S; Matsuyama-M; Kinoshita-Y
Remarkable changes in mitogen responsiveness and rosette-forming capacity of thymocytes with aging due to possible disorders of thymic hormone secretion from thymus epithelial cells of spontaneous
Cell-Mol-Biol. 1987; 33(3): 363-74

Tsichlis-PN
Oncogenesis by Moloney murine leukemia virus.
Anticancer-Res. 1987 Mar-Apr; 7(2): 171-80

Tsoukas-CD; Landgraf-B; Bentin-J; Lamberti-JF; Carson-DA; Vaughan-JH
Structural and functional characteristics of the CD3 (T3) molecular complex on human thymocytes.
J-Immunol. 1987 Jun 1; 138(11): 3885-90

Tsuda-T; Kim-YT; Siskind-GW; DeBlasio-A; Schwab-R; Ershler-W; Weksler-ME
Role of the thymus and T-cells in slow growth of B16 melanoma in old mice.
Cancer-Res. 1987 Jun 15; 47(12): 3097-100

Tufariello-J; DuBois-DC; Almon-RR
Muscle and thymus cell-free glucocorticoid receptor systems are different.
J-Recept-Res. 1986; 6(5-6): 411-24

Turner-CE; Shotton-DM
Isolation and initial biochemical characterisation of caps of two major rat thymocyte glycoproteins: evidence for the involvement of a 205 K Con A binding protein and cytoskeletal components in ca
Cell-Motil-Cytoskeleton. 1987; 8(1): 37-43

Ujiie-T
OK-432-mediated augmentation of antitumor immunity and generation of cytotoxic T lymphocytes.
Jpn-J-Exp-Med. 1987 Apr; 57(2): 103-15

Valesini-G; Barnaba-V; Benvenuto-R; Balsano-F; Mazzanti-P; Cazzola-P
A calf thymus acid lysate improves clinical symptoms and T-cell defects in the early stages of HIV infection: second report.
Eur-J-Cancer-Clin-Oncol. 1987 Dec; 23(12): 1915-9

van-Goor-H; Harms-G; Gerrits-PO; Kroese-FG; Poppema-S; Grond-J
Immunohistochemical antigen demonstration in plastic-embedded lymphoid tissue.
J-Histochem-Cytochem. 1988 Jan; 36(1): 115-20

van-Haelst-UJ; Holland-R
A malignant mediastinal tumor, unclassified, most consistent with thymoma in a 9-year-old girl.
Ultrastruct-Pathol. 1987; 11(5-6): 793-9

Vecsei-L; Faludi-M; Najbauer-J
The effect of "facteur thymique serique" (FTS) on catecholamine and serotonin neurotransmission in discrete brain regions of mice.
Acta-Physiol-Hung. 1987; 69(1): 129-32

Venegas-RJ; Sun-NC
Cardiac tamponade as a presentation of malignant thymoma.
Acta-Cytol. 1988 Mar-Apr; 32(2): 257-62

Villemur-R; Monczak-Y; Rassart-E; Kozak-C; Jolicoeur-P
Identification of a new common provirus integration site in gross passage A murine leukemia virus-induced mouse thymoma DNA.
Mol-Cell-Biol. 1987 Jan; 7(1): 512-22

Vives-J; Sole-J; Suarez-B
Unfractionated human thymocytes have a lower proliferative capacity than CD3-4-8- ones but have a similar capacity for expression of interleukin 2 receptors and production of interleukin 2.
Proc-Natl-Acad-Sci-USA. 1987 Dec; 84(23): 8593-7

von-Domarus-H; Blaha-I
Ectopic thymus in the neck; a case report and review of the literature.
Br-J-Plast-Surg. 1987 Sep; 40(5): 532-5

Vu-ND; Wagner-PD
Effects of proteolysis on the adenosinetriphosphatase activities of thymus myosin.
Biochemistry. 1987 Jul 28; 26(15): 4847-53

Wade-S; Parent-G; Bleiberg-Daniel-F; Maire-B; Fall-M; Schneider-D; Le-Moullac-B; Dardenne-M
Thymulin (Zn-FTS) activity in protein-energy malnutrition: new evidence for interaction between malnutrition and infection on thymic function.
Am-J-Clin-Nutr. 1988 Feb; 47(2): 305-11

Walker-SE
Defective primary and secondary IgG responses to thymic-dependent antigens in autoimmune PN mice.
J-Immunol. 1988 May 15; 140(10): 3426-33

Wang-CR; Chen-GH; Mandy-WJ
Identification of a rabbit class I-like thymocyte-specific antigen.
J-Immunol. 1987 May 15; 138(10): 3352-9

Wang-P; Toyoshima-S; Osawa-T
Physical and functional association of cytosolic inositolphospholipid-specific phospholipase C of calf thymocytes with a GTP-binding protein.
J-Biochem-Tokyo. 1987 Nov; 102(5): 1275-87

Wang-PT; Bigby-M; Sy-MS
Selective down modulation of L3T4 molecules on murine thymocytes by the tumor promoter, phorbol 12-myristate 13-acetate.
J-Immunol. 1987 Oct 1; 139(7): 2157-65

Wang-SS; Chen-ST; Wang-KT; Merrifield-RB
4-Methoxybenzyloxycarbonyl amino acids in solid phase peptide synthesis.
Int-J-Pept-Protein-Res. 1987 Nov; 30(5): 662-7

Warchol-JB; Pilgrim-C; Ginda-WJ; Brelinska-R
Autoradiographic studies on glucose utilization in individual zones of the rat thymus.
Histochemistry. 1988; 88(3-6): 571-3

Warley-A
Concentrations of elements in rat thymocytes measured by X-ray microanalysis.
Cell-Tissue-Res. 1987 Jul; 249(1): 215-20

Warley-A
X-ray microanalysis of cells in suspension and the application of this technique to the study of the thymus gland.
Scanning-Microsc. 1987 Dec; 1(4): 1759-70

Warley-A; Daniel-PM; Guthrie-DL
Diabetes mellitus in the rat: diminution in the number of cells in the thymus and of lymphocytes in the blood.
Q-J-Exp-Physiol. 1987 Oct; 72(4): 609-15

Watanabe-M; Aiso-S; Hibi-T; Watanabe-N; Iwao-Y; Yoshida-T; Asakura-H; Tsuru-S; Tsuchiya-M
Alteration of T cell maturation and proliferation in the mouse thymus induced by serum factors from patients with ulcerative colitis.
Clin-Exp-Immunol. 1987 Jun; 68(3): 596-604

Watanabe-Y; Mitsuyama-M; Koga-T; Handa-T; Yoshikai-Y; Nomoto-K
Protective immunity to Listeria monocytogenes in neonatally thymectomized (NTx) mice: involvement of T cells distinct from those in sham-thymectomized mice.
Immunology. 1988 Apr; 63(4): 649-55

Waymack-JP; Jenkins-M; Warden-GD; Solomkin-J; Law-E Jr; Hummel-R; Miller-A; Alexander-JW
A prospective study of thymopentin in severely burned patients.
Surg-Gynecol-Obstet. 1987 May; 164(5): 423-30

Waymack-JP; Miskell-P; Gonce-SJ; Alexander-JW
Effect of two new immunomodulators on normal and burn injury neutrophils and macrophages.
J-Burn-Care-Rehabil. 1987 Jan-Feb; 8(1): 9-14

Weindruch-R; Naylor-PH; Goldstein-AL; Walford-RL
Influences of aging and dietary restriction on serum thymosin alpha 1 levels in mice.
J-Gerontol. 1988 Mar; 43(2): B40-2

Weiss-A; Dazin-PF; Shields-R; Fu-SM; Lanier-LL
Functional competency of T cell antigen receptors in human thymus.
J-Immunol. 1987 Nov 15; 139(10): 3245-50

Welch-RA; Mutchnick-MG; Weller-FE; Sokol-RJ
Maternal and fetal circulating levels of thymosin alpha 1 during parturition.
Am-J-Reprod-Immunol-Microbiol. 1987 Apr; 13(4): 12

Weller-FE; Mutchnick-MG
Enzyme immunoassay measurement of thymosin beta 4.
J-Immunoassay. 1987; 8(2-3): 203-17

Weller-FE; Mutchnick-MG; Goldstein-AL; Naylor-PH
Enzyme immunoassay measurement of thymosin beta 4 in human serum.
J-Biol-Response-Mod. 1988 Feb; 7(1): 91-6

Welsh-RS; Vyska-K
Phosphopeptides in highly purified calf thymus DNA.
Basic-Appl-Histochem. 1987; 31(3): 281-98

Wen-L; Hanvanich-M; Werner-Favre-C; Brouwers-N; Perrin-LH; Zubler-RH
Limiting dilution assay for human B cells based on their activation by mutant EL4 thymoma cells: total and antimalaria responder B cell frequencies.
Eur-J-Immunol. 1987 Jun; 17(6): 887-92

White-LL; Smith-RA
In vitro sensitization using thymocyte conditioned medium prepared from fetal calf, horse and rabbit sera.
Immunol-Lett. 1987 Jan; 14(2): 87-9

Wiedmeier-SE; Burnham-DK; Singh-U; Daynes-RA
Implantation of thymic epithelial grafts into the anterior chamber of the murine eye: an experimental model for analyzing T-cell ontogeny.
Thymus. 1987; 9(1): 25-44

Wiedmeier-SE; Samlowski-WE; Rasmussen-CJ; Huang-K; Daynes-RA
Effect of ionizing radiation on thymic epithelial cell function. I. Radiation-spared thymic epithelial grafts expedite the recovery of T cell function in lethally irradiated and fetal liver recons
J-Immunol. 1988 Jan 1; 140(1): 21-9

Wilder-RL; Allen-JB; Hansen-C
Thymus-dependent and -independent regulation of Ia antigen expression in situ by cells in the synovium of rats with streptococcal cell wall-induced arthritis. Differences in site and intensity of
J-Clin-Invest. 1987 Apr; 79(4): 1160-71

Willcox-N; Schluep-M; Ritter-MA; Schuurman-HJ; Newsom-Davis-J; Christensson-B
Myasthenic and nonmyasthenic thymoma. An expansion of a minor cortical epithelial cell subset?
Am-J-Pathol. 1987 Jun; 127(3): 447-60

Williams-WW; Falo-LD Jr; Lu-CY; Benacerraf-B; Sy-MS
Effects of in vivo monoclonal anti-I-A antibody treatment in neonatal mice on intrathymic and peripheral class II antigen expression.
Cell-Immunol. 1988 Jan; 111(1): 126-38

Wilson-A; D'Amico-A; Ewing-T; Scollay-R; Shortman-K
Subpopulations of early thymocytes. A cross-correlation flow cytometric analysis of adult mouse Ly-2-L3T4-(CD8-CD4-) thymocytes using eight different surface markers.
J-Immunol. 1988 Mar 1; 140(5): 1461-9

Wilson-A; Ewing-T; Owens-T; Scollay-R; Shortman-K
T cell antigen receptor expression by subsets of Ly-2-L3T4- (CD8-CD4-) thymocytes.
J-Immunol. 1988 Mar 1; 140(5): 1470-6

Wilson-M; Kamin-S; Harvey-W
The role of lipopolysaccharides in endotoxin-induced thymocyte proliferation and chondrocyte collagenase synthesis.
J-Med-Microbiol. 1987 Dec; 24(4): 309-13

Winchurch-RA
Activation of thymocyte responses to interleukin-1 by zinc.
Clin-Immunol-Immunopathol. 1988 May; 47(2): 174-80

Wiranowska-M; Kaido-T; Caspritz-G; Cook-J; Hadden-J
Interleukin-2 and coculture with thymic epithelial cells synergistically induce prothymocyte differentiation and proliferation.
Thymus. 1987; 10(3-4): 231-45

Wirt-DP; Grogan-TM; Nagle-RB; Copeland-JG; Richter-LC; Rangel-CS; Schuchardt-M; Fosse-J; Layton-JM
A comprehensive immunotopographic map of human thymus.
J-Histochem-Cytochem. 1988 Jan; 36(1): 1-12

Wolf-GT; Peterson-KA; Lovett-EJ 3d
In vitro immune modulation by thymosin alpha 1 in patients with head and neck squamous cell carcinoma.
Head-Neck-Surg. 1985 May-Jun; 7(5): 350-6

Wurm-K; Ewert-G; Lohr-G
Sarcoidosis complicated by HTLV III-infection: steroid therapy in combination with thymostimulin.
Sarcoidosis. 1987 Mar; 4(1): 68-70

Wyatt-CR; Magnuson-NS; Perryman-LE
Defective thymocyte maturation in horses with severe combined immunodeficiency.
J-Immunol. 1987 Dec 15; 139(12): 4072-6

Wysocki-H; Wysocki-J; Wierusz-Wysocka-B
The influence of thymus extract on the phagocytosis and the bactericidal capacity of polymorphonuclear neutrophils from children with Down's syndrome.
Ann-NY-Acad-Sci. 1987; 496: 740-2

Xu-CS; Tung-E; Wang-IY; Wang-AC
Identification and characterization of Thy-1 homologues from bovine thymocytes.
Mol-Immunol. 1987 Jul; 24(7): 791-6

Yagi-H; Nishimura-T; Hashimoto-Y
Characteristics of mouse thymocyte-derived, interleukin 2-activated killer cells and their precursors.
Jpn-J-Cancer-Res. 1987 Jul; 78(7): 721-8

Yamada-T; Ohyama-H
Radiation-induced interphase death of rat thymocytes is internally programmed (apoptosis).
Int-J-Radiat-Biol. 1988 Jan; 53(1): 65-75

Yialouris-PP; Evangelatos-GP; Soteriadis-Vlahos-C; Heimer-EP; Felix-AM; Tsitsiloni-OE; Haritos-AA
The identification of prothymosin alpha-like material in vertebrate lymphoid organs by a radioimmunoassay for the N-terminal decapeptide.
J-Immunol-Methods. 1988 Feb 10; 106(2): 267-75

Ying-SY
Inhibins and activins: chemical properties and biological activity.
Proc-Soc-Exp-Biol-Med. 1987 Dec; 186(3): 253-64

Yonuschot-G; Matthews-EK; Corps-AN; Metcalfe-JC
Permeabilization of thymocytes by photon activation of erythrosin.
FEBS-Lett. 1987 Mar 23; 213(2): 401-5

Zepp-F; Cussler-K; Mannhardt-W; Schofer-O; Schulte-Wissermann-H
Intrathymic tolerance induction: determination of tolerance to class II major histocompatibility complex antigens in maturing T lymphocytes by a bone marrow-derived non-lymphoid thymus cell.
Scand-J-Immunol. 1987 Dec; 26(6): 589-601

Zheng-ZM; Lavallee-JT; Bia-FJ; Griffith-BP
Thymic hypoplasia, splenomegaly and immune depression in guinea pigs with neonatal cytomegalovirus infection.
Dev-Comp-Immunol. 1987 Spring; 11(2): 407-18

Zilberman-Y; Howe-LR; Moore-JP; Hesketh-TR; Metcalfe-JC
Calcium regulates inositol 1,3,4,5-tetrakisphosphate production in lysed thymocytes and in intact cells stimulated with concanavalin A.
EMBO-J. 1987 Apr; 6(4): 957-62

Zimecki-M; Wieczorek-Z
Interleukin-1 decreases the level of thymocytes forming rosettes with autologous erythrocytes: a new method of determination of interleukin-1 activity.
Arch-Immunol-Ther-Exp-Warsz. 1987; 35(3): 371-9

Zlotnik-A; Ransom-J; Frank-G; Fischer-M; Howard-M
Interleukin 4 is a growth factor for activated thymocytes: possible role in T-cell ontogeny.
Proc-Natl-Acad-Sci-USA. 1987 Jun; 84(11): 3856-60

BOOK REVIEW

Linscott's Directory of Immunological and Biological Reagents, Fifth Edition, 1988, ISBN 0-9604920-4-6, £57.00 (includes 3 supplements 6 months apart).

Where can you buy motheaten mice, recombinant Interleukin-4, polyclonal sheep anti-kangaroo serum, mouse IgM monoclonal antibody to horse prothymocytes, 29 strains of mycoplasma and a frozen sample of squid semen? The answer to these and many other questions can be found in the latest edition of Linscott's Directory.

It is a dilemma familiar to all laboratory workers wishing to order items such as reagents, animals or cell lines that the necessary catalogues are never there when you need them, and even if they are, it can take a long time searching through them for the relevant information. More importantly, one is often just not aware of all the individual companies from whom particular products can be obtained. Linscott's Directory, by collating all such information, provides an invaluable, time-saving and highly efficient reference source for a wide variety of cellular, molecular, biochemical and immunological reagents. Full details of the products and of the suppliers' addresses, telephone and Telex numbers are given, so that items can be ordered straight from the Directory. Furthermore, the 6-monthly supplements ensure that it is always up-to-date. It is a book that no laboratory should be without.

MARY A. RITTER,
Department of Immunology, Royal Postgraduate Medical School, London.

RECOMMENDED BOOKS

Advanced Immunology. Edited by D. Male, B. Champion and A. Cook. Gower, London/Lippincott, Philadelphia, 1987, 397-445547. £25.00/$39.95.

An Introduction to Immunocytochemistry. Current Techniques and Problems (Revised Edition). Edited by J.M. Polak and S. van Noorden. Oxford University Press (Oxford, England), 1987, 0- 19-856415-5. £5.95.

Animal Models of Immunological Processes. Edited by J. Hay. Academic Press Inc, 1982, ISBN 0-12-333520-5. £46.50.

Antibodies Volume 1 - A practical approach. Edited by D. Catty. IRL Press (Oxford), 1988, Soft 0-047946 85 3; $38.00/£19.00: Hard 0-947946 86 1; $56.00/£28.00.

Basic Immunogenetics (Third Edition). Edited by H.H. Fudenberg. Oxford University Press (Oxford, England), 1984, 0-19-503404-X, £25.00, hardback. 0-19-503405-8, £15.00, paper covers.

Basic Immunology: Immune Mechanisms in Health and Disease. Edited by S. Sell. Elsevier Science Publishers (New York), 1987, 0-444-02238-2. $27.50 (£19.50 outside North America).

Cellular, Molecular and Genetic Approaches to Immunodiagnosis and Immunotherapy. (Proceedings of the VIIIth International Conference on Labelled Antibodies, Tokyo, November, 1985). Edited by K. Kano, S. Mori, T. Sugisaki, Tokyo; M. Torisu, Fukuoka. University of Tokyo Press, Tokyo. ISBN 3-8055-4649-1. $186.75.

Clinical Application of Interferons and their Inducers. Clinical Pharmacology 3 (2nd Edition). Edited by D.A. Stringfellow. Marcel Dekker, 1986, 0-8247-7531-7. £50.95.

Clinical Transplantation - Current Practice and Future Prospects. Edited by G.D.R. Catto. MTP Press Limited (Lancaster, England), 1987, ISBN 0-85200-960-7. Hardbound Dfl. 169.00/£45.00

Complement. Edited by S.K.A. Law and K.B.M. Reid. IRL Press (Oxford, England), 1988, 1-85221-061-3. $11.95/£5.95.

Contemporary Topics in Immunology. Volume 15: Immune Complexes and Human Cancer. Edited by F.A. Salinas and M.G. Hanna, Jr. Plenum Publishing Corporation (New York), 1985, 0-306-41955-6. $49.50 ($59.40 outside US and Canada).

Contemporary Topics in Molecular Immunology. Volume 10: The Interleukins. Edited by S. Gillis and F.P. Inman. Plenum Publishing Corporation (New York), 1985, 0-306-41776-6. $49.50 ($59.40 outside US and Canada).

Culture of Animal Cells (Second Edition). Edited by R. Ian Freshney. Alan R. Liss (New York). 1987, book No.4241. £59.50 (approx).

Current Topics in Aids. Edited by M. Gottlieb, D. Jeffries, D. Mildvan, A. Pinching, T. Quinn and R. Weiss. John Wiley & Sons Limited (West Sussex, England), 1987, 0471-91198-4. £25.00/$49.50.

Essential Immunogenetics. Edited by A.R. Williamson and M.J. Turner. Blackwell Scientific Publications (Oxford), 1988, 0- 632-00236-0. £19.50.

Essential Immunology. Edited by I.M. Roitt. Blackwell Scientific Publications (Oxford), 1988 (6th edition) 0-632-02283- 3. £12.95.

Fundamentals of Immunology (Second Edition). Edited by O.G. Bier, W. Dias da Silva, D. Gotze and I. Mota. Springer- Verlag, 1987. £23.00/$29.95.

Gene Structure and Transcription. Edited by T. Beebee and J. Burke. IRL Press (Oxford, England), 1988, 1-85221-014-1. $11.95/£5.95.

H-2 Antigens - Genes, Molecules, Function. Edited by Chella S. David. Plenum Publishing Corporation (New York), 1988, 0-306-42804-0. $129.50 ($155.40 outside US and Canada).

HLA and Disease Associations. J. Tiwari, P.I. Terasaki. Springer-Verlag (Berlin, New York, London), 1985, Hard cover DM 256, ISBN 3-540-96081-3.

HLA Class II Antigens: A comprehensive review of structure and function. Edited by B.G. Solheim, E. Moller, and S. Ferrone. Springer-Verlag (Berlin, New York, London), 1986, Hard cover DM 226, ISBN 3-540-15316-0.

Hormones and Immunity. Edited by I. Berczi and K. Kovacs. MTP Press Limited (Lancaster, England), 1987, ISBN 0-85200-848-1. Hardbound Dfl. 212.00/£59.00. paper covers.

Immune Mechanisms in Invertebrate Vectors. Edited by A.M. Leckie. Oxford University Press, 1986, 0-19-854004-3. £40.00.

Immune Recognition. Edited by M.J. Owen and J.R. Lamb. IRL Press (Oxford, England), 1988, 1-85221-062-1. $11.95/£5.95.

Immunity in Invertebrates: Cells, Molecules, and Defense Reactions. Edited by M. Brehelin. Proceedings in Life Sciences, Springer-Verlag, 1986, 3-540-15871-5. £34.95.

Immunochemical Techniques, Part H. Edited by G. Di Sabato *et al.* In: *Methods in Enzymology,* Edited by Colowick and Kaplan. Academic Press Inc, 1985, ISBN 0-12-182016-5. £55.50.

Immunochemical Techniques, Part K: *In Vitro* Models of B and T cell Function and Lymphoid Cell Receptors. Edited by G. Di Sabato *et al.* In: *Methods in Enzymology,* Edited by Colowick and Kaplan. Academic Press Inc, 1987, ISBN 0-12-182051-3.

Immunochemistry in Practice. Edited by A.P. Johnstone and R.T. Thorpe. Blackwell Scientific Publications (Oxford), 1987 (2nd edition), 0-632-01723-6. £12.95.

Immunocytochemistry (Third Edition). Edited by L.A. Sternberger. John Wiley & Sons Limited (West Sussex, England). 1985, 0471-86721-7. £50.45/$73.15.

Immunocytochemistry: Modern Methods and Applications. Edited by J.M. Polak and S. Van Noorden. Wright, 1986, 0-7236- 0870-9. £65.00.

Immunology of Endocrine Disease. Edited by A.M. McGregor. MTP Press Limited (Lancaster, England), 1986, ISBN 0-85200-963- 12. Hardbound Dfl. 141.00/£32.50.

Immunology, Immunopathology, and Immunity (Fourth Edition). Edited by S. Sell. Elsevier Science Publishers (New York), 1987, 0-444-01137-4. £37.50/$55.00.

Immunology: A Short Course. Edited by E. Benjamini and S. Leskowitz. Alan R. Liss, New York/Wiley, Chichester, 1988. £15.75/$22.95.

Immunology: A Synthesis. Edited by Edward S. Golub. Sinauer/Blackwell Scientific, 1987. £26.50/$32.95.

Immunology, Immunopathology, and Immunity (Fourth Edition). Edited by S. Sell. Elsevier, 1987. £37.50/$55.00.

Immunology: The Science of Self-Nonself Discrimination. Edited by J. Klein. John Wiley & Sons Limited (West Sussex, England), 1982, 0471-05124-1. £68.80/$99.75 (cloth). 1984, 0471-80762-1. £36.25/$52.55 (paper).

Immunology of the Neonate. Edited by G.R. Burgio *et al.* Springer-Verlag, 1987, 3-540-16339-5. £34.10.

Immunology of Reproduction. Edited by T.G. Wegmann and T.J. Gill. Oxford University Press (Oxford, England), 1983, 0-19- 503096-6. £40.00.

Immunlogy Simplified (Second Edition). Edited by T.R. Bowry. Oxford University Press (Oxford, England), 1984, 0-19- 261340-5. £7.50.

Immunoregulation and Fetal Survival. Edited by T.J. Gill and T.G. Wegmann. Oxford University Press (Oxford, England), 1987, 0-19-503989-0. £40.00.

Interferons and other regulatory cytokines. E. De Maeyer and J. De Maeyer-Guignard. J. Wiley & Sons Ltd. (Sussex, England), 1988, 0471-82887-4. £57.00/$95.00.

Interferons. Their Impact in Biology and Medicine. Edited by J. Taylor-Papadimitriou. Oxford University Press (Oxford, England), 1985, 0-19-261536-X, £22.50, hardback. 0- 19-261481-9, £15.00,

In Vivo Migration of Immune Cells. Edited by W.L. Olszewski, M.D., Ph.D. CRC Press (London) 0-8493-5076-X, 280 pages, September 1987.

Lecture Notes on Immunology. W.G. Reeves. Blackwell Scientific Publications (Oxford), 1987, 0-632-00776-1. £8.50.

Leucocyte Typing II. Edited by E.L. Reinherz, B.F. Haynes, L.M. Nadler, I.D. Bernstein. Springer-Verlag (Berlin, New York, London), 1986, three volumes. Vol.1. Human T Lymphocytes. Hard cover DM 260, ISBN 3-540-96175-5. Vol.2. Human B lymphocytes. Hard cover DM 238, ISBN 3-540-96176-3. Vol.3. Human Myeloid and Hematopoietic Cells, Hard cover DM 194, ISBN 3-540-96177-1.

Leucocyte Typing III. White Cell Differentiation Antigens. Edited by A.J. McMichael. Oxford University Press (Oxford, England), 1987, 0-19-2615521. £80.00.

Linscott's Directory of Immunological and Biological Reagents (Fifth Edition). Edited by W.D. Linscott. Available from Linscott's Directory, P.O. Box 55, East Grinstead RH19 3RX, Surrey. Directory £30, Supplement Pack £15. ISBN 0-9604920-4-6.

Lymphocytes, a practical approach. Edited by G.G. Klaus. In: The Practical Approach Series. Series Editors: D. Rockwood and B.D. Hames. IRC Press (Oxford; Washington), 1987, ISBN 1- 85221-018-4 Hardbound; 1-85221-019-2 softbound.

Lymphokines. Edited by A.Hamblin. IRL Press (Oxford, England), 1988, 1-85221-061-3. $11.95/£5.95.

Lymphokines: A Forum for Immunoregulatory Cell Products. Edited by E. Pick. Academic Press Inc, Volume 12, 1985, ISBN 0-12-43201-0. £24.50. Volume 13, 1987, ISBN 0-12-432013-9. £63.00.

Mechanisms of Lymphocyte Activation and Immune Regulation. Edited by Sudhir Gupta, William E. Paul and Anthony S. Fauci. Plenum Publishing Corporation (New York), 1987, 0-306-42568-8. $55.00 ($66.00 outside US and Canada).

Migration and Homing of Lymphoid Cells. Edited by A. Husband, Ph.D. CRC Press (London) 0-8493-6506-6, 464 pages, May 1988.

Molecular Basis of Lymphokine Action. Edited by D.R. Webb, C.W. Pierce and S. Chen. John Wiley & Sons Ltd. (Sussex, England), 1988, 0-89603139X. £69.96/$101.40.

Molecular Immunology. Edited by B.D. Hames and D.M. Glover. IRL Press (Oxford), 1988, Soft 1-85221-075-3; $30.00/£15.00: Hard 1-85221-079-6; $46.00/£23.00.

Natural History of the Major Histocompatibility Complex. Edited by J. Klein. John Wiley & Sons Limited (West Sussex, England), 1986, 0471-80953-5. £87.15/$126.35.

Progress in Immunology VI. Edited by B. Cinader and R.G. Miller. Academic Press Inc, 1987, ISBN 0-12-174686-0. £57.00 (paperback). ISBN 0-12-174685-2. £144.50 (hardback).

Receptors in Cellular Recognition and Developmental Processes. Edited by R.M. Gorczynski. Academic Press Inc, 1986, ISBN 0-12-290530-X. £41.00.

Recognition of M. Leprae Antigens. Edited by T. Ottenhoff and R. de Vries. MTP Press Limited (Lancaster, England), 1987, ISBN 0-89838-887-2. Hardbound Dfl. 110.00/£37.50.

Regulation of Immune Gene Expression. Edited by M. Feldmann and A. McMichael. Humana Press, 1986, 0-89603-104-7. £66.75.

Soviet Medical Reviews Supplement Series Section D: Immunology. Edited by R.V. Petrov. Gordon and Breach (London), 1988, 3-7186-4801-6.

The Autoimmune Diseases. Edited by N. Rose and I. Mackay. Academic Press Inc, 1985, ISBN 0-12-596921-X. £30.00 (paperback). ISBN 0-12-596920-1. £68.00 (hardback).

The Experimental Foundations of Modern Immunology. Edited by W.R. Clark. Wiley, 1986, 0-471-81508-X. £37.55.

The Rabbit in Contemporary Immunology Research. Edited by S. Dubiski. Longman Scientific & Technicial (Harlow, Essex) 1987, ISBN 0-582-01640-1. £30.00.

The Ruminant Immune System in Health and Disease. Edited by W. Ivan Morrison. Cambridge University Press (Cambridge), 1987, 0-521-32443-2. £55.00.

The T-Cell Receptors. Edited by Tak W. Mak. Plenum Publishing Corporation (New York), 1988, 0-306-42708-7. $42.50 ($51.00 outside US and Canada).

The Year in Immunology 1986-1987. Edited by J.M. Cruse, R.E. Lewis, Jr. Karger, ISBN 3-8055-4616-5. $172.00.

Vertebrate Blood Cells. Edited by A.F. Rowley and N.A. Ratcliffe. Cambridge University Press (Cambridge), 0-521-26032-9. 35.00.

SOME RECENT PhD THESES ON THE THYMUS AND RELATED TOPICS

BRANDT, N. (1988) Qualitative and quantitative ultrastructural studies on the human thymus. University of Kiel, Germany.

PENNINKS, A. (1985) Immunotoxicity of organotin compounds. Rijksuniversiteit of Utrecht. The Netherlands.

LIST OF JOURNALS THAT PUBLISH ARTICLES ON THE THYMUS

Acta Pathologica Microbiologica et Immunologica

Editor in Chief : J. Rygaard M.D.
Manuscripts and editorial matters to :
Editorial Office,
c/o Inger Danielsen,
Secretary,
Johnstrups Alle 6,
DK-1923,
Frederiksberg C,
Denmark.

Advances in Immunology

Editor: Frank J. Dixon,
Scripps Clinic and Research Foundation,
La Jolla,
California,
U.S.A.

American Journal of Pathology

Editor in Chief: Vincent T Marchesi.
Manuscripts to:
The American Journal of Pathology,
P.O. Box 5217,
Hamden,
Connecticut 06518,
U.S.A.

Annual Review of Immunology

Editor: William E. Paul,
National Institutes of Health,
Bethesda,

Maryland,
U.S.A.

Autoimmunity

Editor: Terry Wilkin
Department of Medicine 2,
General Hospital,
Southampton SO9 4XY,
U.K.

Brain, Behaviour, and Immunity

Editor-in-Chief: Robert Ader,
University of Rochester School of Medicine and Dentistry,
New York,
U.S.A.

British Medical Journal

The British Medical Journal,
The British Medical Association,
Tavistock Square,
London WC1H 9JR,
U.K.

Cancer Letters

American, Australasian and Asian Editor: Dr D. Clayson,
Toxicological Research Division,
Bureau of Chemical Safety,
Health Protection Branch,
Health and Welfare Canada,
Tunney's Pasture,
Ontario KIA OL2,
Canada.

Editor for Other Countries: Dr P. Shubik,
Green College,
Oxford OX2 6HG,
UK.

Cell

Editor: Benjamin Lewin
Principal Office:

Cell,
50 Church Street,
Cambridge,
Massachusetts 02138,
U.S.A.

European Editor: Peter W.J. Rigby
European Office:
Laboratory of Eukaryotic Molecular Genetics,
National Institute for Medical Research,
The Ridgway,
Mill Hill,
London NW7 1AA,
U.K.

Cell and Tissue Research

Editors:
Professor D.S. Farner,
Department of Zoology,
University of Washington,
Seattle,
WA 98195,
U.S.A.

Professor D.E. Kelly,
Department of Anatomy and Cell Biology,
University of Southern California,
School of Medicine,
Los Angeles,
CA 90033,
U.S.A.

Professor B. Lofts,
Department of Zoology,
University of Hong Kong,
Hong Kong.

Dr. J.F. Morris,
Department of Human Anatomy,
University of Oxford,
South Parks Road,

Oxford OX1 3QX,
U.K.

Professor A. Oksche,
(Co-ordinating Editor)
Institut für Anatomie und Zytobiologie,
Justus-Liebig-Universität,
Aulweg 123,
d-6300 Giessen,
Federal Republic of Germany.

Professor Berta Scharrer,
Department of Anatomy,
Albert Einstein College of Medicine,
Bronx,
NY 10461,
U.S.A.

Professor L. Vollrath,
Anatomisches Institut,
Johannes Gutenberg-Universität,
Saarstr. 19/21,
D-6500 Mainz,
Federal Republic of Germany.

Cellular Immunology

Editor: Dr. H. Sherwood Lawrence,
Infectious Diseases and Immunology Division,
Department of Medicine,
New York University Medical Centre,
550 First Avenue,
New York,
NY10016,
U.S.A.

Clinical and Experimental Immunology

Editor: Professor J. L. Turk,
Blackwell Scientific Publications Ltd.,
8 John Street,
London WC1N 2ES,
U.K.

Clinical Immunology and Immunopathology

Editors: H. Hugh Fundenberg and Robert T. McCluskey.
Manuscripts to:
Clinical Immunology and Immunopathology,
Editorial Office,
7th. Floor,
1250 Sixth Avenue,
San Diego,
California 92101.
U.S.A.

European Journal of Immunology

Editor in Chief: Dr. B. Kickhofen,
Stubeweg 51,
D-7800 Feiburg,
Federal Republic of Germany.

Human Immunology

Editor in Chief: D. Bernard Amos,
Division of Immunology,
Department of Microbiology and Immunology,
Duke University Medical Center,
Durham,
North Carolina,
U.S.A.

American Editor: Jeffrey R. Dawson,
Division of Immunology,
Department of Microbiology and Immunology,
Duke University Medical Center,
Durham,
North Carolina,
U.S.A.

European Editor: Jon J. van Rood,
Afdeling Immunohaematologie,
Academisch Ziekenhuis,
Leiden,
The Netherlands.

Immunodeficiency Reviews

Editor: Fred S. Rosen,
Harvard Medical School,
The Children's Hospital,
Boston,
U.S.A.

Editor: Maxime Seligmann,
Hospital Saint-Louis,
Paris,
France.

Immunogenetics

Editor: Prof. Dr. Jan Klein,
Abteilung Immungenetik,
Max-Planck-Institut fur Biologie,
Corrensstrasse 42,
D-7400 Tubingen 1,
Federal Republic of Germany.

Immunological Reviews

Editor: Göran Möller,
Department of Immunology,
University of Stockholm,
Biology Building,
F, S-10691,
Stockholm,
Sweden.

Immunology

Editor: Professor M.W. Steward,
Immunology Unit,
Department of Medical Microbiology,
London School of Hygiene and Tropical Medicine,
Keppel Street (Gower Street),
London WC1E 7HT,
U.K.

Immunology and Cell Biology

Editor:
Manuscripts to:
The Editor in Chief,
Immunology and Cell Biology,
Department of Microbiology and Immunology,
University of Adelaide,
Adelaide,
South Australia 5000.

Immunology Letters

Editor-in-Chief: P. Perlmann,
Department of Immunology,
Biology Building,
University of Stockholm,
S-106 91 Stockholm,
Sweden.

Immunology Today

Editor: John Inglis,
Elsevier Publications (Cambridge),
68 Hills Road,
Cambridge CB2 1CA,
U.K.

Immunopharmacology

Editor-in-Chief: J.R. Battisto,
Department of Immunology and Cancer,
The Research Institute,
Cleveland Clinic Foundation,
Cleveland,
Ohio 44106,
U.S.A.

International Reviews of Immunology

Editor: Heinz Kohler,
Roswell Park Memorial Institute,
Buffalo,
New York,
U.S.A.

Editor: Constantin Bona,
Mount Sinai School of Medicine,
New York,
U.S.A.

Journal of Anatomy

Editor: D.B. Moffat
Manuscripts to:
Mrs. R. Davies,
2 Abbotts Walk,
Lechlade Park,
Lechlade,
Gloucestershire GL7 3DB,
U.K.

Journal of Autoimmunity

Editor: J.F. Bach
Inserm U25,
CNRS UA 122,
Hôpital Necker,
161, rue de Sèvres,
75743 Paris,
Cedex 15,
France.

Journal of Clinical Immunology

Editor: Dr Sudhir Gupta,
Division of Basic and Clinical Immunology,
Department of Medicine,
University of California,
College of Medicine,
Irvine,
California 92717,
USA

Journal of Experimental Medicine

Editors: Anthony Cerami, Zanvil A. Cohn, Maclyn McCarthy.
Manuscripts to:
The Editors of The Journal of Experimental Medicine,

The Rockefeller University,
New York 10021,
U.S.A.

Journal of Immunogenetics

Editor: Dr. G. Butcher,
The AFRC Institute of Animal Physiology and Genetics Research,
Babraham,
Cambridge CB2 4AT,
U.K.

Journal of Immunological Methods

Editors: V. Nussenweig,
Department of Pathology,
New York University Medical Center School of Medicine,
550 First Avenue,
New York,
N.Y. 10016,
U.S.A.

M.W. Turner,
Department of Immunology,
Institute of Child Health,
30 Guildford Street,
London WC1N 1EH,
U.K.

Journal of Immunology

Editor in Chief: Dr. Ethan Shevach,
The Journal of Immunology,
9650 Rockville Pike,
Bethesda,
Maryland 20814,
U.S.A.

Journal of Molecular and Cellular Immunology

Editor in Chief: Professor Charles A. Janeway Jr.
Manuscripts to:
J.M.C.I. Editorial Office,
Springer-Verlag,
175 Fifth Avenue,

New York,
N.Y. 10010,
U.S.A.

Journal of Neuroimmunology

Editor for Americas and Asia: Professor Cedric S. Raine,
Department of Pathology (Neuropathology),
Albert Einstein College of Medicine of Yeshiva University,
1300 Maris Park Avenue,
Bronx,
New York 10461,
USA.

Editor for Europe, Australasia and Africa: Dr Peter O. Behan,
Institute of Neurological Sciences,
Southern General Hospital,
1345 Govan Road,
Glasgow G51 4TF,
UK.

Lancet

Editorial Office,
The Lancet,
7 Adam Street,
London WC2N 6AD,
U.K.

Molecular Immunology

Board of Regional Editors:
Professor Keith J. Dorrington,
(Chairman of the Board),
Connaught Research Laboratories,
1755 Steeles Avenue West,
Willowdale,
Ontario.
Canada M2R 3T4.

Professor Gordon L. Ada,
Microbiology Department,
The John Curtin School of Medical Research,
P.O. Box 334,

Canberra City,
A.C.T.,
Australia.

Dr. Michael J. Crumpton,
Imperial Cancer Research Fund,
Lincoln Inn's Fields,
London WC2A 3PX,
U.K.

Dr. Michel Fougereau,
Centre d'Immunologie de Marseille-Luminy,
Case 906,
13288 Marseille Cedex 9,
France.

Dr. Henry Metzger,
Building 10,
National Institutes of Health,
Bethesda,
Maryland 20205,
U.S.A.

Dr. Roald S. Nezlin,
Institute of Molecular Biology,
Academy of Science,
Vavilovstr 32,
Moscow W-334,
USSR 117984.

Nature

Editor: John Maddox,
4 Little Essex Street,
London WC2R 3LF,
U.K.

Neuroimmunomodulation

U.S. Regional Editor: Herbert Novera Spector,
National Institute of Health,
Bethesda,
M.D., U.S.A.

European Regional Editor: Branislav D. Jankovic,
Immunology Research Centre,
Belgrade,
Yugoslavia.

Eastern Europe and USSR Regional Editor: Dr. Ion Baciu,
Faculty of Medicine,
Str. Clinicilol 1,
Rumania.

Reviews on Immunoassay Technology

Editor: S.B. Pal,
Universität Ulm,
Department für Innere Medizin,
Steinhövelstrasse 9,
D-7900 Ulm (Donau),
Federal Republic of Germany.

Scandinavian Journal of Immunology

Managing Editors: M. Harboe, J.B. Natvig and E. Moller.
Manuscripts to:
Scandinavian Journal of Immunology,
Institute for Immunology and Rheumatology,
Fr. Quamsgt. 1,
N-1072 Oslo 1,
Norway.

Science

Editor: Daniel E. Koshland.
Manuscripts to:
Science,
1333 H Street, NW,
Washington, DC 20005,
U.S.A.

Thymus

Editor: J.L. Touraine,
Pavillion P,
Hôpital E. Herriot,

69374 Lyon Cedex 8,
France.

Tissue Antigens

Editor: F. Kissmeyer-Nielsen M.D.,
Tissue Antigens,
Department of Clinical Immunology,
Kommunehospitalet,
DK-8000 Aahus C,
Denmark.

Transplantation

Editors:
R. Batchelor, London, U.K.,
Peter J. Morris, Oxford, U. K.,
E.J. Eichwald, Salt Lake City, Utah, U.S.A.,
David Steinmuller, Ann Arbor, Michigan, U.S.A.,
Anthony P. Monaco, Boston, Massachusetts, U.S.A.,
Mary L. Wood, Boston, Massachusetts, U.S.A.,

European Editoral Office:
Department of Immunology,
Royal Postgraduate Medical School,
Ducane Road,
London W12 OHS,
U.K.

North American Editorial Office:
New England Deaconess Hospital,
185 Pilgrim Road,
Boston,
Massachusetts,
U.S.A.

LISTING OF MONOCLONAL ANTIBODIES REACTIVE WITH THYMOCYTE SURFACE MOLECULES.

Compiled by: Heather M. Ladyman

Royal Postgraduate Medical School, Hammersmith Hospital, Du Cane Road, London W12 0NN.

Cluster of differentiation	Antigen Mol.wt.	Antigen Distribution	Antibody name	Species & Isotype	Source/ref.
CD1a	49	cortical thymocytes,	(420) L119		Bernard (Paris, France)
		langerhans cells. (	432) VIT6b		Knapp (Vienna, Austria)
		Assoc. with B2 microglobulin	(436) 10H39.3		Mawas (Marseille, France)
		Coval. assoc. with CD8, non-coval	(438) NA1/34	M,G2b	McMichael (Oxford, U.K.)
		assoc. with OD1b or OD1c.			(Dako UK., Serotec, UK., Seralab, UK.)
			(431) VIT6		Knapp (Vienna, Austria) (Calbiochem, USA.)
		Formerly T6 antigen.	(423) L404		Bernard (Paris, France)
			(422) L232		Bernard (Paris, France)
			(424) L544		Bernard (Paris, France)
			(427)BL6		Brochier (Montpellier, France) (Serotec, UK.)
			(433) VIT6d		Knapp (Vienna, Austria)
			(440) T6		Reinherz (Boston, USA)
			(435) 10D12.2		Mawas (Marseille, France)
			(425) D47		Boumsell (Paris, France)
			OKT6	M,G1	(Ortho, UK.)
CD1b	45	cortical thymocytes,	(426) WM-25		Bradstock (NSW, Australia)
		Formerly T6 antigen	(429) 7C4/160/4G9		Knapp (Vienna, Austria)
			(437) 4A76	M,G2a	Mawas (Marseille, France) (Serotec, UK.)
			(441) NU-T2		Sagawa (Fukuoka, Japan)
CD1c	43	cortical thymocytes	(421) L161		Bernard (Paris, France)
		subpop. of langerhans cells,	(428) 10C3/213/4D11		Knapp (Vienna, Austria)

		mantle zone in lymphoid organs. Formerly T6 antigen.	(430) 7C6/162/3B10		Knapp (Vienna, Austria)
			(434) M241		Knowles (New York, USA)
CD2	46-50	peripheral blood T cells, thymocytes. Formerly T11 antigen. possible ligand for LFA-3	(445) D66/1		Bernard (Paris, France)
			(446) OCH 217		Columbani (Milano, Italy)
			(447) 9.6		Hansen (Washington, USA)
			(448) 35.1		Martin (London, U.K.)
			(449) 39B21		Mawas (Marseille, France)
			(450) 9-2		Naito
			(451) 9.1		Naito
			(452) F92-3A11	M,G1	Poncelet (Montpellier, France) (Seralab, UK.)
			(453) 95-5-49		Quinones (Bethesda, USA)
			(454) 7Ag		Reinherz (Boston, USA)
			(455) 7E10		Reinherz (Boston, USA)
			(456) MOMO2A6		Reinherz (Boston, USA)
			(457) T11/3PT2H9		Reinherz (Boston, USA)
			(458) T11/3T4-8B5		Reinherz (Boston, USA)
			(459) MT110		Rieber
			(460) MT26		Rieber
			(461) MT910		Rieber
			(462) NU-T1		Sagawa (Fukuoka,Japan)
			(463) TS1/8.1.1		Springer (Boston, USA)
			(464) TS2/18.1.1		Springer (Boston, USA)
			(465) CLB-T11/1		Tetteroo (Amsterdam, Netherlands)
			(466) SA1-11		Yang (New York, USA)
			(467) NU-TER		Yokoyama (Kurume, Japan)
			(91) OR 101	M,G1	Gazit (Israel)
			(84) XIX.8	M,G2b	Bourel (Rennes, France)

Cluster of differentiation	Antigen Mol.wt.	Antigen Distribution	Antibody name	Species, Isotype	Source/ref.
			(109) 9.6	M,G2a	Hansen (Washington, USA)
			(110) 7A9	M,M	Reinherz (Boston, USA)
			T9-10	M,G1	(Dako, UK.)
			AF.10.3	M,G1	(Serotec, UK.)
			39Cl.5	M,G2a	(Serotec, UK.)
			T11	M,G1	(Coulter, UK.)
			OKT11	M,G2a	(Ortho, UK.)
			Anti-Leu5b	M,G2a	(Becton Dickinson, UK.)
CD3	22,26,30	Mature T cells. Constant epitope on the T cell antigen receptor. Formerly T3 antigen	(471) UCHT1		Beverly (London, U.K.)
			(472) X35-3		Bourel (Rennes, France)
			(473) XXIII.141		Bourel (Rennes, France)
			(474) XXIII.46		Bourel (Rennes, France)
			(475) XXIII.87		Bourel (Rennes, France)
			(476) 381		Hansen (Washington, USA)
			(477) VIT3		Knapp (Vienna, Austria)
			(478) VIT3b		Knapp (Vienna, Austria)
			(482) BW264/56		Kurrle (Marburg, FRG)
			(483) 12F6		Lazarovits (Ontario, Canada)
			(484) G19-4.1		Ledbetter (Seattle, USA)
			(485) T3(2ADA)		Reinherz (Boston, USA)
			(486) T3/2T8-2F4		Reinherz (Boston, USA)
			(487) T3/RW2-8C8		Reinherz (Boston, USA)
			(488) TE/RW2-4B6		Reinherz (Boston, USA)

			(489) CLB-T3/3		Tetteroo (Amsterdam, Netherlands)
			(490) CRIS-7		Vilella (Barcelona, Spain)
			(491) YTH12.5		Waldmann (Cambridge, U.K.)
			(492) SK7/LEU4		Warner (California, USA) (Becton Dickinson UK.)
			OKT3		(Ortho UK.)
			SL5	M,G3	(Sera-lab UK.)
			SL7	M,G2a	(Sera-lab UK.)
			X35	M,G2a	(Serotec UK.)
			SPVT3b	M,G2a	(Serotec UK.)
			T3-4B5		(Dako UK.)
CD4	60	T cell subset (helper)	(495) BL4/IOT4	M,G2a	Brochier (Montpellier, France) (Serotec UK.)
		Formerly T4 antigen	(496) LEU3A Evans		(Mountain View, USA)
			(497) VIT4		Knapp (Vienna, Austria)
			(498) BW264/123		Kurrle (Marburg, FRG)
			(499) G1	9-2	Ledbetter (Seattle, USA)
			(500) 66.1		Martin (London, U.K.)
			(501) 13B.8.2	M,G1	Mawas (Marseille, France) (Serotec UK.)
			(502) F101-5		Poncelet (Montpellier, France)
			(503) F101-69(S-T4)	M,G1	Poncelet (Montpellier, France) (Seralab UK.)
			(504) T4		Reinherz (Boston, USA)
			(505) T4/18T3AG		Reinherz (Boston, USA)
			(506) T4/19THY5D7		Reinherz (Boston, USA)
			(507) MT151		Rieber
			(508) MT301		Rieber
			(509) MT321		Rieber

Cluster of differentiation	Antigen Mol.wt.	Antigen Distribution	Antibody name	Species, Isotype	Source/ref.
			(510) 91D6		Szer
			(511) CLB-T4/1		Tetteroo (Amsterdam, Netherlands)
			(512) EDU-2		Vilella (Barcelona, Spain)
			(513) 94B1		Winchester (New York, USA)
			(514) NU-TH/I		Yokoyama (Kurume, Japan)
			T4A		(Coulter UK.)
			T3-10	M,G1	(Dako UK.)
			OKT4		(Ortho UK.)
					OKT4A (Ortho UK.)
			SK3/LEU-3a		(Becton Dickinson UK.)
			T4LA	M,G1	(Seralab UK.)
			BT5.9	M,G3	(Seralab UK.)
			BT5.9 clone 1	M,G1	(Seralab UK.)
			BT5.9 clone 7	M,G2b	(Seralab UK.)
			CLB159	M,G2a	(Jansen Life Sciences, Belgium)
CD5	67	Peripheral blood T cells and thymocytes.	(515) H65		Bai (Beijing, China)
			(516) I247		Bernard (Paris, France)
			(518) UCHT2		Beverly (London, U.K.)
			(520) BL1a/IOT1a		Brochier (Montpellier, France) (Serotec UK.)
			(521) BL1b		Brochier (Montpellier, France)
			(522) WC11.5		Cassano (Gainesville, USA)
			(523) MEM-32		Horejsi (Praha, Czechoslovakia)

			(524) 10.2		Martin (London, U.K.)
			(526) 6-2		Naito
			(527) F93-6D9(S-T1)		Poncelet (Montpellier, France) (Seralab UK.)
			(528) T1		Reinherz (Boston, USA)
			(529) CLB-T1/1		Tetteroo (Amsterdam, Netherlands)
			(530) CRIS1		Vilella (Barcelona, Spain)
			(531) L17F12.LEU1		Warner (California, USA) (Beckton Dickinson UK.)
			(532) TU71		Wernet (Tubingen, FRG)
			(533) Y-2/178		Mason (Oxford, U.K.)
			HuLy-m7	M,M	(Serotec UK.)
			MID5	M,M	(Seralab UK.)
			DK23(DAKO-T1)	M,G1	(Dako UK.)
			OK-CLL		(Ortho UK.)
			(141) YTH26.5	R,G2b	Waldmann (Cambridge, U.K.) (Serotec UK.)
			(124) KS-6	M,G1	Sagawa (Fukuoka,Japan)
CD7	40	Fc{R. Thymocytes and peripheral blood T cells.	(545) 121		Bernard (Paris, France)
			(546) I263		Bernard (Paris, France)
			(547) I458		Bernard (Paris, France)
			(548) 1-3		Dupont (New York, USA)/Naito
			(549) 3A1A		Haynes (Durham, USA)
			(550) 3A1B-4Gb		Haynes (Durham, USA)
			(551) 3A1d-14C2		Haynes (Durham, USA)
			(552) 3A1e-12H7		Haynes (Durham, USA) (Seralab UK.)
			(553) 3A1f-13D1		Haynes (Durham, USA)
			(554) 3A1g-14F12		Haynes (Durham, USA)

Cluster of differentiation	Antigen Mol.wt.	Antigen Distribution	Antibody name	Species, Isotype	Source/ref.
			(555) G3-7		Ledbetter (Seattle, USA)
			(556) T55		Maeda (Tokyo, Japan)
			(557) 8H8.1	M,G2a	Mawas (Marseille, France) (Serotec UK.)
			(558) CL1.3		Morishima (Nagoya, Japan)
			(559) F93-8A6(S-T2)	M,G1	Poncelet (Montpellier, France) (Seralab UK.)
			(560) OKT16		Talle (Ortho UK.)
			(561) CLB-3A1/1		Tetteroo (Amsterdam, Netherlands)
			(562) Tp40		Ueda (Nagoya, Japan)
			(563) 4H9/LEU9		Warner (California, USA) (Beckton Dickinson UK.)
			(564) TU14		Wernet (Tubingen, FRG)
			DK24	M,G2a	(Dako UK.)
			HuLy-m2	M,G2a	(Serotec UK.)
			PAN L7D	M,G2a	(Seralab UK.)
			(142) YTH3.2	R,G2b	Waldmann (Cambridge, U.K.)
CD8	32	T cell subset (cytotoxic/supressor) Formerly T8 antigen.	(565) P218		Bai (Beijing, China)
			(566) L533		Bernard (Paris, France)
			(567) UCHT4		Beverly (London, U.K.)
			(568) BL15		Brochier (Montpellier, France)
			(569) 2D2		Clement
			(570) BMAC8		Dalchau (East Grinstead, U.K.)
			(571) LEU2A		Evans (Mountain View, USA)

(572) C10		Galton (London, U.K.)
(573) T41D8		Girardet
(574) 14		Hogg (London, U.K.)
(575) MEM-31		Horejsi (Praha, Czechoslovakia)
(576) VIT8		Knapp (Vienna, Austria)
(577) M236		Knowles (New York, USA)
(578) BW135/80		Kurrle (Marburg, FRG)
(579) G10-1		Ledbetter (Seattle, USA)
(580) B.9.1.1		Malissen
(581) B.9.11.10	M,G1	Malissen (Serotec UK.)
(582) B.9.2.4		Malissen
(583) B.9.3.1		Malissen
(584) B.9.4.1		Malissen
(585) B.9.7.6		Malissen
(586) B.9.8.6		Malissen
(587) 51.1		Martin (London, U.K.)
(588) 66.2		Martin (London, U.K.)
(589) 4D12.1		Mawas (Marseille, France)
(590) 8E-1.7		Mawas (Marseille, France)
(591) C12/D3		Petranyi (Budapest, Hungary)
(592) F101-87(S-T8a)	M,G1	Poncelet (Montpellier, France) (Seralab UK.)
(593) F80/ID4D11		Poncelet (Montpellier, France)
(594) T8-2T8-19		Reinherz (Boston, USA)
(595) T8/21THY2D3		Reinherz (Boston, USA)
(596) T8/2T8-1C1		Reinherz (Boston, USA)
(598) T8/2T82A1		Reinherz (Boston, USA)
(599) T8/TPE3FP		Reinherz (Boston, USA)

Cluster of differentiation	Antigen Mol.wt.	Antigen Distribution	Antibody name	Species, Isotype	Source/ref.
			(600) MT1014		Rieber
			(601) MT122		Rieber
			(602) MT415		Rieber
			(603) T811		Rieber
			(604) CLB-T8/1		Tetteroo (Amsterdam, Netherlands)
			(605) Tu102		Wernet (Tubingen, FRG)
			(606) Tu68		Wernet (Tubingen, FRG)
			(607) NU-TS/C		Yokoyama (Kurume, Japan)
			MID4	M,M	(Seralab UK.)
			T811	M,G2a	(Seralab UK.)
			MF8	M,M	(Serotec UK.)
			YTC182.20	R,G2b	(Serotec UK.)
			YTC141.1HL	R,G2b	(Serotec UK.)
			DK25(DAKO-T8)	M,G1	(Dako UK.)
			SK1	M,G1	(Beckton Dickinson UK.)
			SK2	M,G2	(Beckton Dickinson UK.)
			OKT8		(Ortho UK.)
			T8A	M,G1	(Coulter UK.)
CD11(a,b & c)	180,165,150	Alpha chains a, b & c of LFA-related molecules - cell adhesion molecules. Formerly LFA-1, Mac-1 and p150/90.			
CD11a	180	LFA-1 alpha chain. Leucocytes.	(383) CC5-07		Monostori (Budapest, Hungary)
			(403) 25.1.2		Mawas (Marseille, France)

			(709) MHM24		M,G McMichael (Oxford, U.K.) (Oxford, U.K.)
			(711) CIMT	M,G1	Pinkington
			(715) 25.31	M,G1	Mawas (Marseille, France) (Serotec UK.)
			(716) 25.5.2	M,G3	Mawas (Marseille, France)
			(717) 2F12	M,G1	Schmidt (Tubingen, West Germany)
			(718) TS1/12.1.1	M,G1	Springer (Boston, USA) .
			(720) TS1/22.1.1.17	M,G1	Springer (Boston, USA)
			(723) TS2/6.1.2	M,G1	Springer (Boston, USA)
			(724) CLD/LFA1/2	M,G2	Tetteroo (Amsterdam, Netherlands)
CD11b	165	Mac-1 alpha chain. Monocytes, granulocytes.	(645) 60.1		Beatty
			(648) 44		Hogg (London, U.K.)
			(649) VIM12		Knapp (Vienna, Austria)
			(651) TM15		Ohto (Saitama, Japan)
			(652) KIM5		Radzun (Kiel, FRG)
			(653) M52L		Reiber
			(654) LM2/1.6.11		Springer (Boston, USA)
			(656) MO1		Todd
			(657) B2.12		Zeijlemaker (Amsterdam, Netherlands)
			(658) CC1.7		Ando
			OKM1		(Ortho UK.)C
			D12		(Becton Dickinson UK.)
			2LPM19c	M,G1	(Dako UK.)
			BRC-GMI	M,G1	(Serotec UK.)
CD11c	150	Alpha chain of p150/90 protein. Monocytes (granulocytes).	(707) 3.9	M,G2a	Hogg (London, U.K.)
			(721) SCHL3		Schwarting (Berlin, FRG)
			(727) L29	M,G1	Lanier (Mountain View, USA)

Cluster of differentiation	Antigen Mol.wt.	Antigen Distribution	Antibody name	Species, Isotype	Source/ref.
			(728) 3KB43		Pulford (Oxford, U.K.)
			(729) KB23		Pulford (Oxford, U.K.)
			(730) KB90	M,G1	Pulford (Oxford, U.K.) (Dako UK.)
			(300) Ki-M1	M,G1	Radzun (Kiel, FRG)
			(256) Bu15	M,G1	Johnson (Birmingham, U.K.)
			S-HCI-		(Becton Dickinson)
			FD11	M,G1	(Seralab UK.)
CD18	95	Common beta chain of LFA-1, Mac-1 and p150/95 (CD11) proteins. Leucocytes.	(361) M232	M,G2a	Bernard (Paris, France)
			(362) 68-5A5	M,G2a	Vilella (Barcelona, Spain)
			(386) BL5/10L1	M,G1	Brochier (Montpellier, France) (Serotec UK.)
			(706) 60.3	M,G2a	Beatty
			(708) MHM23	M,G1	McMichael (Oxford, U.K.) (Oxford, U.K.)
			(710) CLB54	M,G1	Miedema (Amsterdam, Netherlands)
			(712) TS1/18.1		Springer (Boston, USA)
			(719) TS1/18.1.2.23	M,G1	Springer (Boston, USA)
			(725) 1B4	M,G2a	Wright (Southampton, U.K.)
CD25	55	Activated T cells. IL2 receptor low affinity.	(160) 7G7B6	M,G2a	Nelson (Bethesda, USA)
			(175) 4H3		Reinherz (Boston, USA)
			(188) 33B73	R,G2a	Mawas (Marseille, France)
			(769) 2A3		Lanier (Mountain View, USA) (Becton Dickinson UK.)

			(770) 23A9.3		Mawas (Marseille, France)
			(771) 33b3.1		Mawas (Marseille, France)
			(772) 33B73		Mawas (Marseille, France)
			(773) 36A12		Mawas (Marseille, France)
			(774) 39C6.5		Mawas (Marseille, France)
			(776) T1A		Morishima (Nagoya, Japan)
			(778) 4H3		Reinherz (Boston, USA)
			(780) OKT26A		Talle
			(783) Ta60b		Ueda (Nagoya, Japan)
			(784) Tu69		Wernet (Tubingen, FRG)
			(786) SA36-6G		Yang (New York, USA)
			(787) SA36-6H		Yang (New York, USA)
			(782) Ta60a	M,G1	Ueda (Nagoya, Japan) (Seralab UK.)
			IL7.9L	M,G1	(Seralab UK.)
			YTH 906.9HL	R,G2b	(Serotec UK.)
			B1.49.9	M,G2a	(Serotec UK.)
			ACT-1	M,G1	(Dako UK.)
			OKT26a		(Ortho UK.)
			IL-2R1	M,G2a	(Coulter UK.)
CD28	44	T cell subset.	(92) John9.3		Hansen (Washington, USA)
		Possibly cytotoxic T cell precursors.	(125) KOLT-2		Sagawa (Fukuoka,Japan)
CDw29	135	T cell subset.	(139) K20		Bernard (Paris, France)
		T helper cell set for antibody production.	(117) 4B4		Morimoto (Boston, USA) (Coulter UK.)
CD45	220,205	Leucocytes.	(828) 4B2	M,G2	Steinman
	190,180	Leucocyte common antigen (T200).	(815) 9.4	M,G2a	Martin (London, U.K.)

Cluster of differentiation	Antigen Mol.wt.	Antigen Distribution	Antibody name	Species, Isotype	Source/ref.
			(792) Anti-HLe-1	M,G1	Beverly (London, U.K.)
			(797) BMAC1	M,G1	Dalchau (East Grinstead, U.K.)
			(807) CMRF.12		M Hart (Christchurch, New Zealand)
			(805) EO1	M,G2a	Funderud (Oslo,Norway)
			(812) G1-14	M,G1	Ledbetter (Seattle, USA)
			(833a)MEM28		M Horejsi (Praha, Czechoslovakia)
			(398) S-54		M Pesando
			(809) VIT200	M,G2a	Knapp (Vienna, Austria)
			(830) YTH24.5	R,G2b	Waldmann (Cambridge, U.K.) (Serotec UK.)
			(831) YTH54.12	R,G2b	Waldmann (Cambridge, U.K.) (Serotec UK.)
			(808) 71.4	M,G2a	Kraft (Melbourne, Australia)
			(810) 71.5	M,G2a	Kraft (Melbourne, Australia)
			(800) F10-89-4	M,G2a	Dalchau (East Grinstead, U.K.) (Serotec UK.)
			(817) TT52	M,G2a	Ohto (Saitama, Japan)
			(803) 13.4	M,G2a	Fast
			(205) 48		M Bai (Beijing, China)
			(798) BMAC2	M,G1	Dalchau (East Grinstead, U.K.)
			(799) BMAC3	M,G1	Dalchau (East Grinstead, U.K.)
			(820) F101-139		M Poncelet (Montpellier, France) (Seralab UK.)
			(814) G25-1	M,G1	Ledbetter (Seattle, USA)

			(819) S-80		M Pesando
			(833) T29/33		M Trowbridge (San Diego, USA)
			(825) 6RT2	M,G1	Garrido (Granada, Spain)
			(396) 72-5D3	M,G2a	Vilella (Barcelona, Spain)
			(824) MN18		R Ravoet (Brussels, Belgium)
			(818) 1-33	M,G1	Pesando
			(822) 2B11	M,G1	Pulford (Oxford, U.K.)
			(816) PD-7/26/16	M,G1	Mason (Oxford, U.K.)
			(390) TN10A	M,G2a	Morishima (Nagoya, Japan)
			(358) MEM55		M,G Horejsi (Praha, Czechoslovakia)
			(392) 562/10D3/5B4	M,G3	Knapp (Vienna, Austria)
			(359) BL16	M,M	Brochier (Montpellier, France)
			HuLyM4	M,G2a	(Serotec UK.)
			BRC-Lel/GB3	M,G1	(Serotec UK.)
			2B11+PD7/26	M,G1	(Dako UK.)
			2D1		(Becton Dickinson UK.)
CD45R	220,205	Restricted antibodies, recognise high molecular weight chains only. B cells, T cell subsets and myeloid cells.	(137) 111-1C5	M,G1	Vilella (Barcelona, Spain)
			(116) 2H4	M,G1	Morimoto (Boston, USA)
			(796) 3AC5	M,G2a	Clark
			(811) 73.5.17	M,G1	Kraft (Melbourne, Australia)
			(806) CMRF.11		M Hart (Christchurch, New Zealand)
			(804) E7	M,G2a	Franklin
			(801) F8-11-13	M,G1	Dalchau (East Grinstead, U.K.) (Serotec UK.)
			(813) G1-15	M,G1	Ledbetter (Seattle, USA)
			(410) HD3-F5	M,G1	Ando
			(414) KD3	M	Ando
			(832) My11(Leu18)	M,G3	Loken/Lanier (Mountain View, USA)

Cluster of differentiation	Antigen Mol.wt.	Antigen Distribution	Antibody name	Species, Isotype	Source/ref.
					(Becton Dickinson UK.)
			(415) RFB2		M Janossy (London, U.K.)
			(357) RFB5	M,G1	Janossy (London, U.K.)
			(411) S-83		M Pesando
			(827) 10G3		M,G Steinman
			(381) 4KB5		M Pulford (Oxford, U.K.) (Dako UK.)
			(021) RW35-1C5	M,M	Habeshaw
			(351) YTH80.103	R,G2b	Waldmann (Cambridge, U.K.) (Serotec UK.)
	180	Restricted low molecular weight chain of LCA (CD45).	(826) UCHL1	M,G2a	Smith (London, U.K.) (Dako UK.)

Antibody numbers in parentheses are the workshop numbers from the Third International Workshop and Conference of Human Leucocyte Differentiation Antigens, 1986. Ref: Leucocyte Typing III, White Cell Differentiation Antigens, Ed. A.J. McMichael, Oxford University Press, 1987.
Species of origin of antibody and isotype are given where known.
Commercial source of antibody is given where known.

FORTHCOMING MEETINGS

1989

Jan 12-14
3rd Banff Conference on Reproductive Immunology.
Dr Thomas J. Gill III, Department of Pathology, University of Pittsburgh School of Medicine, Pittsburg, Pennsylvannia 15261, USA.

Feb 20-26
Hematopoieis.
UCLA Symposia, 103, Molecular Biology Institute, University of California, Los Angeles, CA 90024-1378, USA.

Feb 21-25
4th International Workshop and Conference on Human Leukocyte Differentiation Antigens.
Prof Dr W. Knapp, c/o Vienna Medical Academy, Alser Strasse 4, A-1090 Vienna, Austria. Telefax: 42 13 83 23. Telex: medak 134743.

April 2-6
Thymus Workshop.
Dr H-J. Schuurman, Division of Immunopathology, Department of Internal Medicine, University Hospital, Catharinjnesingel 101, 3511 GV Utrecht, The Netherlands.

April 19-23
The Molecular Biology of Autoimmune Disease.
Dr A.G. Demaine, Department of Medicine, King's College School of Medicine, Denmark Hill, London, SE5 8RX, UK.

June 1-2
Micronutrient Effects on Immune Functions.
Conference Department, The New York Academy of Sciences, 2 East 63rd Street, New York, NY 10021, USA

July 25-28
The International Conference on Molecular Aspects of Immune Response and Infectious Diseases.
Dr Claudio De Simone, Malattie Infettive, Faculty of Medicine and Surgery, Universita degli Studi dell' Aquila degli Abruzzi, L'Aquila, 67100, Italy.

July 26-29
4th International Congress of Reproductive Immunology.
4th ICRI Secretariat, c/o Prof. L. Metler, Abt. für Geburtshilfe und Gynakologie, University of Kiel, Michaelistrasse 16, D-2300 Kiel 1, Germany.

July 30-Aug 5
7th Internatioanl Congress of Immunology, Berlin.
Congress Bureau, DER Congress, Augsburger Str. 27, D-1000, Berlin 30, Germany

Oct 23-27
6th International Lymphokine Workshop.
Secretariat du Vieme Atelier International sur les Lymphokines, Laboratoire d'immunologie, Batiment Recherche 1, Institut Gustave Roussy, 94805 Villejuif Cedex, France.

1990

Sept 10-12
European Federation of Immunological Sciences,
Edinburgh, Scotland, UK.

1991

June 2-9

European Federation of Immunological Sciences,
Helsinki, Finland.

1992

Aug 23-30
8th International Congress of Immunology,
International Union of Immunological Sciences,
Budapest, Hungary.

INDEX